FORCES OF THE QUANTUM VACUUM

An Introduction to Casimir Physics

FORCES OF THE QUANTUM VACUUM

An Introduction to Casimir Physics

Editors

William Simpson
Weizmann Institute of Science, Israel

Ulf Leonhardt
Weizmann Institute of Science, Israel

World Scientific

NEW JERSEY · LONDON · SINGAPORE · BEIJING · SHANGHAI · HONG KONG · TAIPEI · CHENNAI

Published by

World Scientific Publishing Co. Pte. Ltd.
5 Toh Tuck Link, Singapore 596224
USA office: 27 Warren Street, Suite 401-402, Hackensack, NJ 07601
UK office: 57 Shelton Street, Covent Garden, London WC2H 9HE

Library of Congress Cataloging-in-Publication Data
Forces of the quantum vacuum : an introduction to Casimir physics / editors, William M.R. Simpson,
Ulf Leonhardt, The Weizmann Institute of Science, Israel.
    pages cm
  Includes bibliographical references and index.
  ISBN 978-9814632904 (hardcover : alk. paper) -- ISBN 978-9814632911 (pbk. : alk. paper)
  1. Quantum theory. 2. Casimir effect. I. Simpson, William M. R. II. Leonhardt, Ulf, 1965–
QC680.F67 2015
530.14'33--dc23

                            2014031475

British Library Cataloguing-in-Publication Data
A catalogue record for this book is available from the British Library.

Cover design by Tanya Skrynnik
Photograph of the pier in St Andrews, Scotland, by Spencer Bentley (www.sierrawhiskeybravo.com)

Proofread by Mrs Rachel Parks, St Andrews

For photocopying of material in this volume, please pay a copying fee through the Copyright Clearance Center, Inc., 222 Rosewood Drive, Danvers, MA 01923, USA. In this case permission to photocopy is not required from the publisher.

*William Simpson dedicates this book to his family,
and most especially his brother, John*

Contents

Foreword

Peter Milonni

In 1947 Lamb and Retherford reported the measurement of a difference in the $2s_{1/2}$ and $2p_{1/2}$ energy levels of the hydrogen atom, in contrast to the equal energies calculated using the Dirac equation for the electron in the Coulomb field of the nucleus. This difference comes predominantly from the interaction of the electron with the zero-point electromagnetic field. The energies calculated by standard methods were divergent, owing to the fact that all field frequencies ν and zero-point energies $\frac{1}{2}h\nu$ contribute to the interaction. The Lamb shift not only stimulated the development of new renormalisation techniques; it also inspired, less famously, consideration of "Some observable consequences of the quantum-mechanical fluctuations of the electromagnetic field," the title of a paper published in 1948 by Theodore Welton.

On May 29, 1948, six days before Welton's paper was received by *Physical Review*, Hendrik Casimir communicated to the Royal Netherlands Academy of Arts and Sciences his paper "On the attraction between two perfectly conducting plates." Although he deduced this attraction from the change, caused by the plates, in the (infinite) zero-point energy of the electromagnetic field, it would be false to the facts to think that Casimir took a cue from the Lamb-Retherford work and its implications for the quantum fluctuations and zero-point energy of the field. In reply to a query of mine many years ago, Casimir wrote that he was "not at all familiar with the work of Welton and others" and that he "went [his] own, somewhat clumsy way." He was, however, influenced by a discussion with Niels Bohr concerning the simple expression for the retarded van der Waals interaction that Casimir and Dirk Polder had derived (and submitted for publication in February 1947). Casimir remarked that Bohr "mumbled something about zero-point energy. That was all, but it put me on a new track." In the introduction to his 1948 paper he wrote that he had recently shown that the retarded van der Waals interaction could be derived by consideration of a change in zero-point field energy, and that he would now take the same approach to calculate the force between two perfectly conducting plates.

Casimir's prediction that there is an attraction between the plates due to the change in zero-point electromagnetic energy was, to say the least, a bold one. Although the concept of zero-point energy played a small role early on in the development of quantum theory, the notion that zero-point energy and quantum fluctu-

1

ations of vacuum fields should be taken seriously was not very popular. One reads in Pauli's Nobel Lecture (1945), for example, that "... the zero-point energy of the vacuum derived from the quantised field becomes infinite, a result which is directly connected with the fact that the system considered has an infinite number of degrees of freedom. It is clear that this zero-point energy has no physical reality ..." And thirty years after Casimir's paper, Julian Schwinger called the Casimir force "one of the least intuitive consequences of quantum electrodynamics."

Casimir's papers on the retarded van der Waals interaction and the force between two conducting plates played a role in the interpretation of experiments carried out, most notably in the Soviet Union, in the 1950s. Both papers are cited in the influential paper by Lifshitz (1955) on "The Theory of Molecular Attractive Forces between Solids." Lifshitz calculated the interaction between two dielectric media, in thermal equilibrium and occupying half-spaces with plane-parallel boundaries, considering the interaction "to come about through ... the fluctuating electromagnetic field." Casimir's results were rederived as special cases of Lifshitz's general expression for the force per unit area between the dielectrics. A few years later Sparnaay measured an attractive force between conductors that did "not contradict Casimir's theoretical prediction." It was not until the mid-1990s that the Casimir force was very accurately measured, a few years after the first experimental confirmation of the long-range interaction between a neutral atom and a conducting surface that Casimir and Polder derived in their paper on the retarded van der Waals interaction.

Aside from its basic importance as an interaction between any two bodies, and more generally as a consequence of quantised fields, the Casimir (or Casimir–Lifshitz) effect may take on increasingly practical importance. One reason was recognised more than half a century ago by Feynman (1959) in a talk on the physics involved in making micromachines: "there is the problem that materials stick together by the molecular (van der Waals) attractions." In more recent years considerable efforts have been devoted to controlling this "stiction."

Casimir's 1948 paper — three pages long! — was written with a clarity and a focus on essential physics that was characteristic of papers by many prominent physicists of the period. The more recent literature on Casimir physics, unfortunately, seldom emulates Casimir's clarity, and it often presumes considerable familiarity with quantum electrodynamics, many-body theory, dyadic Green functions, different approaches to "open" quantum systems, and other theoretic tools, let alone sophisticated experimental techniques.

This slender book, in contrast, is a refreshingly readable introduction to Casimir physics. The contributing experts have introduced the concepts and techniques needed for students and nonspecialists to learn the essential physics of van der Waals and Casimir forces, to appreciate what is and what is not well understood, and perhaps to make contributions of their own to this fascinating field.

P. W. Milonni
September 2014

Introduction

WILLIAM SIMPSON

Hendrik Casimir was a Dutch physicist of the last century who made many contributions to Physics. He is most famous for predicting the phenomenon that carries his name. Casimir speculated that two perfect mirrors facing each other in a vacuum would experience an attractive force as a result of vacuum fluctuations present in the cavity, even at zero temperature. This prediction was confirmed in a number of critical experiments decades later, and the field of *Casimir physics* has rapidly proliferated ever since.

For the newcomer, however, it can all seem a little confusing. The number of research papers claiming some relation to the Casimir force and to vacuum energies is legion, and the topics are diverse, capturing the interests of scientists from Cosmology to Atomic Physics, from Chemistry to Quantum Field Theory, and from Biology to Mathematics. Much of this work is simply unfathomable to outsiders, being too specialised or too technical.

In the *Tanakh*, Koheleth complained that 'of the making of many books there is no end,' and we are surely living in an age of unprecedented scholarly output that would set the heads of the wisest teachers of antiquity spinning. However, few *books* have appeared in the field of Casimir physics to date, fewer still seem to be concerned with the uninitiated, and some of them are not books at all but collections of papers. Active researchers today, told to 'publish or perish,' have little time left to unlock their own field of expertise for the next generation knocking on the office door, or to illuminate their working space for curious outsiders staring through clouding window-panes.

We think this is rather troubling and calls for serious reflection, but our immediate response for this particular field is in the form of a collaboration involving eight physicists from four different countries: two of us (the editors) decided what we wanted to put into an introductory book on Casimir forces, all of us wrote something to put into it, and each of us read and criticised the contributions of the others, adjusting his own in response. The result is not a book any one of us would have written, neither is it another atomised collection of review papers, but an organised training manual that seeks, step-by-step, to expound to the reader increasingly sophisticated theories for assimilating the relevant phenomena, and to connect the predictions of theorists with empirical facts, as far as that is possible at the present time.

The book began with a symposium organised by the editors in Stockholm in 2013. We were searching for physicists who were willing to teach as well as display their technical plumage, and we decided to invite a series of speakers to give pedagogical talks on topics close to the foundations of Casimir physics. As I (William) sat by Lake Mälaren the day after our symposium, looking out on 'the town between the bridges' and its brightly coloured buildings, the basic outline of a new book on the Casimir force began to take shape — a book of *physics* concerned with basic principles of nature and empirical facts, without which more exotic speculations or mathematical deductions are in danger of drifting into mere abstraction. But it was a book that I did not believe I had the expertise to write by myself (I was a PhD student at the time). The task of finding and persuading others to 'get on board' proved less difficult than I expected, however, and our collaboration began to take shape through a series of meetings and text messages transmitted that week over the gentle rippling of the Knight Firth.

Before long we had a publisher and the work began in earnest. The difficulty, of course, was maintaining coherence in a manuscript stretched between three continents. Modern mechanisms for the dynamic sharing of our evolving text (like 'Dropbox') proved moderately helpful, along with the pedantic criticisms of its two opinionated editors over e-mail (patiently endured). However, there is no technological substitute for conversation in conducive surroundings. In the summer of 2014, the contributing theorists decided to meet to discuss our evolving manuscript in the coffee houses of Vienna, to agree on some conventions, and to settle our disputes over Sachertorte.

*A 'Vienna Circle' of Casimir Theorists**

Left to right: W.M.R. Simpson, S. Scheel, E. Shahmoon, U. Leonhardt, S.A. Horsley, S.Y. Buhmann

It is important to appreciate that the field of Casimir physics does not at present enjoy a unified and universally accepted theoretical framework, nor do its theorists necessarily agree on what all of its symbolised quantities *mean*: for example, quantum field theory notoriously generates infinite quantities (the quantum electrodynamics we employ in this text is no exception), and this is a delicate problem that has not really been resolved and must be handled with care. In Vienna we adopted this convention: *regularisation* refers to a procedure that makes infinite quantities

---

*The first exercise is to state a complex transformation that turns circles into straight lines.

finite so that we can do arithmetic with them; *renormalisation* refers to a procedure in which we reset the zero of a quantity (typically by subtracting something, and implicitly under regularisation). You will see examples of both in this book, but not everybody in the field uses the same terminology or makes this distinction.

It is also important to emphasise that the field of Casimir physics is evolving rather rapidly, and a second reason for the reluctance among physicists to write books on the subject is that they could quickly become antiquated. We do not doubt that this is the case: the form of macroscopic quantum electrodynamics that marks the summit of this treatise is of fairly recent creation, and it supercedes its predecessors. Undoubtedly, more developments in theory and experiment will soon leave parts of this book behind too. Our intention here is not to write the definitive work on the Casimir effect, however, but to offer an entrance into a fruitful corner of the field. Others will have to judge how successful we have been in writing this book, but we hope to see more collaborative efforts of this kind in the future. It has proven a valuable learning experience for all of us.

The Weizmann Institute of Science, Rehovot
August 2014

Chapter 1

Normal mode quantum electrodynamics: the quantum vacuum and its consequences

STEFAN YOSHI BUHMANN

> Besides, who ever thought any quality to be a heterogeneous aggregate, such as light is discovered to be? But to determine more absolutely what light is, after what manner refracted, and by what modes or actions it produceth in our minds the phantasms of colours, is not so easy.
>
> – Issac Newton, *New Theory about Light and Colours* (1671)

§1. WAVE-PARTICLE DUALITY — WHAT IS LIGHT (LIKE)?

As stated by the United Nations in their proclamation of an International Year of Light in 2015, *'light plays a vital role in our daily lives and is an imperative cross-cutting discipline of science in the 21st century'* [1]. But what exactly is this important entity?

Light exhibits a whole range of interesting properties (like colour, energy, momentum, speed, polarisation and behaviours (like reflection, refraction, diffraction, interference, absorption, emission). Some of them are accessible by our immediate sensory experience while others can only be revealed via careful experimentation. A possible, but unsatisfactory definition of light would state that it is nothing more or less than an object which has these properties and shows these behaviours. However, this descriptive characterisation in terms of seemingly unrelated attributes does not explain anything, nor does it provide us with any intuition.

1.1 Light as a particle

As a first step towards greater insight into the nature of light, we could ask the alternative question: What is light like? Does it behave like anything else that we know of from our everyday experience? *Newton* tried to find an answer to this question by studying *refraction* in prisms [2]. He concluded that light is like a stream of *corpuscles*: tiny, not necessarily indivisible particles which travel along straight trajectories at finite speed [3]. The particle analogue is able to explain how obstacles give rise to a shadow and how light is reflected (as the corpuscles bounce off a mirror) and refracted (as they change direction upon being attracted towards a medium).

1.2   Light as a wave

Newton's contemporary *Huygens* put forward a very different analogy. He stated that light is like a wave: a continuous oscillation moving through space [4]. According to the principle now bearing his name, the propagation of such a wave can be understood by regarding it as made up from spherical waves. In this way, reflection and refraction can be understood by thinking of spherical waves emanating from an illuminated surface. As with Newton's particle analogy, the wave model provides us with an intuition of how light may behave in a given situation.

However, as Newton objected, waves can go around corners whereas light seemingly doesn't. Or does it? As *Young* demonstrated with his famous *double-slit experiment* [5], light does exhibit the typical wave behaviour of diffraction and interference. Two coherently illuminated slits in a screen are the sources of two spherical light waves. Their superposition gives rise to an oscillatory pattern in the space behind the slits. This wave behaviour is analogous to the patterns of ripples created by two stones dropped into the surface of a lake.

1.3   The photon

It seemed as though the argument was settled in favour of Huygens' wave analogy. But this was not the end of the story. Based on observations of the *photoelectric effect, Einstein* concluded that light consists of discrete quanta of fixed energy which can only be emitted or absorbed as a whole [6]. It appears Newton was partly correct.

So is light like a particle or is it like a wave? The answer is light is a bit of both—and much more. In short, one could say that light is emitted and absorbed like a particle, but propagates like a wave. This integration of the two diametrically opposing descriptions of light as discrete particles on the one hand, and as a continuous wave on the other, has surprising consequences. To see this, let us return to Young's double-slit experiment and imagine what happens when we gradually reduce the light coming through the slits. We will begin to see that the continuous wave pattern formed behind the slits is actually composed of discrete spots. One can even reduce the light source so much that single light quanta, or particles, pass through the slits one at a time [7]. If we record them on a screen, each of them will leave a single bright spot. Yet after recording many such events, these spots will mysteriously form the continuous interference pattern of the wave model again. How do the light particles accomplish this? They are all mutually independent and each may only leave a single bright spot, which has to contribute to the overall interference pattern. There can only be one conclusion: the place where each particle leaves its mark has to be random in a very specific way. The probability distribution of each particle on the screen has to be governed by the same global interference pattern. In this way, a continuous pattern may evolve out of single, discrete particles.

We have captured the essence of light as described by *quantum electrodynamics (QED)*: light is an electromagnetic wave which is quantised into particles of discrete energy called *photons*. The photons are randomly distributed in such a way that the wave emerges from their collective behaviour.

In the next two sections, we substantiate these ideas by presenting the mathematical formalism of quantum electrodynamics. We will start from the wave behaviour of light by considering *classical electrodynamics* in free space. The particle behaviour will be introduced by progressing to the corresponding *quantum electrodynamics*.

Exercise 1.1
Try to come up with as many processes or phenomena involving light as you can. Which of them can be understood within particle or wave models or both?

---

§2. CLASSICAL ELECTRODYNAMICS IN FREE SPACE
— WHAT DO PHOTONS LOOK LIKE?

As shown by *Maxwell* [8], the electric and magnetic fields are governed by a consistent set of four equations. In free space and in the absence of charges or currents, the *Maxwell equations* assume the form

$$\mathbf{\nabla} \cdot \mathbf{E} = 0, \tag{2.1}$$

$$\mathbf{\nabla} \cdot \mathbf{B} = 0, \tag{2.2}$$

$$\mathbf{\nabla} \times \mathbf{E} + \dot{\mathbf{B}} = 0, \tag{2.3}$$

$$\mathbf{\nabla} \times \mathbf{B} - \frac{1}{c^2} \dot{\mathbf{E}} = 0. \tag{2.4}$$

The last two of these equations can be combined to give

$$\mathbf{\nabla} \times (\mathbf{\nabla} \times \mathbf{E}) + \frac{1}{c^2} \ddot{\mathbf{E}} = 0. \tag{2.5}$$

The double curl can be expanded via the rule $\mathbf{\nabla} \times (\mathbf{\nabla} \times \mathbf{E}) = \mathbf{\nabla}(\mathbf{\nabla} \cdot \mathbf{E}) - \nabla^2 \mathbf{E}$. Invoking Maxwell equation (2.1), we arrive at the *wave equation*

$$\nabla^2 \mathbf{E} - \frac{1}{c^2} \ddot{\mathbf{E}} = 0. \tag{2.6}$$

Exercise 2.1
Derive the wave equation from the Maxwell equation.

We have thus combined the coupled equations for the electromagnetic field into Eqs. (2.1) and (2.6) for the electric field alone. But what does this field look like and how does it evolve in time?

To answer this question, we need to solve the wave equation. We observe that it naturally splits into two terms, one acting only on the spatial coordinates and one acting only on the time coordinate. We may therefore use *separation of variables* and assume that the solution can be separated into purely position- and time-dependent factors accordingly:

$$E(r,t) = \sum_j [u_j(r)a_j(t) + u_j^*(r)a_j^*(t)]. \tag{2.7}$$

To obtain a solution of the wave equation, we then have to ensure that the position- and time-dependent functions solve two separate eigenvalue equations: the functions $u_j(r)$ must be solutions to

$$\nabla \times [\nabla \times u_j(r)] - \frac{\omega_j^2}{c^2} u_j(r) = 0, \tag{2.8}$$

or equivalently, the wave equation

$$\nabla^2 u_j(r) + \frac{\omega_j^2}{c^2} u_j(r) = 0 \tag{2.9}$$

together with

$$\nabla \cdot u_j(r) = 0. \tag{2.10}$$

They determine the spatial structure, form or 'mode' of the electromagnetic field and hence commonly referred to as mode functions, or simply *modes*. The time-dependent functions $a_j(t)$ must obey

$$\ddot{a}_j(t) + \omega_j^2 a_j(t) = 0. \tag{2.11}$$

They are the amplitudes of the associated modes. The spatial and temporal equations and their solution are linked by the yet undetermined eigenvalues ω_j.

Exercise 2.2
Show that the decomposition (2.7) solves the wave equation provided that the eigenvalue equations hold.

Note that it is often mathematically convenient to allow for complex-valued mode functions. For instance, *complex modes* take a particularly simple form in free space or for waves which are not linearly polarised. As an additional benefit, the above complex mode expansion with amplitudes a_j and a_j^* is closely analogous to the corresponding quantised version, to be introduced in the following Section 3. The physical fields $E(r,t)$ and $B(r,t)$ constructed with the complex modes must of course be real. This has been ensured by including the second term in Eq. (2.7).

To proceed towards a more explicit form for the electromagnetic field, we must solve the eigenvalue equations. Let us first concentrate on the mode functions which govern its spatial profile. They must be solutions to the wave equation (2.9) together with Eq. (2.10). In addition, the structure of the electromagnetic field obviously depends on the geometry in which it is contained. Are we dealing with an unbounded region of free space or is the field trapped by a cavity of perfectly conducting walls? What is the shape of the walls? Are there additional bodies or objects providing an obstacle for the field? In mathematical terms, the mode functions depend on the boundary conditions for the electromagnetic field as imposed by the present environment. We will give explicit solutions for two very simple geometries: an unbounded region of free space and a cuboid cavity bounded by perfectly conducting walls.

2.1 Normal modes in free space

An infinite region of empty space causes two problems: first, the absence of a finite boundary surface renders it difficult to identify a complete set of modes as unique solutions to a boundary value problem; secondly, we will later want to count photons by integrating over the volume which they occupy. This will lead to a divergence for an infinite volume. To circumvent this problem, we use a trick first introduced by *Rayleigh* [9]: we contain the field inside a fictitious cubic box of length L and volume $V = L^3$. We impose *periodic boundary conditions* but the field is not contained: when impinging on a surface of the box, it simply re-emerges from the opposing surface. One could say that the volume is finite, but unbounded.

The simplest solutions to the wave equation (2.9) inside the box are plane, *propagating waves*:

$$u_j(r) = u_j e^{ik \cdot r}, \quad r \in [-L/2, L/2]^3. \tag{2.12}$$

One can easily verify this by direct substitution. We find that the wave vector k must fulfil the *dispersion relation*

$$k^2 = \frac{\omega_j^2}{c^2}. \tag{2.13}$$

In addition, Eq. (2.10) implies that

$$k \cdot u_j = 0. \tag{2.14}$$

This means that the electric field is *transverse*: it oscillates in a direction which is perpendicular to its propagation. For any given propagation direction $e_k = k/k$, we can choose two transverse *polarisation vectors* $e_{k\lambda}$ ($\lambda = 1, 2$) of unit length such that the three vectors form a complete orthonormal basis:

$$e_k \otimes e_k + \sum_{\lambda=1,2} e_{k\lambda} \otimes e_{k\lambda} = 1 \tag{2.15}$$

($[a \otimes b]_{ij} = a_i b_j$: *dyadic product*, 1: unit tensor). For each wave vector, we hence have two transverse modes

$$u_{k\lambda}(r) = \frac{1}{\sqrt{V}} e_{k\lambda} e^{ik \cdot r}. \tag{2.16}$$

The *mathematical normalisation* factor $1/\sqrt{V}$ ensures that

$$\int_V d^3r\, \boldsymbol{u}_{k\lambda}(\boldsymbol{r}) \cdot \boldsymbol{u}^*_{k\lambda}(\boldsymbol{r}) = 1. \tag{2.17}$$

Finally, the periodic boundary conditions state that the vector potentials on opposite faces of the cubic volume must coincide. This restricts the allowed wave vectors to discrete values

$$\boldsymbol{k} = \frac{2\pi}{L}\,\boldsymbol{n}, \quad \boldsymbol{n} \in \mathbb{Z}^3. \tag{2.18}$$

We have hence found all eigensolutions to the wave equation,

$$\boldsymbol{u}_{n\lambda}(\boldsymbol{r}) = \frac{1}{\sqrt{V}}\,\boldsymbol{e}_{n\lambda}e^{2\pi i \boldsymbol{n}\cdot\boldsymbol{r}/L}, \quad \boldsymbol{n} \in \mathbb{Z}^3, \lambda = 1, 2 \tag{2.19}$$

with associated eigenvalues

$$\omega_n = c\sqrt{\frac{2\pi}{L}}\,\sqrt{n_x^2 + n_y^2 + n_z^2}\,. \tag{2.20}$$

They are orthonormal

$$\int_V d^3r\, \boldsymbol{u}_{n\lambda}(\boldsymbol{r}) \cdot \boldsymbol{u}^*_{n'\lambda'}(\boldsymbol{r}) = \delta_{nn'}\delta_{\lambda\lambda'} \tag{2.21}$$

and complete: they can be used to construct any periodic transverse electric field

$$\boldsymbol{E}(\boldsymbol{r}) = \sum_{n,\lambda}\left[\boldsymbol{u}_{n\lambda}(\boldsymbol{r})a_{n\lambda} + \boldsymbol{u}^*_{n\lambda}(\boldsymbol{r})a^*_{n\lambda}\right] \tag{2.22}$$

with expansion coefficients

$$a_{n\lambda} = \frac{1}{2}\int_V d^3r\, \boldsymbol{u}^*_{n\lambda}(\boldsymbol{r}) \cdot \boldsymbol{E}(\boldsymbol{r}). \tag{2.23}$$

For any concrete calculation, we have to make sure that the boundary conditions introduced by our fictitious box do not lead to unphysical artefacts. To avoid this, the dimensions of the box must be much larger than any length scale of the system under consideration. We can then take the limit $L \to \infty$, where the wave vector takes continuous values. This results in a continuous set of modes

$$\boldsymbol{u}_{k\lambda}(\boldsymbol{r}) = \frac{1}{\sqrt{(2\pi)^3}}\,\boldsymbol{e}_{k\lambda}e^{i\boldsymbol{k}\cdot\boldsymbol{r}}, \quad \boldsymbol{k} \in \mathbb{R}^3, \lambda = 1, 2 \tag{2.24}$$

with eigenvalues

$$\omega_k = ck. \tag{2.25}$$

Here, we have redefined the normalisation constant according to

$$\frac{1}{\sqrt{V}} \mapsto \frac{1}{\sqrt{(2\pi)^3}}, \tag{2.26}$$

such that the orthonormality relations take the form

$$\int d^3r\, \boldsymbol{u}_{\boldsymbol{k}\lambda}(\boldsymbol{r}) \cdot \boldsymbol{u}^*_{\boldsymbol{k}'\lambda'}(\boldsymbol{r}) = \delta(\boldsymbol{k} - \boldsymbol{k}')\delta_{\lambda\lambda'}. \tag{2.27}$$

As shown by Eq. (2.18), a single mode occupies a volume $d^3k = (2\pi)^3/V$ in $\boldsymbol{k}$-space. In the continuum limit, sums over $\boldsymbol{n}$ hence transform to integrals over $\boldsymbol{k}$ according to

$$\sum_{\boldsymbol{n}} \mapsto \frac{V}{(2\pi)^3} \int d^3k. \tag{2.28}$$

Now the *completeness relation* reads

$$\int d^3k \sum_{\lambda} \boldsymbol{u}_{\boldsymbol{k}\lambda}(\boldsymbol{r}) \otimes \boldsymbol{u}^*_{\boldsymbol{k}\lambda}(\boldsymbol{r}') = \boldsymbol{\delta}^{\perp}(\boldsymbol{r} - \boldsymbol{r}'), \tag{2.29}$$

where

$$\boldsymbol{\delta}^{\perp}(\boldsymbol{r}) = \frac{1}{(2\pi)^3} \int d^3k \left(\mathbf{1} - \frac{\boldsymbol{k} \otimes \boldsymbol{k}}{k^2} \right) e^{i\boldsymbol{k}\cdot\boldsymbol{r}} \tag{2.30}$$

is the *transverse delta function*. It plays a similar role as the delta function, but in the space of transverse functions:

$$\int d^3r\, \boldsymbol{\delta}^{\perp}(\boldsymbol{r} - \boldsymbol{r}') \cdot \boldsymbol{a}(\boldsymbol{r}') = \boldsymbol{a}(\boldsymbol{r}) \quad \Leftrightarrow \quad \nabla \cdot \boldsymbol{a} = 0. \tag{2.31}$$

The continuous, propagating modes can thus be used to construct arbitrary, but not necessarily periodic, transverse electric fields.

Exercise 2.3

Show that the continuous mode functions in free space satisfy the completeness relation (2.29).

2.2 Normal modes inside a cuboid cavity

Next, we consider a cuboid cavity of linear dimensions (L_x, L_y, L_z) and volume $V = L_x L_y L_z$ which is bounded by perfectly conducting walls. By adapting the geometric parameters, this model can also be used to study QED effects near a single plate or between two parallel plates. The *perfect-conductor boundary conditions* require the parallel component of the electric field to vanish on the cavity walls:

$$E_y(x, y, z) = E_z(x, y, z) = 0 \quad \text{for } x = 0, L, \tag{2.32}$$
$$E_x(x, y, z) = E_z(x, y, z) = 0 \quad \text{for } y = 0, L, \tag{2.33}$$
$$E_x(x, y, z) = E_y(x, y, z) = 0 \quad \text{for } z = 0, L. \tag{2.34}$$

As a result, the normal modes are not propagating waves as in free space, but *standing waves*. Their wave vector takes discrete values

$$\boldsymbol{k} = \pi \left(n_x/L_x, n_y/L_y, n_z/L_z \right) \quad \text{with } \boldsymbol{n} \in \mathbb{N}^3 \tag{2.35}$$

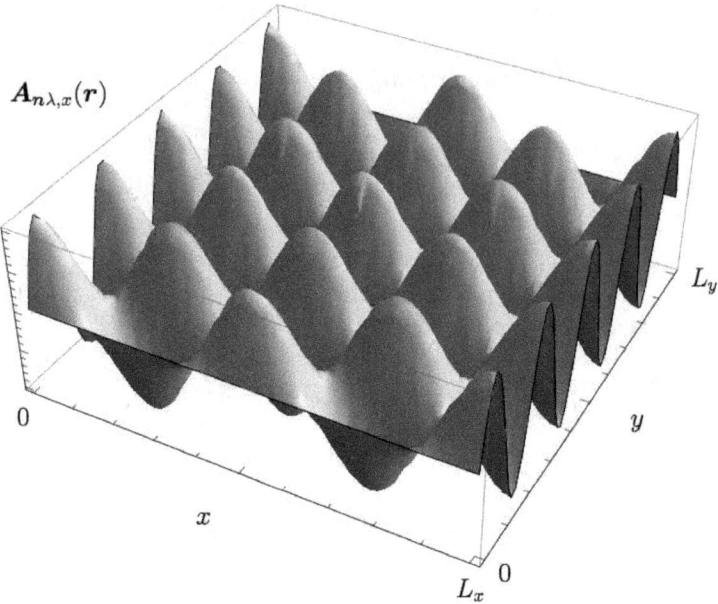

Figure 1.1: Normal modes $\boldsymbol{u_{n\lambda}(r)}$ inside a cuboid cavity of dimensions (L_x, L_y, L_z) enclosed by perfectly conducting walls: the boundary conditions require the modes to be standing waves where integer multiples of half the wavelength in each direction must coincide with the respective cavity dimension. We show the spatial profile of one such mode with these integers taking the values $\boldsymbol{n} = (4, 9, 0)$. Note that due to the vanishing of n_z, the wave vector of the mode lies in the xy-plane. As the boundary conditions at the xz- and yz-walls of the cavity prevent the mode from having any z-component, only one polarisation $(\lambda = 1)$ exists, which also lies in the xz-plane.

with associated eigenfrequencies

$$\omega_{\boldsymbol{n}} = c \sqrt{\left(\frac{\pi n_x}{L_x}\right)^2 + \left(\frac{\pi n_y}{L_y}\right)^2 + \left(\frac{\pi n_z}{L_z}\right)^2}. \tag{2.36}$$

Choosing orthogonal polarisation unit vectors $\boldsymbol{e_{n\lambda}}$ $(\lambda = 1, 2)$ as before, a complete set of modes can be given as

$$\boldsymbol{u_{n\lambda}(r)} = \sqrt{\frac{8}{V}} \begin{pmatrix} e_{n\lambda,x} \cos(\pi n_x x/L_x) \sin(\pi n_y y/L_y) \sin(\pi n_z z/L_z) \\ e_{n\lambda,y} \sin(\pi n_x x/L_x) \cos(\pi n_y y/L_y) \sin(\pi n_z z/L_z) \\ e_{n\lambda,z} \sin(\pi n_x x/L_x) \sin(\pi n_y y/L_y) \cos(\pi n_z z/L_z) \end{pmatrix},$$

$$\boldsymbol{r} \in [0, L_x] \times [0, L_y] \times [0, L_z]. \tag{2.37}$$

Exercise 2.4
Verify explicitly that the above normal modes in a cuboid cavity satisfy the wave equation (2.9) and the transversality condition (2.10).

The modes vanish identically if two or more of n_x, n_y, n_z are zero. If exactly one n_i vanishes, then the wave vector lies in a plane parallel to one set of opposing cavity walls. The boundary conditions imposed by the other walls then requires the mode to be polarised in this plane as well, so that only one polarisation exists.

The mathematical normalisation constant $\sqrt{8/V}$ ensures that the modes satisfy the orthonormality conditions (2.21). As in the free-space case, the modes form a complete set such that every transverse electric field satisfying the boundary conditions can be expanded according to Eqs. (2.22) and (2.23). As an illustration, the mode $n = (4, 9, 0)$ is shown in Fig. 1.1.

2.3 Normal modes: general remarks

The spatial profile of the electric field, and also of the magnetic field, is governed by the mode functions. The modes can always be chosen to be orthonormal:

$$\int_V d^3r\, u_j(r) \cdot u_{j'}^*(r) = \delta_{jj'}. \tag{2.38}$$

They form a complete set such that the electric field at any given instance of time can be expanded as

$$E(r, t) = \sum_j \left[u_j(r)a_j(t) + u_j^*(r)a_j^*(t) \right] \tag{2.39}$$

with coefficients

$$a_j(t) = \frac{1}{2} \int_V d^3r\, u_j^*(r) \cdot E(r, t). \tag{2.40}$$

As required by the wave equation, the modes always oscillate in space. The details of these oscillations depend on the geometry, boundary conditions and chosen set of modes. For instance, the normal modes in the presence of perfectly conducting spheres or cylinders are most conveniently chosen to be spherical or cylindrical waves, respectively. Depending on the utilised modes, the mode label j can have continuous components (such as the wave number in an unbounded direction, $\int d^3k$) or discrete elements (such as the wave number in a bounded direction $\sum_k$, polar or azimuthal wave numbers in spherical or cylindrical geometries, $\sum_{l=0}^{\infty} \sum_{m=-l}^{l}$, or polarisation, $\sum_{\lambda=1,2}$).

It should be noted that — unlike what may be suggested by our examples of free space and a cuboid cavity — normal modes for the electric field are not always necessarily transverse. An example of longitudinal modes, which have a non-zero component along their propagation direction, are some of the wave-guide modes discussed in Section 13.2. In order to satisfy the condition $\nabla \cdot E = 0$, these modes must then have a spatially varying profile in the plane perpendicular to their propagation direction.

2.4   Electromagnetic waves

Let us next turn our attention to the time-dependence of the electromagnetic field. It is contained in the coefficients $a_j(t)$. Solving the respective dynamical equation (2.11), we find

$$a_j(t) = a_j(0)e^{-i\omega_j t}. \tag{2.41}$$

Combining the results of this section, we find that the electric field is given by

$$\boldsymbol{E}(\boldsymbol{r},t) = \sum_j \left[\boldsymbol{u}_j(\boldsymbol{r})a_j(0)e^{-i\omega_j t} + \boldsymbol{u}_j^*(\boldsymbol{r})a_j^*(0)e^{i\omega_j t} \right]. \tag{2.42}$$

Recalling the Faraday law (2.3), we can use this solution to also obtain the magnetic field

$$\boldsymbol{B}(\boldsymbol{r},t) = \sum_j \frac{1}{i\omega_j} \left[\boldsymbol{\nabla} \times \boldsymbol{u}_j(\boldsymbol{r})a_j(0)e^{-i\omega_j t} - \boldsymbol{\nabla} \times \boldsymbol{u}_j(\boldsymbol{r})a_j^*(0)e^{i\omega_j t} \right]. \tag{2.43}$$

Starting from the Maxwell equations, we have hence found general solutions for the intertwined electric and magnetic fields. These solutions exhibit all the classical properties of the physical entity known to us as 'light': first, their spatio-temporal behaviour is that of waves. For *monochromatic light* containing only modes of a single frequency, the field profiles oscillate globally as time progresses. For instance, a single free-space mode $\boldsymbol{u}_{\boldsymbol{k}\lambda}(\boldsymbol{r})$ with amplitude $a_{\boldsymbol{k}\lambda}(0) = ae^{i\phi}$ corresponds to a propagating plane wave

$$\boldsymbol{E}(\boldsymbol{r},t) = \boldsymbol{E}_0 \cos(\boldsymbol{k} \cdot \boldsymbol{r} - \omega t + \phi) \tag{2.44}$$

where $\boldsymbol{E}_0 = 2a\boldsymbol{e}_{\boldsymbol{k}\lambda}/\sqrt{(2\pi)^3}$ and $\omega = ck$. Wave fronts as defined by a constant phase $\boldsymbol{k} \cdot \boldsymbol{r} - \omega t + \phi$ obviously propagate with the speed of light, $\omega/k = c$. Similarly, a single mode of a cuboid cavity corresponds to a standing wave (recall Fig. 1.1) whose amplitude periodically oscillates in time. For *non-monochromatic light*, the different oscillation frequencies associated with the constituent modes result in a nontrivial temporal evolution of the wave profile. For instance, light pulses of finite length are superpositions of modes from a range of frequencies. In dispersing media with a frequency-dependent refractive index, such pulses will disperse: their profile will become flatter and broader as they travel through space. Furthermore, the wave solutions to the Maxwell equations undergo reflection and transmission when impinging on a surface; they form interference patterns when incident on holes, slits or gratings; they diffract when met by small obstacles; and they attenuate when spreading out into larger regions of space.

We have thus found the classical electrodynamics answer to the question: what is light? Light is an electromagnetic wave that consists of intertwined electric and magnetic fields.

2.5   Energy of the electromagnetic field

Having introduced light as an electromagnetic wave, we next address the question: what is the energy of this wave? The energy of the electromagnetic field is manifest in its ability to do work on electrically charged matter. As a simple example, let us consider an ensemble of point charges q_α at positions $\boldsymbol{r}_\alpha$ and with velocities $\dot{\boldsymbol{r}}_\alpha$. They give rise to a microscopic charge density

$$\rho(\boldsymbol{r}) = \sum_\alpha q_\alpha \delta(\boldsymbol{r} - \boldsymbol{r}_\alpha) \tag{2.45}$$

and a microscopic current density

$$\boldsymbol{j}(\boldsymbol{r}) = \sum_\alpha q_\alpha \dot{\boldsymbol{r}}_\alpha \delta(\boldsymbol{r} - \boldsymbol{r}_\alpha). \tag{2.46}$$

The charge and current densities act as sources for the electromagnetic field, so that the Maxwell equations (2.1) and (2.4) take their well-known forms

$$\boldsymbol{\nabla} \cdot \boldsymbol{E} = \frac{\rho}{\varepsilon_0}, \tag{2.47}$$

$$\boldsymbol{\nabla} \times \boldsymbol{B} - \frac{1}{c^2}\dot{\boldsymbol{E}} = \mu_0 \boldsymbol{j} \tag{2.48}$$

as *Gauss' law* and *Ampère's law*, respectively. More important for our current question, the electromagnetic field acts on these charges by exerting Lorentz forces

$$\boldsymbol{F}_\alpha = q_\alpha[\boldsymbol{E}(\boldsymbol{r}_\alpha) + \dot{\boldsymbol{r}}_\alpha \times \boldsymbol{B}(\boldsymbol{r}_\alpha)]. \tag{2.49}$$

The power generated by these forces is

$$\dot{W} = \sum_\alpha \boldsymbol{F}_\alpha \cdot \dot{\boldsymbol{r}}_\alpha = \sum_\alpha q_\alpha \dot{\boldsymbol{r}}_\alpha \cdot \boldsymbol{E}(\boldsymbol{r}_\alpha). \tag{2.50}$$

Note that the magnetic forces do no work as they are perpendicular to the particle velocities. Using the above definition of the current density, the power can be expressed as

$$\dot{W} = \int_V \mathrm{d}^3 r \, \boldsymbol{j} \cdot \boldsymbol{E} \tag{2.51}$$

where V is an arbitrary volume containing the particles. We want to express this rate of work in terms of the electromagnetic field alone. To this end, we eliminate the current density by means of the Ampère law,

$$\dot{W} = \int_V \mathrm{d}^3 r \left[\frac{1}{\mu_0} \boldsymbol{E} \cdot \boldsymbol{\nabla} \times \boldsymbol{B} - \varepsilon_0 \boldsymbol{E} \cdot \dot{\boldsymbol{E}} \right]. \tag{2.52}$$

Using the vector identity $\boldsymbol{\nabla} \cdot (\boldsymbol{a} \times \boldsymbol{b}) = \boldsymbol{b} \cdot (\boldsymbol{\nabla} \times \boldsymbol{a}) - \boldsymbol{a} \cdot (\boldsymbol{\nabla} \times \boldsymbol{b})$ and invoking the *Faraday law* (2.3), we find

$$\dot{W} = \int_V \mathrm{d}^3 r \left[-\frac{1}{\mu_0} \boldsymbol{B} \cdot \dot{\boldsymbol{B}} - \varepsilon_0 \boldsymbol{E} \cdot \dot{\boldsymbol{E}} - \frac{1}{\mu_0} \boldsymbol{\nabla} \cdot (\boldsymbol{E} \times \boldsymbol{B}) \right]. \tag{2.53}$$

Finally invoking the divergence theorem $\int_V \mathrm{d}^3 r \, \boldsymbol{\nabla} \cdot \boldsymbol{f} = \int_{\partial V} \mathrm{d}\boldsymbol{a} \cdot \boldsymbol{f}$, we arrive at the *Poynting theorem*:

$$\dot{\mathcal{E}} + \int_{\partial V} \mathrm{d}\boldsymbol{a} \cdot \boldsymbol{S} = -\int_V \mathrm{d}^3 r \, \boldsymbol{j} \cdot \boldsymbol{E}. \tag{2.54}$$

Here,

$$\mathcal{E} = \frac{1}{2} \int_V \mathrm{d}^3 r \left[\varepsilon_0 \, \mathbf{E}^2 + \frac{1}{\mu_0} \, \mathbf{B}^2 \right] \tag{2.55}$$

is the energy of the electromagnetic field contained in volume V and the second term above describes the energy flowing out of the volume as governed by the Poynting vector

$$\mathbf{S} = \frac{1}{\mu_0} \mathbf{E} \times \mathbf{B}. \tag{2.56}$$

Exercise 2.5
Derive Poynting's theorem explicitly by starting from the rate of work (2.50).

Poynting's theorem is an energy conservation law: the change of field energy within the volume plus energy flowing out of the volume equals minus the work done on the charged particles. Note that the simple form (2.55) for the field energy is only valid in free space. It can be easily adapted to a non-dispersive, non-absorbing medium of relative dielectric *permittivity* ε and *permeability* μ by making the replacements $\varepsilon_0 \mapsto \varepsilon_0 \varepsilon$ and $\mu_0 \mapsto \mu_0 \mu$. For the dispersing and absorbing media discussed in Chapters 3 and 4, the permittivity and permeability are frequency-dependent and they possess a positive imaginary part. In these cases, simple generalisations of Poynting's theorem exist only for almost monochromatic electromagnetic fields [11].

As stated in the previous section, the particle behaviour of light becomes manifest most prominently in its emission and absorption. In the next section, we will see how this particle nature of light can be incorporated into our model, leading to quantum electrodynamics.

§3. NORMAL MODE QUANTUM ELECTRODYNAMICS — HOW TO COUNT PHOTONS?

As stated by Einstein in his explanation of the photoelectric effect [6], the electromagnetic field is packaged into quanta. Our discussion of the Young experiment [5] has shown that the integration of such particle-like behaviour into a continuous wave picture requires the particles to exhibit a degree of randomness. In this section, we are going to introduce the quantum theory of light which gives a mathematical description of this random behaviour.

3.1 Basic elements of quantum theory

As in any standard quantum theory, we are going to encode the statistics of light in a normalised *state* vector, $\langle \psi | \psi \rangle = 1$, and represent any physical property or *observable* by a linear, hermitian operator $\hat{O} = \hat{O}^\dagger$. Being hermitian, the operator has real eigenvalues O_j which represent the possible outcomes of a measurement of the observable:

$$\hat{O} |O_j\rangle = O_j |O_j\rangle. \tag{3.1}$$

The associated eigenvectors $|O_j\rangle$ form a complete orthonormal basis, so that they can be used to expand the operator:

$$\hat{O} = \sum_j O_j |O_j\rangle\langle O_j|. \tag{3.2}$$

A measurement of the observable yields a result O_j with a probability

$$p(O_j) = |\langle\psi|O_j\rangle|^2. \tag{3.3}$$

Repeating the measurement many times, we thus find a quantum average

$$\langle\hat{O}\rangle = \langle\psi|\hat{O}|\psi\rangle. \tag{3.4}$$

The quantum state $|\psi\rangle$ hence contains all the information about the statistics by determining for an arbitrary observable $\hat{O}$ the probabilities $p(O_j)$ and hence also the average $\langle\hat{O}\rangle$ and fluctuations

$$\Delta\hat{O} = \hat{O} - \langle\hat{O}\rangle. \tag{3.5}$$

The quantum statistics encoded in the state vector $|\psi\rangle$ is fundamentally different from any classical statistics. In particular, measuring an observable $\hat{O}$ may influence the subsequent measurement of another observable $\hat{P}$. Mathematically speaking, this happens whenever the two observables do not possess a common basis of eigenvectors. Or equivalently, when their *commutator*

$$[\hat{O}, \hat{P}] = \hat{O}\hat{P} - \hat{P}\hat{O} \tag{3.6}$$

is non-zero.

Exercise 3.1
Show that two observables $\hat{O}$ and $\hat{P}$ possess a common basis of eigenvectors if their commutator vanishes.

As a consequence, the fluctuations of the two observables cannot become arbitrarily small: they are bounded from below by the *Heisenberg uncertainty relation* [10]

$$\langle(\Delta\hat{O})^2\rangle\langle(\Delta\hat{P})^2\rangle \geq \tfrac{1}{2}|\langle[\hat{O}, \hat{P}]\rangle|^2. \tag{3.7}$$

In many situations, quantum and classical randomness arise simultaneously. For instance, we may not know the quantum state of a system. Instead, we just know that it may be in a given state $|\psi_k\rangle$ with a given classical probability P_k. The average of an observable $\hat{O}$ is then a combination of classical and quantum averages:

$$\langle\hat{O}\rangle = \sum_k P_k\langle\hat{O}\rangle_k \tag{3.8}$$

with $\langle\hat{O}\rangle_k = \langle\psi_k|\hat{O}|\psi_k\rangle$. Alternatively, we may encode mixed classical-quantum statistics in a *density matrix*

$$\hat{\rho} = \sum_k P_k|\psi_k\rangle\langle\psi_k|, \tag{3.9}$$

which must be normalised, $\text{tr}\,\hat{\rho} = 1$. The above operator average can then equivalently be written as

$$\langle \hat{O} \rangle = \text{tr}(\hat{O}\hat{\rho}). \tag{3.10}$$

As a special case, a pure state $|\psi\rangle$, which exhibits a purely quantum statistics, has a density matrix $\hat{\rho} = |\psi\rangle\langle\psi|$.

Exercise 3.2
Show that the quantum average of an observable $\hat{O}$ is given by $\langle\psi|\hat{O}|\psi\rangle$ for a system in a pure state $|\psi\rangle$ and by $\text{tr}(\hat{\rho}\hat{O})$ for a system in a mixed state $\hat{\rho}$.

3.2 The quantised electromagnetic field

After this recapitulation of the essence of quantum theory in general, let us return our attention to electromagnetism. In order to formulate a quantum theory of light, we must replace the classical electromagnetic fields presented in the previous section by quantum observables as described by operators. Replacing the amplitudes a_j and a_j^* in Eq. (2.39) by operators $\hat{a}_j$ and $\hat{a}_j^\dagger$ the quantised *electric field* reads

$$\hat{\boldsymbol{E}}(\boldsymbol{r}) = \sum_j \left[\boldsymbol{E}_j(\boldsymbol{r})\hat{a}_j + \boldsymbol{E}_j^*(\boldsymbol{r})\hat{a}_j^\dagger \right] \tag{3.11}$$

while the quantised *magnetic field* assumes the corresponding form

$$\hat{\boldsymbol{B}}(\boldsymbol{r}) = \sum_j \frac{1}{i\omega_j} \left[\boldsymbol{\nabla} \times \boldsymbol{E}_j(\boldsymbol{r})\hat{a}_j - \boldsymbol{\nabla} \times \boldsymbol{E}_j^*(\boldsymbol{r})\hat{a}_j^\dagger \right]. \tag{3.12}$$

These are our first examples of *quantum fields*: operators which depend continuously on position. The position dependence is contained in mode functions $\boldsymbol{E}_j(\boldsymbol{r})$, while the operator nature resides in the amplitude operators $\hat{a}_j$ and $\hat{a}_j^\dagger$. Note that the mode functions $\boldsymbol{u}_j(\boldsymbol{r})$ used in the previous section have a dimension $1/\sqrt{\text{Volume}}$. The classical amplitudes a_j then have to possess the right physical dimension to ensure that $\boldsymbol{E}(\boldsymbol{r})$ has the dimension of an electric field strength. As we will see, the quantum amplitudes $\hat{a}_j$ on the other hand are typically dimensionless. We have thus had to replace the mode functions $\boldsymbol{u}_j(\boldsymbol{r})$ by rescaled modes

$$\boldsymbol{E}_j(\boldsymbol{r}) = c_j \boldsymbol{u}_j(\boldsymbol{r}) \tag{3.13}$$

which carry the correct physical dimension of an electric field strength. The appropriate scaling constants c_j will be determined below.

We next ask ourselves the question: what are the properties of the operators $\hat{a}_j$ and $\hat{a}_j^\dagger$ which describe the random behaviour of light? The key to the answer lies in Einstein's conclusion from the photoelectric effect: the electromagnetic field of frequency ω is quantised into photons which each carry an energy $\hbar\omega$. When counting the number of photons, we expect to find a certain integer number $n = 0, 1, 2, 3 \ldots$ with a probability $p(n)$. Note the difference between the well-known 'first quantisation' of non-relativistic quantum mechanics, where the position and momentum of a given particle is random, and this 'second quantisation' of quantum field theory, where even the number of particles (photons) is uncertain.

3.3   Quantum statistics and Fock states

Let us first concentrate on characterising the quantum statistics of a single mode j. We require a *number operator* $\hat{n}$ whose eigenvalues n, the possible measurement outcomes, are non-negative integers:

$$\hat{n}|n\rangle = n|n\rangle. \tag{3.14}$$

The associated eigenstates $|n\rangle$ are known as *Fock states* [12]. One can easily show that the operator

$$\hat{n} = \hat{a}^\dagger \hat{a} \tag{3.15}$$

has the required eigenvalues, provided that the amplitude operators satisfy the commutation relations

$$\left[\hat{a}, \hat{a}\right] = \left[\hat{a}^\dagger, \hat{a}^\dagger\right] = 0, \qquad \left[\hat{a}, \hat{a}^\dagger\right] = 1. \tag{3.16}$$

This is a consequence of two simple properties of the eigenstates and eigenvalues which follow immediately from the eigenvalue equation and the commutation relations: First, all eigenvalues are non-negative, $n \geq 0$. Secondly, the operators $\hat{a}^\dagger$ and $\hat{a}$ raise or lower the eigenvalues by exactly one,

$$\hat{a}^\dagger|n\rangle = \sqrt{n+1}\,|n+1\rangle, \tag{3.17}$$
$$\hat{a}|n\rangle = \sqrt{n}\,|n-1\rangle. \tag{3.18}$$

Exercise 3.3

Use the commutation relations (3.16) to show that the eigenvalues of the photon number operator (3.15) are non-negative and that $\hat{a}^\dagger$ and $\hat{a}$ act as creation and annihilation operators in accordance with Eqs. (3.17) and (3.18).

They are therefore known as photon *creation* and *annihilation operators*. To ensure that a repeated application of the annihilation operator does not to lead negative eigenvalues, we must conclude the existence of an eigenstate $|0\rangle$ with vanishing eigenvalue,

$$\hat{a}|0\rangle = 0. \tag{3.19}$$

This state is the *quantum vacuum* state where the number of photons is zero,

$$\hat{n}|0\rangle = 0|0\rangle. \tag{3.20}$$

The quantum vacuum is the state referred to in the title of this book. Its properties will be discussed in more detail in the next section. Combining the above properties of creation and annihilation operators and quantum vacuum, we conclude that the number operator $\hat{n}$ has the required non-negative integer eigenvalues $n = 0, 1, 2, 3 \ldots$ The associated *Fock states* [12] can be obtained by repeated application of the creation operator,

$$|n\rangle = \frac{1}{\sqrt{n!}}\left(\hat{a}^\dagger\right)^n|0\rangle. \tag{3.21}$$

The constant $1/\sqrt{n!}$ ensures that these orthogonal states are normalised,

$$\langle m|n\rangle = \delta_{mn}. \tag{3.22}$$

The Fock states form a basis and they can be used to determine the probability of finding n photons when a given manifestation of light is described by a state $|\psi\rangle$,

$$p(n) = |\langle n|\psi\rangle|^2. \tag{3.23}$$

We next generalise the quantum statistics to the multi-mode case. Photons in different modes are independent of each other: making a measurement to count the number of photons in a given mode j should have no influence on a subsequent measurement of the number of photons in a different mode j'. Mathematically, this means that operators belonging to different modes commute, so that the multi-mode *bosonic commutation relations* read

$$[\hat{a}_j, \hat{a}_{j'}] = [\hat{a}_j^\dagger, \hat{a}_{j'}^\dagger] = 0, \qquad [\hat{a}_j, \hat{a}_{j'}^\dagger] = \delta_{jj'}. \tag{3.24}$$

Recall that the mode labels may include continuous and discrete components, so that $\delta_{jj'}$ may represent the delta function or the ordinary Kronecker delta. For instance, we have

$$[\hat{a}_{n\lambda}, \hat{a}_{n'\lambda'}] = [\hat{a}_{n\lambda}^\dagger, \hat{a}_{n'\lambda'}^\dagger] = 0, \qquad [\hat{a}_{n\lambda}, \hat{a}_{n'\lambda'}^\dagger] = \delta_{nn'}\delta_{\lambda\lambda'} \tag{3.25}$$

for the normal modes of a finite free-space volume and cuboid cavity, but

$$[\hat{a}_{\boldsymbol{k}\lambda}, \hat{a}_{\boldsymbol{k}'\lambda'}] = [\hat{a}_{\boldsymbol{k}\lambda}^\dagger, \hat{a}_{\boldsymbol{k}'\lambda'}^\dagger] = 0, \qquad [\hat{a}_{\boldsymbol{k}\lambda}, \hat{a}_{\boldsymbol{k}'\lambda'}^\dagger] = \delta(\boldsymbol{k} - \boldsymbol{k}')\delta_{\lambda\lambda'} \tag{3.26}$$

in infinite free space. The *quantum vacuum* $|0\rangle = |\{0_j\}\rangle$ of the multi-mode electromagnetic field is the state where each mode contains no photons:

$$\hat{a}_j|0\rangle = 0. \tag{3.27}$$

Excited *Fock states* of the electromagnetic field can be obtained by repeated application of creation operators. For instance, a two-mode excited state with photon 1 in mode j and photon 2 in a different mode j' reads

$$|jj'\rangle = \hat{a}_j^\dagger\hat{a}_{j'}^\dagger|0\rangle. \tag{3.28}$$

As the two operators commute, this state coincides with the state $|j'j\rangle = \hat{a}_{j'}^\dagger\hat{a}_j^\dagger|0\rangle$ where photon 1 is in mode j' and photon 2 is in mode j. Photons are therefore *bosons*: indistinguishable particles whose wave function is symmetric under exchange. In particular, this implies that more than one photon can occupy a single mode.

3.4 Energy of the electromagnetic field

To complete our field quantisation, we have to ensure that each photon carries an energy $\hbar\omega_j$ as noted by Einstein [6]. We have already replaced the mathematical normalisation of the modes (2.38) with a physical normalisation

$$\int_V d^3r\, \boldsymbol{E}_j(\boldsymbol{r}) \cdot \boldsymbol{E}_{j'}^*(\boldsymbol{r}) = c_j c_{j'}^* \, \delta_{jj'}, \tag{3.29}$$

recalling Eq. (3.13). The constants c_j have to be determined such that Einstein's observation holds. We start from the total energy (2.55) of the electromagnetic field

$$\hat{H}_{\mathrm{F}} = \frac{1}{2}\int d^3r \left[\varepsilon_0 \,\hat{\boldsymbol{E}}^2(\boldsymbol{r}) + \frac{1}{\mu_0}\,\hat{\boldsymbol{B}}^2(\boldsymbol{r}) \right], \tag{3.30}$$

where the electric and magnetic fields are now quantum operators. Using the above expansions, we find

$$\begin{aligned}
\hat{H}_{\mathrm{F}} = \; & \frac{1}{2}\sum_{jj'}\int d^3r \left\{ \varepsilon_0 \boldsymbol{E}_j(\boldsymbol{r}) \cdot \boldsymbol{E}_{j'}^*(\boldsymbol{r}) \big(\hat{a}_j \hat{a}_{j'}^\dagger + \hat{a}_{j'}^\dagger \hat{a}_j\big) \right. \\
& \left. + \frac{1}{2\mu_0 \omega_j \omega_{j'}} [\boldsymbol{\nabla}\times\boldsymbol{E}_j(\boldsymbol{r})] \cdot [\boldsymbol{\nabla}\times\boldsymbol{E}_{j'}^*(\boldsymbol{r})]\big(\hat{a}_j\hat{a}_{j'}^\dagger + \hat{a}_{j'}^\dagger\hat{a}_j\big) \right\}.
\end{aligned} \tag{3.31}$$

We evaluate the magnetic part of the field energy by use of partial integration (where terms on the surface of V vanish due to the boundary conditions) and the wave equation (2.9). Exploiting the orthonormality of the modes, the field energy then assumes the form

$$\hat{H}_{\mathrm{F}} = 2\varepsilon_0 \sum_j |c_j|^2 \big(\hat{n}_j + \tfrac{1}{2}\big). \tag{3.32}$$

Note that terms proportional to $\hat{a}_j\hat{a}_{j'}$ or $\hat{a}_j^\dagger\hat{a}_{j'}^\dagger$ vanish after combining electric and magnetic field energies.

Exercise 3.4
Derive the above expression for the field energy by starting from Eq. (3.30) and using the mode expansions (3.11) and (3.12) for the electric and magnetic fields.

The energy of the electromagnetic field is proportional to the photon number, as suggested by Einstein. We require that the addition of a photon should add an amount $\hbar\omega_j$ to the *field energy*:

$$\hat{H}_{\mathrm{F}} = \sum_j \big(\hat{n}_j + \tfrac{1}{2}\big)\hbar\omega_j. \tag{3.33}$$

By comparing the two expressions, we find that the constants c_j have to take the values

$$c_j = \mathrm{i}\sqrt{\frac{\hbar\omega_j}{2\varepsilon_0}}. \tag{3.34}$$

The phase factor i is a common convention.* Applying this physical normalisation to the examples from the previous section, the *modes* read

$$\boldsymbol{E}_{n\lambda}(\boldsymbol{r}) = i\sqrt{\frac{\hbar\omega_n}{2\varepsilon_0 V}}\, \boldsymbol{e}_{n\lambda} e^{2\pi i n \cdot \boldsymbol{r}/L} \qquad (3.35)$$

for a finite free-space volume,

$$\boldsymbol{E}_{k\lambda}(\boldsymbol{r}) = i\sqrt{\frac{\hbar\omega_k}{16\pi^3\varepsilon_0}}\, \boldsymbol{e}_{k\lambda} e^{i\boldsymbol{k}\cdot\boldsymbol{r}} \qquad (3.36)$$

in the limit of infinite *free space* and

$$\boldsymbol{E}_{n\lambda}(\boldsymbol{r}) = i\sqrt{\frac{4\hbar\omega_n}{\varepsilon_0 V}} \begin{pmatrix} e_{n\lambda,x}\cos(\pi n_x x/L_x)\sin(\pi n_y y/L_y)\sin(\pi n_z z/L_z) \\ e_{n\lambda,y}\sin(\pi n_x x/L_x)\cos(\pi n_y y/L_y)\sin(\pi n_z z/L_z) \\ e_{n\lambda,z}\sin(\pi n_x x/L_x)\sin(\pi n_y y/L_y)\cos(\pi n_z z/L_z) \end{pmatrix} \qquad (3.37)$$

inside a *cuboid cavity*.

We have constructed an explicit quantisation scheme for the electromagnetic field: the electric and magnetic fields are operator fields which are given by the above expansions in terms of classical mode functions and associated bosonic operators. The bosonic commutation relations can be used to evaluate field commutators and quantum averages. The latter represent expectation values of repeated field measurements. Normal mode quantum electrodynamics is an example of a quantum field theory.

3.5 Quantum dynamics of the electromagnetic field

The energy of the electromagnetic field (3.33) also plays the role of a *field Hamiltonian*, generating the *Heisenberg equations* of motion

$$\dot{\hat{a}}_j = \frac{1}{i\hbar}[\hat{a}_j, \hat{H}_F] = -i\omega_j \hat{a}_j. \qquad (3.38)$$

Their solution

$$\hat{a}_j(t) = e^{-i\omega_j t}\hat{a}_j \qquad (3.39)$$

implies that the time-dependent *electric* and *magnetic fields* in the *Heisenberg picture* are given by

$$\hat{\boldsymbol{E}}(\boldsymbol{r},t) = \sum_j \left[\boldsymbol{E}_j(\boldsymbol{r})e^{-i\omega_j t}\hat{a}_j + \boldsymbol{E}_j^*(\boldsymbol{r})e^{i\omega_j t}\hat{a}_j^\dagger \right], \qquad (3.40)$$

$$\hat{\boldsymbol{B}}(\boldsymbol{r},t) = \sum_j \frac{1}{i\omega_j} \left[\boldsymbol{\nabla} \times \boldsymbol{E}_j(\boldsymbol{r})e^{-i\omega_j t}\hat{a}_j - \boldsymbol{\nabla} \times \boldsymbol{E}_j^*(\boldsymbol{r})e^{i\omega_j t}\hat{a}_j^\dagger \right]. \qquad (3.41)$$

The electromagnetic fields oscillate in space and time just like a wave in classical electrodynamics. Yet the energy contained in this wave is quantised into well-defined packages, as sketched in Fig. 1.2. They exhibit a random number statistics according to the boson model. Quantum electrodynamics has thus provided us with a model of light which integrates wave and particle behaviours into a coherent whole.

---

*The quantisation of the electromagnetic field is often based on the modes for the vector potential. When defining the latter without phase factor, the electric-field modes acquire an i.

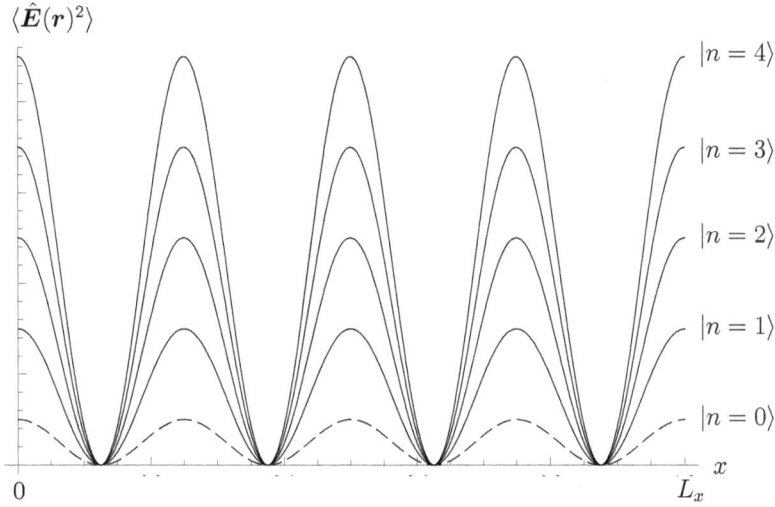

Figure 1.2: Quantum statistics of the electromagnetic field inside a cuboid cavity of dimensions (L_x, L_y, L_z) enclosed by perfectly conducting walls: we display position-dependent amplitudes of the electric field associated with the $\boldsymbol{n} = (4, 9, 0)$ mode for different photon numbers.

§4. The quantum vacuum — What are virtual photons?

The introduction of particle behaviour into the electromagnetic field via bosonic operators has had an unexpected side-effect: the energy (3.33) contained in the electromagnetic field in the presence of n photons is not $n\hbar\omega$, but $(n + \frac{1}{2})\hbar\omega$. Even in the *quantum vacuum* $|0\rangle$ where every possible mode is occupied by zero photons, we find an average energy

$$E_0 = \langle \hat{H}_F \rangle = \sum_j \tfrac{1}{2}\hbar\omega_j. \tag{4.1}$$

This infinite expression corresponds to the energy of half a photon in each mode. What is the significance of this vacuum energy? To answer this question, we must turn our attention to the random nature of the quantum electromagnetic field. The *electric* and *magnetic field* operators have a non-vanishing *commutator* which can be evaluated from the bosonic commutation relations by using the completeness relation of the modes. In infinite free space, we find

$$\left[\hat{\boldsymbol{E}}(\boldsymbol{r}), \hat{\boldsymbol{B}}(\boldsymbol{r}')\right] = \frac{i\hbar}{\varepsilon_0} \boldsymbol{\nabla} \times \boldsymbol{\delta}(\boldsymbol{r} - \boldsymbol{r}') \tag{4.2}$$

where $\boldsymbol{\delta}(\boldsymbol{r} - \boldsymbol{r}') = \mathbf{1}\delta(\boldsymbol{r} - \boldsymbol{r}')$ is the delta tensor and $[\boldsymbol{a} \times \mathrm{T}]_{ij} = \epsilon_{ikl}a_k T_{lj}$ is the cross product of a vector with a tensor.

Exercise 4.1
Starting from the expansions (3.11) and (3.12) in terms of normal modes (3.36), show that the electric and magnetic fields in infinite free space satisfy the above commutation relation.

According to the *Heisenberg uncertainty relation* [10], the values of these non-commuting electric and magnetic fields cannot be determined to arbitrary precision at the same time. In particular, we cannot ascertain that these fields simultaneously vanish. Instead, both of these fields are random quantities. They vanish on average in the vacuum state, but they exhibit *quantum fluctuations* around this average. Since both fields equally contribute to the energy of the electromagnetic field, it is obvious that this energy can be minimised, but never caused to vanish completely.

4.1   Vacuum fluctuations

Although the total energy associated with the vacuum fluctuations diverges, we can identify a finite *spectral energy density* $\rho_0(\omega)$. This is the energy per unit volume of field fluctuations within a given frequency interval $[\omega, \omega + d\omega]$,

$$\left\langle \frac{\varepsilon_0}{2}\, \hat{\boldsymbol{E}}^2(\boldsymbol{r}) + \frac{1}{2\mu_0}\, \hat{\boldsymbol{B}}^2(\boldsymbol{r}) \right\rangle = \int_0^\infty d\omega\, \rho_0(\omega). \tag{4.3}$$

To find its value, we substitute the normal mode expansions into the above expression, noting that only terms $\hat{a}\hat{a}^\dagger$ contribute when the field is in its vacuum state, and evaluate products of the normal modes (3.36) in infinite free space as

$$\boldsymbol{E}_{\boldsymbol{k}\lambda}(\boldsymbol{r}) \cdot \boldsymbol{E}^*_{\boldsymbol{k}\lambda}(\boldsymbol{r}) = \frac{\hbar\omega_k}{16\pi^3\varepsilon_0}, \tag{4.4}$$

$$\nabla \times \boldsymbol{E}_{\boldsymbol{k}\lambda}(\boldsymbol{r}) \cdot \nabla \times \boldsymbol{E}^*_{\boldsymbol{k}\lambda}(\boldsymbol{r}) = \frac{\hbar\omega_k k^2}{16\pi^3\varepsilon_0}. \tag{4.5}$$

Carrying out the trivial sum over polarisations and transforming the wave vector integral according to $\int d^3k = 4\pi \int_0^\infty dk\, k^2 = 4\pi \int_0^\infty d\omega\, \omega^2/c^3$, we find

$$\rho_0(\omega) = \frac{\hbar\omega^3}{2\pi^2 c^3}. \tag{4.6}$$

What would happen if we attempted to measure the electromagnetic field in its vacuum state? Suppose that we could prepare the vacuum state by completely shielding a region of space from all external electromagnetic fields and removing all possible sources such as charges or currents from within. If we were to place an idealised detector in this region which is perfectly sensitive to electric fields in a given frequency interval $[\omega, \omega + d\omega]$, then a series of measurements would give the following results: the average of these measurements would vanish,

$$\langle \hat{\boldsymbol{E}}(\boldsymbol{r}) \rangle = \boldsymbol{0}, \tag{4.7}$$

while their variance would be governed by the *vacuum fluctuations*

$$\langle \hat{\boldsymbol{E}}^2(\boldsymbol{r}) \rangle_{[\omega,\omega+d\omega]} = \frac{\hbar\omega^3}{2\pi^2\varepsilon_0 c^3}\, d\omega. \tag{4.8}$$

Exercise 4.2
Calculate the vacuum fluctuations $\langle \hat{\boldsymbol{E}}^2(\boldsymbol{r}) \rangle_{[\omega,\omega+\mathrm{d}\omega]}$ of the electric field in infinite free space in an infinitesimal frequency interval by using the mode expansion (3.11).

The vacuum fluctuations of the electromagnetic field constitute the light that remains after we have tried to 'turn off' the light as far as we can. They are commonly referred to as *'virtual photons'*. A virtual photon carries an energy of $\frac{1}{2}\hbar\omega$. It is *virtual* in the following sense: it cannot be converted to other forms of energy. This distinguishes virtual photons from the real photons contained in an n-photon state $|n\rangle$. The latter can be absorbed, releasing their energy to the absorbing system. This crucial difference aside, virtual photons share most of the properties of their cousins, the *real photons*. For instance, they are reflected and transmitted when incident on surfaces and they affect charged or polarisable objects via the Lorentz force. As will be argued in the remainder of this chapter, virtual photons may be seen as responsible for a range of physically observable effects, such as the Lamb shift of atomic energy levels in free space, the spontaneous decay of an excited atom, the Casimir–Polder force between an atom and a metal or dielectric body, and the Casimir force between two such bodies.

4.2   Thermal fluctuations

Before considering the observable consequences of virtual photons in detail, it is worth pointing out that they are closely related to the much more tangible thermal photons. To see this, we assume that the electromagnetic field is not in its ground state, but in thermal equilibrium with finite temperature T. This *thermal state* of the electromagnetic field is characterised by a *density matrix*

$$\hat{\rho}_T = \frac{e^{-\hat{H}_\mathrm{F}/(k_\mathrm{B}T)}}{\mathrm{tr}\left[e^{-\hat{H}_\mathrm{F}/(k_\mathrm{B}T)}\right]}. \tag{4.9}$$

Let us first concentrate on a single mode of the electromagnetic field whose occupation at thermal equilibrium is governed by the respective density matrix

$$\hat{\rho}_T = \frac{e^{-(\hat{n}+1/2)\hbar\omega/(k_\mathrm{B}T)}}{\mathrm{tr}\left[e^{-(\hat{n}+1/2)\hbar\omega/(k_\mathrm{B}T)}\right]}. \tag{4.10}$$

The probability of finding n photons is given by the *Boltzmann distribution* [13]

$$
\begin{aligned}
p(n) &= \langle n|\hat{\rho}_T|n\rangle = \frac{\langle n|e^{-(\hat{n}+1/2)\hbar\omega/(k_\mathrm{B}T)}|n\rangle}{\sum_{k=0}^{\infty}\langle k|e^{-(\hat{n}+1/2)\hbar\omega/(k_\mathrm{B}T)}|k\rangle} \\
&= \frac{e^{-n\hbar\omega/(k_\mathrm{B}T)}}{\sum_{k=0}^{\infty} e^{-k\hbar\omega/(k_\mathrm{B}T)}} = e^{-n\hbar\omega/(k_\mathrm{B}T)}\left[1 - e^{-\hbar\omega/(k_\mathrm{B}T)}\right],
\end{aligned}
\tag{4.11}
$$

where we have used the geometric sum

$$\sum_{k=0}^{\infty} q^k = \frac{1}{1-q}. \tag{4.12}$$

Note that the vacuum energy does not influence this probability, as the respective factor $e^{-(1/2)\hbar\omega/(k_B T)}$ cancels. The average *thermal photon number* is then given by

$$n(\omega) = \langle \hat{n} \rangle = \sum_{n=0}^{\infty} \langle n | \hat{n}\hat{\rho}_T | n \rangle = \sum_{n=0}^{\infty} n p(n) = \frac{1}{e^{\hbar\omega/(k_B T)} - 1}. \tag{4.13}$$

It is governed by *Bose–Einstein statistics* [14, 15].

Exercise 4.3
Use the geometric sum (4.12) to calculate the average thermal photon number.

With this result, we can now find the spectral energy density of the electromagnetic field at temperature T

$$\left\langle \frac{\varepsilon_0}{2} \hat{E}^2(r) + \frac{1}{2\mu_0} \hat{B}^2(r) \right\rangle = \int_0^\infty d\omega \, \rho(\omega). \tag{4.14}$$

Using the multi-mode thermal density matrix (4.9), we find that each mode is independently governed by Bose–Einstein statistics. Terms $\hat{a}_j^\dagger \hat{a}_{j'}$ and $\hat{a}_j \hat{a}_{j'}^\dagger$ thus lead to contributions $n(\omega_j)$ and $n(\omega_j) + 1$, respectively. Following analogous steps as in the vacuum case, we arrive at a *spectral energy density*

$$\rho(\omega) = \frac{\hbar\omega^3}{\pi^2 c^3} \left[n(\omega) + \tfrac{1}{2} \right]. \tag{4.15}$$

Exercise 4.4
Use the normal mode expansions (3.11) and (3.12) to show that the spectral energy density of the electromagnetic field at temperature T is given by the above equation.

The energy density of the *thermal photons* alone reads

$$\rho_T(\omega) = \frac{\hbar\omega^3}{\pi^2 c^3} n(\omega) = \frac{\hbar\omega^3}{\pi^2 c^3} \frac{1}{e^{\hbar\omega/(k_B T)} - 1}. \tag{4.16}$$

This is the famous *Planck spectrum* for *blackbody radiation*. Its asymptotes are the *Rayleigh–Jeans law* [9, 16]

$$\rho_T(\omega) = \frac{k_B T \omega^2}{\pi^2 c^3} \tag{4.17}$$

for small frequencies $\hbar\omega \ll k_B T$ and the *Wien law* [17]

$$\rho_T(\omega) = \frac{\hbar\omega^3 e^{-\hbar\omega/(k_B T)}}{\pi^2 c^3} \tag{4.18}$$

in the limit of high frequencies $\hbar\omega \gg k_B T$. In his alternative derivation of the blackbody spectrum [19], Planck introduced the notion of *zero-point energy* for the first time.

We have thus seen that both virtual and thermal photons contribute to the spectral energy density of the electromagnetic field, where an inclusion of the latter leads to a factor $2n(\omega) + 1$, compare Eqs. (4.6) and (4.15). Similarly, virtual and thermal photons both give rise to fluctuations of the electromagnetic field. Generalising the above result (4.8) to finite temperature by including the factor $2n(\omega) + 1$, these fluctuations read

$$\langle \hat{E}^2(r) \rangle_{[\omega,\omega+d\omega]} = \frac{\hbar\omega^3}{\pi^2\varepsilon_0 c^3} \left[n(\omega) + \tfrac{1}{2} \right] d\omega. \qquad (4.19)$$

In the limit of high temperatures, the thermal photons dominate and we obtain

$$\langle \hat{E}^2(r) \rangle_{[\omega,\omega+d\omega]} = \frac{k_B T \omega^2}{\pi^2\varepsilon_0 c^3} \, d\omega \qquad (4.20)$$

for $k_B T \gg \hbar\omega$. As evident from the absence of the Planck constant $\hbar$ which governs all quantum fluctuations, the fluctuations of the electric field are a purely classical effect in this case.

§5. QUANTUM VACUUM EFFECTS IN FREE SPACE — WHAT CAN UNBOUNDED VIRTUAL PHOTONS DO TO MATTER?

As seen, virtual photons are an unexpected consequence of the particle nature of the electromagnetic field: the field in its vacuum state fluctuates, leading to a finite spectral energy density even in the absence of real photons. The virtual photons associated with the vacuum fluctuations carry one half the energy of the real photons. However, this energy cannot be absorbed by another system such as an atom or a surface. So how can we find evidence for the existence of virtual photons? In this section, we will consider the simplest case of virtual photons in free space.

5.1   Atom–field interactions

Let us first probe virtual photons by means of an *atom*: an electrically neutral system of particles (nucleus and electrons) with charges q_α, masses m_α, positions $\hat{r}_\alpha$ and canonically conjugate momenta $\hat{p}_\alpha$. The positions and momenta satisfy the *canonical commutation relations*

$$[\hat{r}_\alpha, \hat{r}_\beta] = [\hat{p}_\alpha, \hat{p}_\beta] = 0, \qquad [\hat{r}_\alpha, \hat{p}_\beta] = i\hbar\delta_{\alpha\beta}\mathbf{1}. \qquad (5.1)$$

In the absence of external electromagnetic fields, the internal dynamics of the particles is governed by an *atomic Hamiltonian*

$$\hat{H}_A = \sum_\alpha \frac{\hat{p}_\alpha^2}{2m_\alpha} + \sum_{\alpha<\beta} \frac{q_\alpha q_\beta}{4\pi\varepsilon_0 |\hat{r}_\alpha - \hat{r}_\beta|}. \qquad (5.2)$$

Here, the first term represents the kinetic energy of the particles while the second term is their *Coulomb interaction*. For an atom of given species, we can find the internal energy eigenstates $|n\rangle$ with energies E_n, so that the Hamiltonian can be expanded in the form

$$\hat{H}_A = \sum_n E_n |n\rangle\langle n|. \qquad (5.3)$$

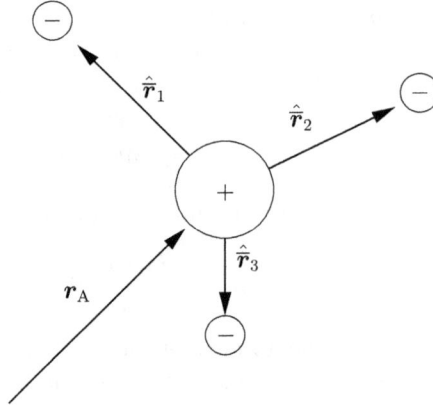

Figure 1.3: Atomic polarisation: the polarisation is non-vanishing along the lines connecting the nucleus at (approximately) the centre of mass r_A and the electrons at relative positions $\hat{\bar{r}}_\alpha$ ($\alpha = 1, 2, 3$). On these lines, it takes the values of the dipole moments correponding to the respective electrons.

We are interested in the effect of our quantum electromagnetic field on the atom. According to the *multipolar coupling scheme* [21,22], the atom–field coupling

$$\hat{H}_{AF} = -\int d^3r \, \hat{\boldsymbol{P}}_A(\boldsymbol{r}) \cdot \hat{\boldsymbol{E}}(\boldsymbol{r}) \tag{5.4}$$

is mediated via the *atomic polarisation*

$$\hat{\boldsymbol{P}}_A(\boldsymbol{r}) = \sum_\alpha q_\alpha \hat{\bar{\boldsymbol{r}}}_\alpha \int_0^1 d\sigma \, \delta(\boldsymbol{r} - \boldsymbol{r}_A - \sigma \hat{\bar{\boldsymbol{r}}}_\alpha). \tag{5.5}$$

The polarisation is illustrated in Fig. 1.3. It is defined in terms of the atomic *centre of mass* r_A and *relative coordinates* $\hat{\bar{\boldsymbol{r}}}_\alpha = \hat{\boldsymbol{r}}_\alpha - \boldsymbol{r}_A$; and it takes non-zero values along the lines connecting the nucleus and the electrons. In most cases of practical interest, the electric field is almost constant on the length scale corresponding to the size of the atom. We can then apply the multipole expansion giving the multipolar coupling scheme its name: to leading order in the relative coordinates $\hat{\bar{\boldsymbol{r}}}_\alpha$, the atom–field coupling is approximated by

$$\hat{H}_{AF} = -\hat{\boldsymbol{d}} \cdot \hat{\boldsymbol{E}}(\boldsymbol{r}_A) \tag{5.6}$$

where

$$\hat{\boldsymbol{d}} = \sum_{\alpha \in A} q_\alpha \hat{\bar{\boldsymbol{r}}}_\alpha = \sum_{\alpha \in A} q_\alpha \hat{\boldsymbol{r}}_\alpha \tag{5.7}$$

is the atomic *electric dipole moment*. For obvious reasons, this is known as the *electric dipole approximation*.

Exercise 5.1
Expand the multipolar coupling Hamiltonian (5.4) to leading order in the relative coordinates $\hat{\boldsymbol{r}}_\alpha$ and hence obtain the electric-dipole interaction (5.6).

By virtue of the electric-dipole interaction, an electromagnetic field can induce internal atomic transitions from state $|n\rangle$ to state $|m\rangle$. The probabilities for such transitions are governed by the dipole matrix elements $\boldsymbol{d}_{mn} = \langle m|\hat{\boldsymbol{d}}|n\rangle$. Possible dipole transitions from a given state $|n\rangle$ to different internal states are constrained by the *Thomas–Reiche–Kuhn sum rule* [23–25]:

$$\frac{1}{2\hbar}\sum_m \omega_{mn}(\boldsymbol{d}_{nm} \otimes \boldsymbol{d}_{mn} + \boldsymbol{d}_{mn} \otimes \boldsymbol{d}_{nm}) = \sum_\alpha \frac{q_\alpha^2}{2m_\alpha}\,\boldsymbol{1}. \tag{5.8}$$

In particular, this rule shows that the specific combination of dipole matrix elements and transition frequencies on the left hand side does not depend on the state $|n\rangle$.

Exercise 5.2
Calculate the commutator $[\hat{\boldsymbol{r}}_\alpha, \hat{H}_A]$ in two alternative ways, by writing the atomic Hamiltonian in the forms (5.2) and (5.3). Hence show that

$$\sum_{\alpha\in A} \frac{q_\alpha}{m_\alpha}\,\langle m|\hat{\boldsymbol{p}}_\alpha|n\rangle = i\omega_{mn}\boldsymbol{d}_{mn}.$$

Use this result to prove the Thomas–Reiche–Kuhn sum rule.

In the multipolar coupling scheme, the atomic polarisation not only interacts with the electric field, but also with itself. This self-interaction is given by [21, 22]

$$\hat{H}_{\mathrm{P}} = \frac{1}{2\varepsilon_0} \int \mathrm{d}^3 r\, \hat{\boldsymbol{P}}_{\mathrm{A}}^\perp(\boldsymbol{r}) \cdot \hat{\boldsymbol{P}}_{\mathrm{A}}^\perp(\boldsymbol{r}) \tag{5.9}$$

where

$$\boldsymbol{a}^\perp(\boldsymbol{r}) = \int \mathrm{d}^3 r'\, \boldsymbol{\delta}^\perp(\boldsymbol{r} - \boldsymbol{r}') \cdot \boldsymbol{a}(\boldsymbol{r}') \tag{5.10}$$

is the transverse part of a vector field, recall definition (2.30) of the transverse delta function. In electric-dipole approximation, the polarisation self-interaction simplifies to

$$\hat{H}_{\mathrm{P}} = \frac{\mu_0}{6\pi^2 c}\,\hat{\boldsymbol{d}}^2 \int_0^\infty \mathrm{d}\omega\,\omega^2. \tag{5.11}$$

It is evident that this self-interaction is infinite, as the frequency integral diverges.

Exercise 5.3
Derive the polarisation self-interaction in electric dipole approximation by expanding to leading order in the relative particle coordinates and using the explicit form of the transverse delta function ($\omega = ck$).

The total Hamiltonian describing the quantum electromagnetic field and a single atom is given by

$$\hat{H} = \hat{H}_F + \hat{H}_A + \hat{H}_{AF} + \hat{H}_P$$
$$= \sum_j \left(\hat{n}_j + \tfrac{1}{2}\right)\hbar\omega_j + \sum_n E_n |n\rangle\langle n| - \hat{\boldsymbol{d}} \cdot \hat{\boldsymbol{E}}(\boldsymbol{r}_A) + \frac{\mu_0}{6\pi^2 c}\,\hat{\boldsymbol{d}}^2 \int_0^\infty d\omega\,\omega^2. \quad (5.12)$$

5.2   Lamb shift

As first discovered by *Lamb* and *Retherford* [20], the interaction of an atom with the quantum electromagnetic field leads to an observable shift of the atomic energy levels which may be understood in terms of virtual photons. To see this, we start from an atom in an energy eigenstate $|n\rangle$ and consider its interaction with the vacuum electromagnetic field in free space $|0\rangle$. The resulting atomic energy shift can be calculated using second order perturbation theory,

$$\Delta_2 E_\psi = \sum_{\phi\neq\psi} \frac{\langle\psi|\hat{H}_{AF}|\phi\rangle\langle\phi|\hat{H}_{AF}|\psi\rangle}{E_\psi - E_\phi}. \quad (5.13)$$

The electric dipole term above couples our initial state $|\psi\rangle = |n\rangle|0\rangle$ to intermediate states $|\phi\rangle = |m\rangle|1_{\boldsymbol{k}\lambda}\rangle$ containing one photon of wave vector $\boldsymbol{k}$ and polarisation λ. The energy denominator thus reads $E_\psi - E_\phi = E_n - (E_m + \hbar\omega_k) = -\hbar(\omega_{mn} + \omega_k)$ with $\omega_{mn} = (E_m - E_n)/\hbar$. The energy shift thus takes the concrete form

$$\Delta_2 E_n = -\sum_m \int d^3k \sum_{\lambda=1,2} \frac{\langle 0|\langle n|\hat{\boldsymbol{d}}\cdot\hat{\boldsymbol{E}}(\boldsymbol{r}_A)|m\rangle|1_{\boldsymbol{k}\lambda}\rangle\langle 1_{\boldsymbol{k}\lambda}|\langle m|\hat{\boldsymbol{d}}\cdot\hat{\boldsymbol{E}}(\boldsymbol{r}_A)|n\rangle|0\rangle}{\hbar(\omega_{mn} + \omega_k)}. \quad (5.14)$$

Read from right to left, it can be graphically represented by the *Feynman diagram* shown in Fig. 1.4: The atom in state $|n\rangle$ emits a photon $|1_{\boldsymbol{k}\lambda}\rangle$ while making a transition to state $|m\rangle$. It then reabsorbs this photon upon returning to its original state $|n\rangle$. In particular when the atom is in its ground state $(n = 0)$, it is evident that the transitions and the emitted photon must be virtual. The atom is originally in its lowest energy eigenstate, so that it does not have the energy required for emitting a real photon let alone making a transition to an excited state at the same time. The energy for this process is borrowed from the quantum vacuum, to which it must be returned in the end.

The matrix element for virtual-photon emission can be found by using the field expansion (3.11) and the bosonic commutation relations (3.24),

$$\langle 1_{\boldsymbol{k}\lambda}|\langle m|\hat{\boldsymbol{d}}\cdot\hat{\boldsymbol{E}}(\boldsymbol{r}_A)|n\rangle|0\rangle = \boldsymbol{d}_{mn}\cdot\boldsymbol{E}_{\boldsymbol{k}\lambda}^*(\boldsymbol{r}_A). \quad (5.15)$$

Here, $\boldsymbol{d}_{mn} = \langle m|\hat{\boldsymbol{d}}|n\rangle$ is the electric dipole matrix element for internal atomic transitions. Similarly, the reabsorption of the virtual photon is described by

$$\langle 0|\langle n|\hat{\boldsymbol{d}}\cdot\hat{\boldsymbol{E}}(\boldsymbol{r}_A)|m\rangle|1_{\boldsymbol{k}\lambda}\rangle = \boldsymbol{d}_{nm}\cdot\boldsymbol{E}_{\boldsymbol{k}\lambda}(\boldsymbol{r}_A). \quad (5.16)$$

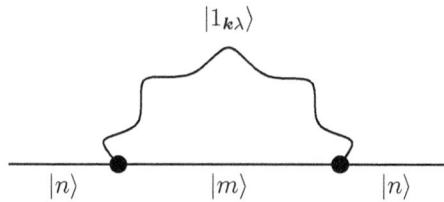

Figure 1.4: Feynman diagram for the Lamb shift: an atom in state $|n\rangle$ is presented by the solid line at the far right of the diagram. It interacts with the quantum electric field by emitting a virtual photon $|1_{k\lambda}\rangle$ of wave vector $\boldsymbol{k}$ and polarisation λ (wavy line) while making a transition to an intermediate state $|m\rangle$ (central solid line). Finally, it interacts a second time by reabsorbing the virtual photon and returning to its original state (solid line on the left).

With these results, the energy shift takes the form

$$\Delta_2 E_n = -\sum_m \int \mathrm{d}^3 k \sum_{\lambda=1,2} \frac{\boldsymbol{d}_{nm} \cdot \boldsymbol{E}_{k\lambda}(\boldsymbol{r}_A) \otimes \boldsymbol{E}^*_{k\lambda}(\boldsymbol{r}_A) \cdot \boldsymbol{d}_{mn}}{\hbar(\omega_{mn} + \omega_k)}. \tag{5.17}$$

Next, we use the normal modes in infinite free space (3.36) to find

$$\Delta_2 E_n = -\frac{1}{16\pi^3 \varepsilon_0} \sum_m \int \mathrm{d}^3 k\, \omega_k \sum_{\lambda=1,2} \frac{\boldsymbol{d}_{nm} \cdot \boldsymbol{e}_{k\lambda} \otimes \boldsymbol{e}_{k\lambda} \cdot \boldsymbol{d}_{mn}}{\omega_{mn} + \omega_k}. \tag{5.18}$$

To evaluate the sum over the polarisation unit vectors, we make use of their completeness (2.15) and introduce spherical coordinates $\int \mathrm{d}^3 k = \int_0^\infty \mathrm{d}k\, k^2 \int \mathrm{d}\Omega$:

$$\int \mathrm{d}\Omega \sum_{\lambda=1,2} \boldsymbol{e}_{k\lambda} \otimes \boldsymbol{e}_{k\lambda} = \frac{8\pi}{3}\, \mathbf{1}. \tag{5.19}$$

Exercise 5.4
Carry out the angular integral $\int \mathrm{d}\Omega$ to obtain the above sum over polarisation unit vectors.

Changing the integration variable to $\omega = ck$, we obtain

$$\Delta_2 E_n = -\frac{\mu_0}{6\pi^2 c} \sum_m |\boldsymbol{d}_{nm}|^2 \int_0^\infty \mathrm{d}\omega\, \frac{\omega^3}{\omega_{mn} + \omega}. \tag{5.20}$$

So far, we have only considered the second-order energy shift due to the atom–field coupling Hamiltonian $\hat{H}_{AF}$. It is proportional to the square of the dipole matrix elements which are a measure of the atom–field coupling strength. To obtain a consistent perturbative approximation, we have to include the first-order energy shift due to $\hat{H}_P$ which is also quadratic in the dipole operator:

$$\Delta_1 E_n = \langle n|\hat{H}_P|n\rangle = \frac{\mu_0}{6\pi^2 c} \sum_m |\boldsymbol{d}_{nm}|^2 \int_0^\infty \mathrm{d}\omega\, \omega^2. \tag{5.21}$$

Combining the two contributions, we find

$$\Delta E_n = \Delta_1 E_n + \Delta_2 E_n = \frac{\mu_0}{6\pi^2 c} \sum_m \omega_{mn} |\boldsymbol{d}_{nm}|^2 \int_0^\infty d\omega \, \frac{\omega^2}{\omega_{mn} + \omega} . \tag{5.22}$$

The total energy shift is strongly divergent. This has two main causes which are generic to vacuum-QED effects: First, our result exhibits an *ultraviolet divergence* due to our assumption that virtual photons of arbitrarily high frequencies contribute. Secondly, it contains *self energies*: physically unobservable, constant energy shifts. We will identify and address these two types of divergences one after another by applying two procedures called regularisation and renormalisation. They are essential tools used in quantum electrodynamics and other quantum field theories for obtaining finite, meaningful results. We will encounter more examples when applying them in Section 7 to calculate a finite Casimir force.

Regularisation. We will first use *regularisation* to address the ultraviolet divergence. As noted above, the frequency integral diverges because it has no upper bound: the atom interacts with virtual photons of all frequencies in the same way. However, *Bethe* has pointed out that the non-relativistic quantum electrodynamics used in our derivation only applies up to energies smaller than the electron rest energy mc^2 [26]. At such energies, virtual electron–positron pairs will begin to contribute to the quantum vacuum, fundamentally altering the atom–field interaction and requiring a relativistic quantum electrodynamics description which is well beyond the scope of this book. Conceding our ignorance of how virtual photons of such high energies might interact with an atom, we introduce a *cut-off regularisation*, limiting the frequency integral to the range $\hbar\omega \le mc^2$ where our non-relativistic theory holds,

$$\Delta E_n = \frac{\mu_0}{6\pi^2 c} \sum_m \omega_{mn} |\boldsymbol{d}_{nm}|^2 \int_0^{mc^2/\hbar} d\omega \, \frac{\omega^2}{\omega_{mn} + \omega} . \tag{5.23}$$

The result is a very large, but finite energy shift.

Renormalisation. The energy shift contains two large, but physically unobservable self-energies which we are going to remove by means of *renormalisation*: by redefining the zero of the energy scale.* The first unobservable contribution to the energy shift is the state-independent *self-energy of the atom*. Atomic eigenenergies can only be measured relative to each other, so that a constant shift applied to all eigenstates of an atom is irrelevant. A tool for identifying the state-independent contribution is the Thomas–Reiche–Kuhn sum rule (5.8). Taking its trace, we find

$$\frac{1}{\hbar} \sum_m \omega_{mn} |\boldsymbol{d}_{nm}|^2 = \sum_\alpha \frac{3q_\alpha^2}{2m_\alpha} . \tag{5.24}$$

---

*Note that this simple notion of renormalisation as a shift of the zero on the energy scale differs from the use of the term in many quantum field theories. In the latter context, it commonly refers to a self-consistent removal of divergences by distinguishing between bare and effective quantities and identifying the latter with the finite observables.

Upon substituting this result, the energy shift can be written in the form

$$\Delta E_n = -\frac{\mu_0}{6\pi^2 c}\sum_m \omega_{mn}^2 |\boldsymbol{d}_{nm}|^2 \int_0^{mc^2/\hbar} d\omega\, \frac{\omega}{\omega_{mn}+\omega}$$
$$+\frac{\hbar\mu_0}{4\pi^2 c}\sum_\alpha \frac{q_\alpha^2}{m_\alpha}\int_0^{mc^2/\hbar} d\omega\,\omega. \tag{5.25}$$

Exercise 5.5
Use the Thomas–Reiche–Kuhn sum rule to separate the energy shift (5.23) into its state-dependent and independent parts as given by the above equation.

We can now perform our first renormalisation of the energy shift by discarding the state-independent self-energy of the atom:

$$\Delta E_n \mapsto -\frac{\mu_0}{6\pi^2 c}\sum_m \omega_{mn}^2 |\boldsymbol{d}_{nm}|^2 \int_0^{mc^2/\hbar} d\omega\, \frac{\omega}{\omega_{mn}+\omega}. \tag{5.26}$$

The second self-energy contained in our energy shift is that of the individual electrons. It has its origins in the preparation of the initial setup: our starting point is an atom in free space which interacts with the virtual photons of the quantum vacuum. In order to prepare such a situation, one would have to first place the nucleus and all the electrons contained in the atom into the quantum vacuum. Due to the ubiquitous nature of virtual photons, they will interact with the individual atomic particles.

Our innocent-looking starting point of an atom in free space has led to an energy shift which contains both the individual *electron self-energies* and the Lamb shift that we are interested in. The latter is the effect of virtual photons on the atom as a whole. Luckily, the former can be identified by considering the energy shift of an unbound collection of charged particles. Free electrons have a continuous spectrum so that the energy separation ω_{mn} between neighbouring states become arbitrarily small. The energy shift of an unbound system of charged particles can therefore be obtained from the above result (5.26) by performing the limit $\omega_{mn}/\omega \to 0$:

$$\Delta E_{n,\text{electron}} = -\frac{\mu_0}{6\pi^2 c}\sum_m \omega_{mn}^2 |\boldsymbol{d}_{nm}|^2 \int_0^{mc^2/\hbar} d\omega. \tag{5.27}$$

Our energy shift hence reads

$$\Delta E_n = \frac{\mu_0}{6\pi^2 c}\sum_m \omega_{mn}^3 |\boldsymbol{d}_{nm}|^2 \int_0^{mc^2/\hbar} d\omega\, \frac{1}{\omega_{mn}+\omega} + \Delta E_{n,\text{electron}}. \tag{5.28}$$

We apply a second renormalisation by discarding the free-electron self energy:

$$\Delta E_n \mapsto \frac{\mu_0}{6\pi^2 c}\sum_m \omega_{mn}^3 |\boldsymbol{d}_{nm}|^2 \int_0^{mc^2/\hbar} d\omega\, \frac{1}{\omega_{mn}+\omega}. \tag{5.29}$$

(a) Absorption   (b) Stimulated emission   (c) Spontaneous emission

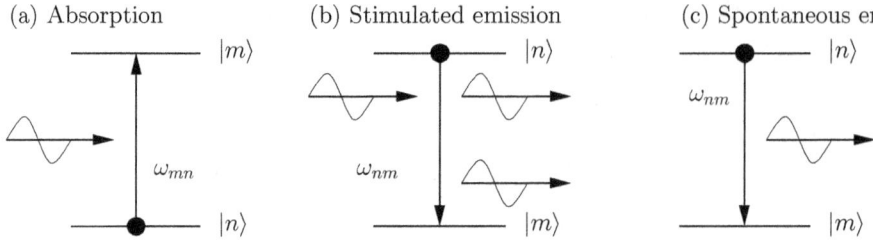

Figure 1.5: Basic atom–photon interactions. (a) Absorption: an atom absorbs a thermal photon while making a transition to a higher energy eigenstate. (b) Stimulated emission: a thermal photon causes an atom to decay to a lower state while emitting an additional, identical photon. (c) Spontaneous emission: an excited atom spontaneously decays to a lower state while emitting a photon.

Carrying out the frequency integration, we find Bethe's famous result for the *Lamb shift* [26]:

$$\Delta E_n = \frac{\mu_0}{6\pi^2 c} \sum_m \omega_{mn}^3 |d_{nm}|^2 \ln\left(\frac{mc^2}{\hbar\omega_{mn}}\right). \tag{5.30}$$

Note that a regularisation via a relativistic high-frequency cut-off is not needed in the presence of matter with a realistic, frequency-dependent response. For instance, material mirrors become transparent at high frequencies and cease to interact with virtual photons long before the breakdown of non-relativistic quantum electrodynamics.

   The coupling of an atom to the quantum vacuum thus leads to a state-dependent shift of the atomic energy levels. This shift lifts the degeneracy between the $^2S_{1/2}$ and $^2P_{1/2}$ states. The perturbative calculation using Feynman diagrams has allowed us to interpret the Lamb shift as a consequence of virtual photons. Virtual photons hence affect individual atoms in free space by changing their internal level structure.

5.3   Spontaneous decay

In addition to changing the atomic energy levels, virtual photons also influence the population of these levels. To see this, let us first recall what real photons can do to an atom in an energy eigenstate $|n\rangle$. According to *Einstein* [27], thermal radiation with spectral energy density $\rho_T(\omega)$ can be *absorbed* by an atom which makes a transition to a higher-lying state $|m\rangle$, see Fig. 1.5(a). The corresponding transition rate is given by $B_{nm}\rho_T(\omega_{mn})$ where

$$B_{nm} = \frac{\pi|d_{nm}|^2}{3\varepsilon_0\hbar^2} \tag{5.31}$$

is the *Einstein B-coefficient*. Alternatively, thermal radiation can stimulate decay to a lower-lying state with a rate $B_{nm}\rho_T(\omega_{nm})$. The process is analogous to *stimulated emission* in a laser, where atoms coherently add photons to the laser signal while making transitions to lower lying energy eigenstates. As illustrated in Fig. 1.5(b), a thermal photon causes the atom to decay while emitting an additional, identical photon.

Virtual photons obviously cannot be absorbed by an atom. Stimulated emission on the other hand only requires the presence of a photon which is not absorbed in the process. Can virtual photons cause an atom to decay to a lower energy eigenstate in the same way then? Yes, they can—and the corresponding process is *spontaneous decay*: the atom decays to a lower energy eigenstate while emitting a real photon, see Fig. 1.5(c). The spontaneous *decay rate* is given by the *Einstein A-coefficient*

$$A_{nm} = \frac{\omega_{nm}^3 |\boldsymbol{d}_{nm}|^2}{3\pi\varepsilon_0\hbar c^3}. \tag{5.32}$$

In a naïve first attempt to derive this rate, let us assume that spontaneous decay is literally decay stimulated by virtual photons. Replacing the thermal energy density $\rho_T(\omega)$ with the energy density (4.6) of the quantum vacuum, we find

$$B_{nm}\rho_0(\omega_{nm}) = \frac{\omega_{nm}^3 |\boldsymbol{d}_{nm}|^2}{6\pi\varepsilon_0\hbar c^3}. \tag{5.33}$$

This is only half the correct Einstein A-coefficient. We conclude that virtual photons possibly contribute to the spontaneous decay of an atom, but there must be an additional process responsible for the other half.

To find the complete answer, we need to solve the coupled atom–field dynamics where initially, the atom is in an excited energy eigenstate $|n\rangle$ and the field is in its vacuum state $|0\rangle$. For the weak atom–field interaction present in free space, one can then apply the *Markov approximation* [28] which assumes that the behaviour of atom and field only depends on their interaction at a given instant. Memory effects due to their coupled past evolution are thus discarded. One finds that the probabilities $p_n(t)$ for the atom to be in state $|n\rangle$ at time t are governed by the equations

$$\dot{p}_n(t) = -\sum_{m<n} A_{nm}p_n(t) + \sum_{m>n} A_{mn}p_m(t) \tag{5.34}$$

where the rates are indeed the full Einstein A-coefficients.

The calculation reveals that two processes are potentially responsible for spontaneous decay [29]. First, as described above, the *vacuum fluctuations* or virtual photons disturb the atom, causing it to decay. Secondly, the fluctuations of the atom's electric dipole moment create an electromagnetic field which then acts back on the atom. This second contribution to spontaneous decay is known as *radiation reaction*. It contributes the missing half to the Einstein A-coefficient. However, the precise answer as to how much vacuum fluctuations versus radiation reaction each contribute to the decay rate sensitively depends on the chosen ordering of the atomic and field operators in the interaction Hamiltonian (5.6). As these atom and field operators commute at equal times, we can equivalently express the interaction

(a)   (b)

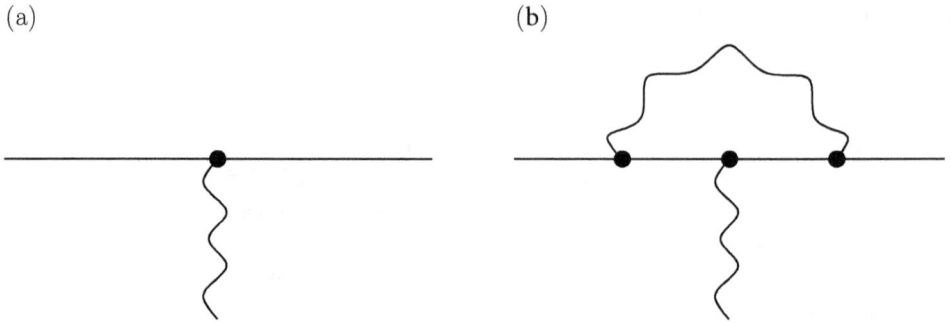

Figure 1.6: Feynman diagrams relevant for the modification of the electron magnetic moment. (a) An electron, as represented by the right solid line interacts with an applied magnetic field (wavy line) and emerges in a new state (left solid line). (b) An electron incoming from the right emits a virtual photon (curved wavy line); it then interacts with an applied magnetic field (straight wavy line) and finally reabsorbs the virtual photon.

in *normal ordering* (with field annihilation operators to the right and creation operators to the left),

$$\hat{H}_{\mathrm{AF}} = -\sum_j \left[\hat{\boldsymbol{d}} \cdot \boldsymbol{E}_j(\boldsymbol{r}_{\mathrm{A}})\hat{a}_j + \hat{a}_j^\dagger \boldsymbol{E}_j^*(\boldsymbol{r}_{\mathrm{A}}) \cdot \hat{\boldsymbol{d}} \right], \tag{5.35}$$

anti-normal ordering (with annihilation operators to the left and creation operators to the right),

$$\hat{H}_{\mathrm{AF}} = -\sum_j \left[\hat{a}_j \boldsymbol{E}_j(\boldsymbol{r}_{\mathrm{A}}) \cdot \hat{\boldsymbol{d}} + \hat{\boldsymbol{d}} \cdot \boldsymbol{E}_j^*(\boldsymbol{r}_{\mathrm{A}})\hat{a}_j^\dagger \right], \tag{5.36}$$

or an arbitrary interpolation between these extremes. Different orderings must lead to the same result for the total decay rate, but suggest alternative interpretations in terms of either virtual photons, radiation reaction or both.

5.4   Electron magnetic moment

As mentioned, virtual photons do not only affect atoms as a whole, but also the individual electrons contained inside them. One prominent consequence of this interaction is a modification of the *electron magnetic moment*. The electron spin $\hat{\boldsymbol{s}}$ induces a magnetic moment

$$\hat{\boldsymbol{m}} = \frac{g\mu_{\mathrm{B}}}{\hbar}\,\hat{\boldsymbol{s}}, \tag{5.37}$$

where $\mu_{\mathrm{B}} = e\hbar/(2m)$ is the Bohr magneton. The *electron g-factor* takes the value $g = 2$ in the absence of virtual photons. In non-relativistic quantum mechanics, the magnetic moment couples to any magnetic field via the *Pauli interaction*

$$\hat{H}_{\mathrm{eF}} = -\hat{\boldsymbol{m}} \cdot \hat{\boldsymbol{B}}(\boldsymbol{r}_{\mathrm{e}}). \tag{5.38}$$

This can be graphically represented by the Feynman diagram of Fig. 1.6(a): the electron (represented by the incoming straight line from the right) interacts with

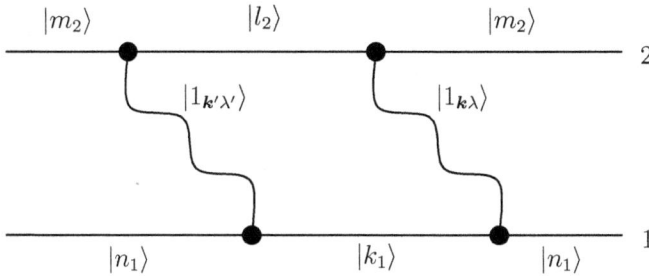

Figure 1.7: Feynman diagram contributing to the van der Waals potential of two atoms: Atom 1 in initial state n_1, represented by the lower solid line incoming from the right, emits a virtual photon $|1_{k\lambda}\rangle$ with wave vector k and polarisation λ (right-hand wavy line) making a transition to an intermediate state k_1. The photon is subsequently absorbed by atom 2 (upper solid line) upon making a transition $|m_2\rangle \rightarrow |l_2\rangle$. The atoms then exchange a second virtual photon $|1_{k'\lambda'}\rangle$ and return to their respective initial states.

the magnetic field (depicted by the wavy line) and emerges from this interaction in a new state (outgoing straight line towards the left).

As predicted by *Schwinger* [30] and first measured by Kusch and Foley [31], virtual photons lead to a modification of the Pauli interaction. The leading process responsible is depicted in Fig. 1.6(b): the electron emits a virtual photon (represented by the curved wavy line), interacts with the magnetic field and then reabsorbs the virtual photon. The effect of this process can be calculated by means of perturbation theory in close analogy to our Lamb shift evaluation. The resulting modification of the effective Pauli interaction can be interpreted as a change in the electron g-factor or a change in its magnetic moment [32].

5.5   Van der Waals potential

So far, we have considered the emission and reabsorption of virtual photons by a single object, be it an atom or an electron. As we found, they change the internal structure and properties of such an object. To make the virtual-photon concept more tangible, we next show that virtual photons can even propagate from one atom to another.

We consider two atoms 1 and 2 in their respective internal-energy eigenstates $|n_1\rangle$ and $|m_2\rangle$ in free space. Each of these atoms will interact with the electromagnetic field via an electric-dipole coupling of the type (5.6), allowing it to emit and absorb virtual photons while changing their internal state. The fourth-order energy shift arising from these interactions contains processes where photons emitted by one atom are absorbed by the other. One possible sequence of events is shown in the Feynman diagram of Fig. 1.7: atom 1 emits a virtual photon $|1_{k\lambda}\rangle$ making a transition from its original state $|n_1\rangle$ to an intermediate state $|k_1\rangle$. Next, atom 2 absorbs the same photon and makes a transition from its original state $|m_2\rangle$ to a new state $|l_2\rangle$.

Subsequently, atom 1 emits a second virtual photon $|1_{k'\lambda'}\rangle$ while returning to its original state. The photon is again absorbed by atom 2 which also makes a transition to its original state. As in the case of the Lamb shift, the virtual-photon emission for two ground state atoms is virtual: the virtual photons can only exist as intermediate states and they have to be returned to the quantum vacuum in the end.

Alternative sequences of virtual two-photon exchanges between the two atoms are also possible and described by different Feynman diagrams. To count them, we note that atom 1 can be involved in any 2 of the 4 different emission/absorption events, leaving atom 2 with the other two. Furthermore, there are 2 possibilities regarding the pairing of the four interactions by the two virtual photons. This leaves a total of $\binom{4}{2} \times 2 = 12$ distinct Feynman diagrams in which two virtual photons are exchanged between the two atoms [33]. Summing up all these diagrams, we obtain an energy shift $\Delta_4 E(r_{12})$ which depends on the separation of the two atoms. This is the attractive *van der Waals potential* $U(r_{12})$ [34] as first calculated by Casimir and Polder [35]. It will be discussed in detail in Section 13.

Hence, we have seen that virtual photons in free space couple to charged particles, leading to a number of fundamental effects: the Lamb shift and spontaneous decay of an isolated atom, a modification of the electron magnetic moment and the van der Waals interaction between two atoms.

With the exception of spontaneous decay, the processes responsible for these effects can be visualised by Feynman diagrams. In this picture, virtual photons are represented by internal wavy lines which can be emitted and absorbed by one or several objects. These events do not have to obey energy conservation, but all virtual photons have to be reabsorbed eventually.

§6. QUANTUM VACUUM EFFECTS NEAR BOUNDARIES — CAN MIRRORS CHANGE THE EFFECT OF VIRTUAL PHOTONS ON MATTER?

As seen in Section 2, the normal modes of the electromagnetic field strongly depend on the boundary conditions set by material surfaces. The boundaries determine whether the modes are propagating waves (as in free space) or standing waves (for instance inside a cuboid cavity). They dictate whether the wave vectors are continuous (as in unbounded free space) or discrete (as for a finite free-space region or inside a cuboid cavity). Boundary effects are most pronounced in the vicinity of a surface. For instance, a perfectly conducting mirror enforces the parallel component of the electric field to vanish for each mode.

In other words, the environment dictates the shape of the virtual and real photons which can exist. As seen in the previous section, virtual photons interact with particles such as atoms or electrons, giving rise to a number of fundamental effects such as the Lamb shift, spontaneous decay, the electron magnetic moment and the van der Waals interaction. If the presence of boundaries changes the spatial profile of these virtual photons, then it must also have an effect on the resulting quantum-vacuum effects.

6.1 Casimir–Polder potential

We have derived the Lamb shift in free space as the second-order energy shift due to the atom–photon interaction. The result (5.17) was given in terms of the normal modes in free space. The effect of boundaries on this shift can then considered quite easily by simply replacing the free-space modes by the respective bounded modes,

$$\Delta_2 E_n = -\sum_m \sum_j \frac{d_{nm} \cdot E_j(r_A) \otimes E_j^*(r_A) \cdot d_{mn}}{\hbar(\omega_{mn} + \omega_j)}. \tag{6.1}$$

The effect of this innocent-looking change on the energy shift is very profound: The mode sums in the presence of a boundary sensitively depend on the atomic position. As a consequence, the energy shift depends on the atomic position as well. This leads to two observable effects: First, it means that the transition frequencies of an atom in a bounded environment differ from their free-space value. This was experimentally confirmed by Heinzen and Feld who performed spectroscopic measurements of atoms placed between two spherical mirrors [37].

Secondly, a position-dependent energy can be interpreted as a potential: the famous *Casimir–Polder potential* [35] as first measured by Hinds and co-workers [36]. It will be considered in detail in Section 14. In our virtual-photon picture, the associated Casimir–Polder force arises as follows. The atom emits a virtual photon, which is reflected off the present boundaries and then reabsorbed (recall Fig. 1.4). As a consequence of the reflection, it exerts a force on the atom. It turns out that this force is attractive for a ground-state atom interacting with a perfectly conducting wall.

6.2 Purcell effect

In our simplified picture, the spontaneous decay of an excited atom in state $|n\rangle$ to a lower-lying level $|m\rangle$ is stimulated by virtual photons. The decay rate is proportional to the vacuum energy density $\rho_0(r_A, \omega_{nm})$ of the electromagnetic field at the atomic position and the respective atomic transition frequency. Again, boundaries influence the available modes and hence their energy-density in a fundamental way. The resulting environment-dependence of spontaneous decay has been named after its discoverer *Purcell* [38].

The *Purcell effect* manifests itself in two ways. First, boundaries forming a confined space such as our cuboid cavity restrict the frequencies of allowed modes to certain discrete values. Depending on whether or not the atomic transition frequency matches the frequency of such a mode, the decay rate can be strongly enhanced [39] or suppressed [40].

Secondly, the decay rate of an atom in the presence of boundaries is also a local probe of the spatial profile of the vacuum field fluctuations. In close analogy to the Lamb shift, the decay rate becomes a position-dependent quantity. Depending on where an ensemble of excited atoms are situated relative to the present boundaries, they may thus release their energy sooner or later. Note however, that the spontaneous decay rate describes the average dynamics of such an ensemble. The exact timing of spontaneous emission for an individual atom is inherently random.

6.3   Electron magnetic moment

The electron magnetic dipole moment is influenced by boundaries in close analogy with the Lamb shift. Recalling the Feynman diagram from Fig. 1.6(b), the electron emits a virtual photon and interacts with an external magnetic field before reabsorbing the photon. In the presence of boundaries, the virtual photon can be reflected off a surface prior to its reabsorption. This leads to a position-dependent magnetic moment of the electron [41].

6.4   Van der Waals potential

As shown in Fig. 1.7 the van der Waals potential between two ground-state atoms is due to the exchange of two virtual photons between them. A nearby mirror can reflect either one or even both of these atoms. This leads to a modification of the interaction potential, as each atom interacts not only with the other atom directly, but in addition with the other atom's mirror image, Fig. 1.8(a). The effect of the plate depends on the alignment of the atoms relative to the plate. The potential is maximally enhanced for perpendicular alignment, Fig. 1.8(b). In the limit of one atom approaching the plate, the enhancement factor relative to free space is 8/3. For parallel alignment, the mirror reduces the interaction to 2/3 of its free-space value in the on-surface limit, Fig. 1.8(c).

§7. THE CASIMIR FORCE
— CAN VIRTUAL PHOTONS EXERT A PRESSURE?

As seen in the previous two sections, virtual photons interact with particles such as atoms or electrons. By calculating the respective energies, we have shown how virtual photons can change the internal structure of the particles and even introduce an interaction between them. But what are the effects of virtual photons on macroscopic objects? To answer this question, we need to leave free-space QED behind and instead consider virtual photons confined between material boundaries. The simplest such confinement is the cuboid cavity of perfectly conducting plates introduced in Section 2. We are going to show that the confined virtual photons lead to an observable macroscopic force between the confining plates, the famous Casimir force [42].

7.1   Perfectly conducting plates

Following Casimir's seminal work, we will derive the Casimir force from the vacuum energy of the electromagnetic field. As shown in Section 4, this energy is the sum (4.1) over the vacuum energies of all possible modes or virtual photons. In the

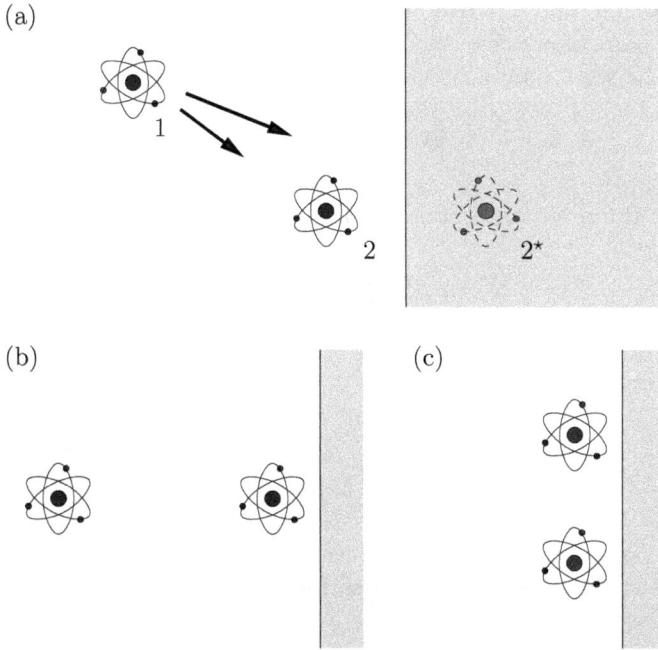

Figure 1.8: Mirror-induced modification of the van der Waals potential. (a) Atom 1 interacts with atom 2, leading to the free-space van der Waals interaction. A nearby mirror leads to an image 2* of atom 2 which also interacts with atom 1. This additional interaction results in a deviation of the interatomic potential from its free-space value which depends on the alignment of the atoms with respect to the mirror. (b) Perpendicular alignment: the line connecting the atoms is perpendicular to the mirror surface; the van der Waals potential is enhanced. (c) Parallel alignment: the line connecting the atoms is parallel to the mirror; the potential is reduced.

case of the cuboid cavity, the frequency of an individual mode is given by (2.36), so that the associated vacuum energy reads

$$E_{n\lambda} = \tfrac{1}{2}\hbar\omega_n = \frac{\hbar c}{2}\sqrt{\left(\frac{\pi n_x}{L_x}\right)^2 + \left(\frac{\pi n_y}{L_y}\right)^2 + \left(\frac{\pi n_z}{L_z}\right)^2}. \tag{7.1}$$

Summing over all modes as characterised by their frequencies and polarisations, we find a total vacuum energy

$$E_0 = \frac{\hbar c}{2} \sum_{n\in\mathbb{N}^3} \sum_{\lambda} \sqrt{\left(\frac{\pi n_x}{L_x}\right)^2 + \left(\frac{\pi n_y}{L_y}\right)^2 + \left(\frac{\pi n_z}{L_z}\right)^2}. \tag{7.2}$$

We want to use this energy to derive the Casimir force between two perfectly conducting plates, for instance the two opposing plates at $z = 0$ and $z = L_z$. To that end, we assume that the two plates in question are much closer to one another than the other pairs of opposing boundaries, $L_z \ll L_x, L_y$. In the directions parallel to the plates, the wave numbers then take quasi-continuous values

$$k_x = \frac{\pi n_x}{L_x} \in [0, \infty), \qquad k_y = \frac{\pi n_y}{L_y} \in [0, \infty). \tag{7.3}$$

Approximating the respective mode sums by integrals according to

$$\sum_{n_x, n_y = 0}^{\infty} \mapsto \frac{L_x L_y}{\pi^2} \int_0^{\infty} dk_x \int_0^{\infty} dk_y, \tag{7.4}$$

the vacuum energy takes the form

$$E_0 = \frac{\hbar c L_x L_y}{2\pi^2} \int_0^{\infty} dk_x \int_0^{\infty} dk_y \sum_{n_z = 0}^{\infty} \sum_{\lambda} \sqrt{k_x^2 + k_y^2 + \left(\frac{\pi n_z}{L_z}\right)^2}. \tag{7.5}$$

The sum over polarisations is trivial, because the mode energies are polarisation-independent. Recalling that two polarisations exist for each $n_z = 1, 2 \ldots$ but only one for $n_z = 0$, we have

$$E_0 = \frac{\hbar c L_x L_y}{\pi^2} \int_0^{\infty} dk_x \int_0^{\infty} dk_y \sum_{n_z = 0}^{\infty}{}' \sqrt{k_x^2 + k_y^2 + \left(\frac{\pi n_z}{L_z}\right)^2}, \tag{7.6}$$

where the prime indicates that the $n_z = 0$ term carries a weight of $\frac{1}{2}$. Finally, we introduce polar coordinates in the $k_x k_y$ plane, $\int_0^{\infty} dk_x \int_0^{\infty} dk_y = (\pi/2) \int_0^{\infty} dk_{\parallel} \, k_{\parallel}$. Denoting the area of the plates by $A = L_x L_y$ and the distance between them by $L = L_z$, we find a Casimir energy

$$E_0(L) = \frac{\hbar c A}{2\pi} \int_0^{\infty} dk_{\parallel} \, k_{\parallel} \sum_{n=0}^{\infty}{}' \sqrt{k_{\parallel}^2 + \left(\frac{\pi n}{L}\right)^2}. \tag{7.7}$$

Just like the Lamb shift, the Casimir energy is infinite. A number of different approaches can be used to regularise the Casimir energy, which have different physical interpretations, but lead to the same result for the observable Casimir force. We will here follow Casimir's original approach. Alternative regularisation schemes are discussed in the Appendix.

Cut-off regularisation. In close analogy to the Lamb-shift case from Section 5, the Casimir energy contains the two generic divergences of vacuum QED: the *ultraviolet divergence* and self-energies. The former arises because virtual photons of arbitrarily large energies have been included. As in the case of the Lamb shift, one might argue that non-relativistic QED breaks down at energies larger than mc^2. However, as already noted by Casimir himself, our assumption of perfectly conducting plates becomes unrealistic at much smaller frequencies already. Real metal plates become transparent to electromagnetic waves at frequencies which exceed the plasma frequency ω_P of the conduction electrons. At such high frequencies, virtual photons cease to interact with the plates and therefore cannot contribute to the Casimir energy. To account for this *high-frequency transparency*, we introduce a *cut-off function* $\mathrm{e}^{-\lambda_P k}$ which exponentially suppresses contributions from virtual photons whose wavelength $k = \sqrt{k_\parallel^2 + k_z^2}$ exceeds the plasma wavelength $\lambda_P = 2\pi c/\omega_P$:

$$E_0(L) = \frac{\hbar c A}{2\pi} \int_0^\infty \mathrm{d}k_\parallel \, k_\parallel \sum_{n=0}^{\infty}{}' \mathrm{e}^{-\lambda_P \sqrt{k_\parallel^2 + (\pi n/L)^2}} \sqrt{k_\parallel^2 + \left(\frac{\pi n}{L}\right)^2} . \tag{7.8}$$

The Casimir energy with *cut-off regularisation* is finite for any given λ_P.

Renormalisation. The regularised Casimir energy contains a large, physically unobservable *self-energy* which arises as soon as we prepare our original setup: when introducing the plates into the quantum vacuum, each of them individually interacts with the virtual photons. The corresponding self-energy can be identified in the limit $L \to \infty$. In this limit, the mode sum in the z direction also becomes quasi-continuous and we have

$$E_0(L \to \infty) = \frac{\hbar c A L}{2\pi^2} \int_0^\infty \mathrm{d}k_\parallel \, k_\parallel \int_0^\infty \mathrm{d}k_z \, \mathrm{e}^{-\lambda_P \sqrt{k_\parallel^2 + k_z^2}} \sqrt{k_\parallel^2 + k_z^2} . \tag{7.9}$$

We *renormalise* the Casimir energy by subtracting this self-energy. The resulting genuine two-plate energy reads

$$E_0(L) \;\mapsto\; \frac{\hbar c A}{2\pi} \int_0^\infty \mathrm{d}k_\parallel \, k_\parallel \left[\sum_{n=0}^{\infty}{}' \mathrm{e}^{-\lambda_P \sqrt{k_\parallel^2 + (\pi n/L)^2}} \sqrt{k_\parallel^2 + \left(\frac{\pi n}{L}\right)^2} \right.$$
$$\left. - \frac{L}{\pi} \int_0^\infty \mathrm{d}k_z \, \mathrm{e}^{-\lambda_P \sqrt{k_\parallel^2 + k_z^2}} \sqrt{k_\parallel^2 + k_z^2} \right] . \tag{7.10}$$

Assuming that $L \gg \lambda_P$, we can use the *Euler–Maclaurin formula*

$$\sum_{n=0}^{\infty}{}' f(n) - \int_0^\infty \mathrm{d}x \, f(x) = -\tfrac{1}{12} f'(0) + \tfrac{1}{720} f'''(0) + \dots \tag{7.11}$$

to express the result as a Taylor series in λ_P/L. To leading order, one finds

$$E_0(L) = -\frac{\pi^2 \hbar c A}{720 L^3} . \tag{7.12}$$

Exercise 7.1

Use the Euler–Maclaurin formula (7.11) to calculate the leading-order approximation to the Casimir energy (7.8) in the limit $L \gg \lambda_P$.

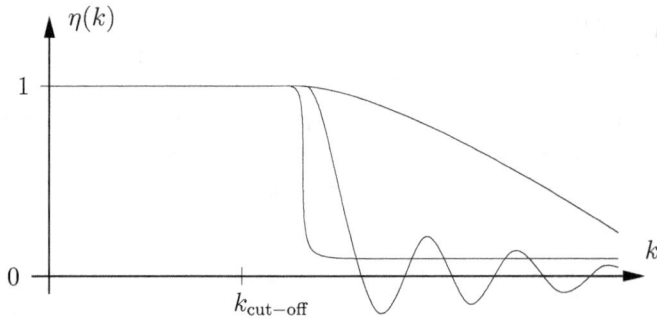

Figure 1.9: Cut-off independence: a range of cut-off functions can be used to regularise the Casimir energy. The illustrated functions $\eta(k)$ take the value unity for wave numbers smaller than a given cut-off $k_{\text{cut-off}}$ and go to zero for values much larger than $k_{\text{cut-off}}$. They hence restrict the modes contributing to the Casimir energy to those with small wave numbers. Independent of the particular profile and the chosen value for the cut-off $k_{\text{cut-off}}$, all cut-off functions lead to the same Casimir energy between two perfectly conducting plates.

By calculating the total vacuum energy of virtual photons confined between two perfectly conducting plates, we have thus found an energy which depends on the separation of the plates. Taking the derivative with respect to this separation, the potential gives rise to a *Casimir force* per unit area [42]

$$\frac{F(L)}{A} = -\frac{E_0'(L)}{A} = -\frac{\pi^2 \hbar c}{240 L^4} . \tag{7.13}$$

The presence of $\hbar$ in this result shows that the Casimir force is a pure quantum effect which vanishes in the classical limit. The speed of light c indicates that retardation is also relevant.

Cut-off independence. Note that in order to obtain a finite Casimir force, we have had to account for the fact that the plates become transparent at some cut-off frequency ω_P. Our result is hence only valid at separations which are much larger than the respective wavelength λ_P. However, the asymptotic result itself does not explicitly depend on the plasma wavelength as long as $L \gg \lambda_P$. One can show that it does not even depend on the shape of the cut-off function: the Casimir force between two perfectly conducting plates is *cut-off independent*.

We are going to illustrate this by considering a large class of *cut-off functions* $\eta(k)$. In order to be useful cut-off functions, they must be equal to unity for small values of k and fall off for large k:

$$\eta(k) = \begin{cases} 1, & k \le k_{\text{cut-off}}, \\ 0, & k \gg k_{\text{cut-off}}. \end{cases} \tag{7.14}$$

where $k_{\text{cut-off}}$ is the cut-off wave number. Possible examples of cut-off functions fulfilling these criteria are shown in Fig. 1.9. Note that Casimir's exponential cut-off function fulfils our requirements approximately for sufficiently large λ_P.

Introducing a cut-off function, the Casimir energy (7.7) is regularised to

$$E_0(L) = \frac{\hbar c A}{2\pi} \int_0^\infty \mathrm{d}k_\| \, k_\| \sum_{n=0}^{\infty}{}' \eta(k_\|^2 + \pi^2 n^2/L^2)\sqrt{k_\|^2 + \left(\frac{\pi n}{L}\right)^2} \tag{7.15}$$

and hence the renormalised Casimir energy reads

$$E_0(L) \mapsto \frac{\hbar c A}{2\pi} \int_0^\infty \mathrm{d}k_\| \, k_\| \left[\sum_{n=0}^{\infty}{}' \eta\left(\sqrt{k_\|^2 + \pi^2 n^2/L^2}\right)\sqrt{k_\|^2 + \left(\frac{\pi n}{L}\right)^2} \right.$$
$$\left. -\frac{L}{\pi} \int_0^\infty \mathrm{d}k_z \, \eta\left(\sqrt{k_\|^2 + k_z^2}\right)\sqrt{k_\|^2 + k_z^2} \right]. \tag{7.16}$$

This expression can once more be evaluated by means of the Euler–Maclaurin formula (7.11). To make the dependence on n more explicit, we make the substitution $x = k_\|^2 L^2/\pi^2 + n^2$, $\mathrm{d}k_\| \, k_\| = \pi^2/(2L^2)\mathrm{d}x$. The function $f(n)$ of the Euler–Maclaurin formula then reads

$$f(n) = \frac{\pi^2 \hbar c A}{4L^3} \int_{n^2}^\infty \mathrm{d}x \, \eta(\pi\sqrt{x}/L)\sqrt{x}. \tag{7.17}$$

Exploiting the fact that $\eta'(0) = \eta''(0) = 0$, one then easily finds

$$f'(0) = f''(0) = 0, \quad f'''(0) = -\frac{\pi^2 \hbar c A}{L^3}. \tag{7.18}$$

Substituting this into the Euler–Maclaurin formula (7.11), we recover the Casimir energy (7.12).

Exercise 7.2

Starting from the regularised and renormalised Casimir energy (7.16), use the Euler–Maclaurin formula (7.11) to show that the Casimir energy is cut-off independent and given by Eq. (7.12).

The Casimir force between two perfectly conducting plates hence requires a finite cut-off in order to produce a convergent result, but it is cut-off independent: the asymptotic result in the limit $L \gg k_{\text{cut-off}}$ depends neither on the chosen value for the cut-off $k_{\text{cut-off}}$ nor on the specific cut-off profile. Cut-off independence results from the fact that modes with low frequencies give the dominant contribution to the renormalised Casimir energy, as they are most strongly altered by the presence of the perfectly conducting boundaries. The cut-off independence is the underlying reason why the alternative regularisation schemes presented in the Appendix all lead to the same Casimir energy.

Cut-off independence should not be taken for granted: for instance, the Casimir stress in the presence of inhomogeneous media [43] appears to be *cut-off dependent**

---

*For further discussion of this problem, see Sections 39.2 and 39.3.

and cut-off dependent contributions also arise in piston geometries [44]. When such contributions do not cancel, one has to choose that cut-off value and profile which most closely corresponds to the behaviour of the actual physical system of interest.

The cut-off independence of results should not be mistaken for a measure of physical significance. For instance, as we will see in Section 21, the Casimir force between two realistic metal or dielectric plates depends on their reflectivity. The high-frequency transparency of such plates then imposes an implicit natural cut-off, leading to a finite Casimir force. This physically observable force is far from cut-off independent, but depends sensitively on the intricate details of the plates' material response to electromagnetic fields. After all, why should one expect that the Casimir force between two objects would not depend on their physical properties, such as the plasma frequency?

Sign of the Casimir force. We have found the Casimir force from the total energy of the vacuum electromagnetic field between two perfectly conducting plates. It has turned out that the separation-dependent part of this energy is negative and that it grows in magnitude as the plates come closer together. The corresponding Casimir force is hence attractive. The sign of the force emerges in a rather unintuitive way from two partially cancelling terms in Eq. (7.10), the vacuum energies at finite and infinite separations. It would be nice to have a simpler and more intuitive way of predicting whether the Casimir force in a given situation is attractive or repulsive.

A very tempting interpretation of the Casimir force involves virtual photons. Real photons exert a pressure on metallic plates. So what if we assume that virtual photons can do the same? In analogy to the gas pressure emerging from particles impinging onto two plates, the momentum of virtual photons would lead to a photon pressure. To calculate the net pressure, we would have to compare the number of photons incident from the space between the plates to those incident from the outside. The possible virtual photons between the plates are constrained to discrete wave lengths while those on the outside are not. We conclude that the pressure from the outside dominates, leading to the correct attractive force. We have thus deduced the sign of the force from counting and comparing numbers of virtual photons between and beyond the plates. Unfortunately, this simplistic picture is unreliable, because it may lead to wrong conclusions about the sign of the Casimir force [45]. To see this, we consider a variation of the original Casimir problem which was first introduced by Boyer [46].

7.2   Conducting and permeable plates

Boyer replaced one of Casimir's perfectly conducting plates by an infinitely magnet-ically permeable one ($\mu \to \infty$). To study this problem, we once more start from a cuboid cavity where now five of the cavity walls are perfectly conducting, but the wall at $z = L_z$ is infinitely permeable. This modified plate will enforce the parallel component of the magnetic field $\boldsymbol{B}$ to vanish. By virtue of the Faraday law (2.3), this implies a vanishing of the parallel component of $\nabla \times \boldsymbol{E}$.* The wave vector along the z-direction is then constrained to discrete values

$$k_z = \frac{\pi\left(n_z + \frac{1}{2}\right)}{L_z} \quad \text{with } n_z \in \mathrm{N} \tag{7.19}$$

and the eigenfrequencies of the normal modes inside the modified cavity read

$$\omega_n = c \sqrt{\left(\frac{\pi n_x}{L_x}\right)^2 + \left(\frac{\pi n_y}{L_y}\right)^2 + \left[\frac{\pi\left(n_z + \frac{1}{2}\right)}{L_z}\right]^2}. \tag{7.20}$$

instead of Eq. (2.36). The only difference with respect to the original Casimir problem is the constant shift $+\frac{1}{2}$. However, this innocent-looking term is able to fundamentally change the resulting force. Following Casimir's procedure based on a cut-off regularisation, we now find a positive Casimir energy (Ex. 1.1)

$$E_0(L) = +\frac{7\pi^2 \hbar c A}{5760 L^3}. \tag{7.21}$$

This leads to a repulsive *Casimir force* per unit area [46]

$$\frac{F(L)}{A} = +\frac{7\pi^2 \hbar c}{1920 L^4}. \tag{7.22}$$

which is smaller in magnitude than Casimir's attractive result by a factor 7/8.

Boyer's result is a counter-example to our heuristic rule of deriving the sign of the Casimir force from virtual-photon counting: in his configuration, there are still more virtual photons impinging on the plates from the outside. However, the Casimir pressure is not attractive as predicted from this reasoning. The resolution to the apparent discrepancy lies in the fact that the momentum of virtual photons has a direction. Rather than simply counting their number, one must therefore determine the associated vacuum radiation pressure in the direction normal to the plates from the *Maxwell stress tensor* [47]. This procedure yields the correct sign and magnitude for the Casimir force. Unfortunately, however, it does not provide us with a simple intuition as to why the force is attractive or repulsive in any particular case.

7.3   Thermal effects

In our considerations of the Casimir force so far, we have consistently considered the electromagnetic field to be in its vacuum state. The force is then entirely due to virtual photons. As stated in Section 4, this is only true in the absolute ground state. At finite temperature, virtual photons are complemented by thermal photons which also contribute to the Casimir force.

---

*This is an example of a Neumann boundary condition where the derivative of our mode functions is specified. In contradistinction, the boundary condition of a perfectly conducting plate constrains the mode functions themselves and is known as a Dirichlet boundary condition.

To calculate the Casimir force at finite temperature, also known as thermal Casimir force, we again start from the total energy (3.33) of the electromagnetic field. Assuming the field to be in a thermal state $\hat{\rho}_T$ of uniform temperature T, one can easily evaluate the quantum average of this energy by means of Eq. (4.13):

$$E = \sum_j \left[n(\omega_j) + \tfrac{1}{2} \right] \hbar \omega_j \tag{7.23}$$

$$\equiv E_T + E_0 \tag{7.24}$$

Having already evaluated the ground-state Casimir force E_0, we are here going to concentrate on the additional contribution

$$E_T = \sum_j n(\omega_j) \hbar \omega_j. \tag{7.25}$$

due to thermal photons. Substituting the normal modes of the cuboid cavity, it assumes the form

$$E_T = \sum_{n \in \mathbb{N}^3} \sum_\lambda n(\omega_n) \hbar \omega_n \tag{7.26}$$

with eigenfrequencies (2.36):

$$\omega_n = c \sqrt{ \left(\frac{\pi n_x}{L_x} \right)^2 + \left(\frac{\pi n_y}{L_y} \right)^2 + \left(\frac{\pi n_z}{L_z} \right)^2 }. \tag{7.27}$$

In general, this expression can only be evaluated numerically (Ex. 1.2). However, analytic results can be obtained for small and large temperatures. To that end, we note that the thermal photon number (4.13) approximates to

$$n(\omega) = \frac{1}{e^{\hbar\omega/(k_B T)} - 1} = \begin{cases} e^{-\hbar\omega/(k_B T)} & \text{for } k_B T \ll \hbar\omega \\ \dfrac{k_B T}{\hbar\omega} & \text{for } k_B T \gg \hbar\omega \end{cases} \tag{7.28}$$

in these two limits. However, these approximations cannot immediately be used, because the Casimir energy is a sum over modes of various frequencies. At any given temperature, some of these modes will be governed by the low-temperature limit with very small thermal photon numbers, some will exhibit large thermal photon numbers according to the high-temperature limit and others will have an intermediate thermal population. To apply the correct approximation, we have to determine which modes dominate the Casimir energy.

For a plate separation L, the allowed modes within the cavity have frequencies of the order $\omega_n = \pi c/L$ or higher. The low-temperature limit is hence realised when $k_B T \ll hc/(2L)$. In this case, all relevant thermal photon numbers are exponentially suppressed and the Casimir force is well approximated by its zero-temperature limit (7.13). However, the above condition can equivalently be expressed as $L \ll hc/(2k_B T)$. In other words, the low-temperature limit is not uniformly valid, but applies only up to a certain maximum distance which depends on the given temperature. For instance, we have $hc/(2k_B T) = 48\,\mu$m at room temperature ($T = 300$ K). A small-temperature limit is hence always simultaneously a small-distance limit. One must bear this in mind when neglecting the effect of thermal photons on the Casimir force.

Similarly, the opposite, high-temperature limit $k_B T \gg hc/L$ is simultaneously a large-distance limit, $L \gg hc/(k_B T)$. In this case, thermal photons dominate the Casimir energy. Using the above approximation for the photon number together with the Euler-Maclaurin formula, one finds (Ex. 1.2) [48]

$$E(L) = -\frac{\zeta(2)k_B T A}{8\pi L^2} \qquad (7.29)$$

where $\zeta(2) \approx 1.2$. This corresponds to a *thermal Casimir force* per unit area of

$$\frac{F(L)}{A} = -\frac{\zeta(2)k_B T}{4\pi L^3} \,. \qquad (7.30)$$

Due to the influence of thermal photons, the Casimir force hence decreases less rapidly at large distances. The result in the high-temperature limit does not depend on the Planck constant $\hbar$ associated with quantum fluctuations. This shows that the Casimir force is an entirely classical effect in this limit.

§8. THE QUANTUM VACUUM REVISITED
—SO HOW REAL ARE VIRTUAL PHOTONS AFTER ALL?

In the previous sections, we have discussed a host of quantum effects: the Lamb shift, spontaneous decay, a modification of the electron magnetic moment, the van der Waals interaction, and most famously, the Casimir force. All of these effects have been experimentally verified. So we have clear evidence for the existence of vacuum energy and virtual photons. Or do we? We will discuss this question separately for vacuum energy and virtual photons.

8.1 Vacuum energy

The *vacuum energy* is a constant contribution to the Hamiltonian of the electromagnetic field (3.33). Being a constant, it commutes with any field operator and hence it does not affect the Heisenberg equations of motion (3.38) for the electromagnetic field. As in classical mechanics, one could dismiss such a global energy offset as meaningless, because only changes in energy are physically observable.

One exception is gravity as described by general relativity [49]: according to Einstein's field equations, any energy acts as a source for gravity. In particular, the vacuum energy of the electromagnetic field in free space, if it exists, would have an impact on the expansion of the universe (see Section 39.4 for more details). However, there is one problem: the vacuum energy is infinite. It can be made finite in analogy to our regularisation of the Casimir energy (Section 7). Here, we note that our current theory is likely to become invalid at energies where quantum gravity phenomena might arise. The relevant energy scale is $m_P c^2$ where $m_P = \sqrt{\hbar c/G}$ is the *Planck mass*. The latter is defined such that the gravitational energy $G m_P^2/r$ of two Planck-mass particles at distance r is equal to the energy $\hbar \omega = \hbar c/r$ of a photon whose inverse wave number is r. Conceding our ignorance of what the quantum vacuum might look like at such scales, we can obtain a finite vacuum energy by introducing a cut-off $m_P c^2$. However, the resulting vacuum energy is roughly 120 orders of magnitude larger than the value required to explain observations of the expanding universe [50]! This huge discrepancy is known as the *hierarchy problem.* Unless it is resolved, we do not have an answer as to whether the vacuum energy of the electromagnetic field is 'real', in the sense that it gives rise to gravity.

8.2   Changes in vacuum energy

On the other hand, changes in the vacuum energy of the electromagnetic field are indeed observable. The most prominent example is the *Casimir energy*, which we have derived in the previous section as the separation-dependent part of the vacuum energy in the presence of two perfectly conducting plates. The Casimir energy is a potential energy which leads to a conservative force. It can reversibly be converted into kinetic energy by allowing the plates to move under the influence of the force. The varying Casimir energy in a movable geometric configuration is hence an observable entity. Incidentally, this energy is consistent with general relativity, because it satisfies the equivalence principle [51]: its inertial and gravitational masses coincide. However, whether this energy is really a consequence of vacuum energy is a matter of interpretation: as shown by Schwinger, DeRaad and Milton [52], one can alternatively derive the Casimir force from interactions of charges inside the plates without any reference to the energy or fluctuations of the vacuum electromagnetic field (see also Ref. [53]).

A different example of a varying vacuum energy is the Lamb shift. As shown in Section 5, the interaction of an atom with the vacuum electromagnetic field leads to a shift in its internal energy. Again, this effect is only observable as an energy change. Rather than varying the separation of two plates, we consider here a variation with respect to the internal state of the atom, in other words: the distance of the atom's bound electrons from the nucleus. As the vacuum energy is different for different internal states, it leads to an observable contribution to the transition frequency between two such states. As shown by Milonni and co-workers, this may be interpreted as evidence for the existence of vacuum fluctuations, but alternative derivations of the same effect exist that make no reference to vacuum fluctuations [29]. Instead, they attribute the Lamb shift to radiation reaction: quantum fluctuations of the atomic electric dipole moment give rise to an electromagnetic field which then acts back on the atom.

8.3   Virtual photons

So what about virtual photons: what are they and do they exist? The most explicit manifestations of virtual photons that we have encountered so far featured on the Feynman diagrams for calculating the Lamb shift (Fig. 1.4), the modified electron magnetic moment (Fig. 1.6) and the van der Waals interaction (Fig. 1.7). In all of these cases, they were depicted by internal lines which began and ended with a field–matter interaction vertex. These lines are graphical representations of intermediate states in perturbation theory where the electromagnetic field caries a single-quantum excitation. We have read the Feynman diagrams and their corresponding perturbative contributions intuitively as sequences of events where virtual photons are emitted and reabsorbed and atoms change their internal states in the process. However, this interpretation should not be taken too literally: after all, each Feynman diagram is just a graphical shorthand for a perturbative approximation. The corresponding exact observable quantum effect is the sum of an infinite hierarchy of such diagrams. Virtual photons are hence simply a suggestive way of representing excited intermediate field states in perturbation theory.

The evidence regarding vacuum energy and virtual photons thus stands as follows: the relevance of the absolute vacuum energy is currently an open problem. Changes in vacuum energy lead to observable quantum effects. However, alternative explanations of these effects exist which make no reference to vacuum energy. Virtual photons feature in Feynman diagrams which are graphical representations of a perturbative approximation. One might say that they are just as real or not as the mathematical terms which they represent, providing a useful intuitive picture.

§9. NORMAL MODE QED: DISCUSSION
— ARE PHOTONS REALLY THAT SIMPLE?

In this chapter, we have introduced many important quantum effects which arise when the electromagnetic field is in its vacuum state. We have based our considerations on idealised normal mode QED which is a model for the quantised electromagnetic field in the presence of perfectly conducting bodies. We have used the energy of the vacuum electromagnetic field interacting with matter as our point of departure.

9.1   Advantages

This approach has a number of advantages over the more sophisticated methods presented in the remainder of this book:

- *It is simple.* In simple perfect-conductor geometries, the normal modes are simple analytic expressions. The calculations shown in this chapter are often the simplest (and the historic) way to calculate the respective quantum effects.

- *It is based on first principles.* Normal mode QED can be constructed from first principles by canonical quantisation of the electromagnetic field. It is free of heuristic model assumptions or effective approximations such as the linear response description underlying Lifshitz theory.

- *Its underlying model is clearly defined.* Normal mode QED typically starts with the assumption that all present bodies are perfect conductors. This means that the parallel electric field vanishes exactly on their surface, full stop. No limiting procedures of infinite conductivity are needed. These may often lead to ambiguities when combined with other idealisations such as vanishing temperature or instantaneous interactions.

- *It provides a qualitative intuition for quantum-vacuum effects.* Normal mode QED can provide us with a rough idea of which quantum effects are to be expected in a given scenario, together with a feeling for their sign (e.g., attractive vs. repulsive forces, enhancement vs. reduction of the electron magnetic moment).

- *Its quantitative predictions are often not too far off.* Quantum effects involving ground-state atoms and/or metal bodies at separations much larger than the metallic plasma wavelength are often remarkably well approximated by the normal mode result. They provide a useful point of reference and often an upper limit for a more accurate result based on a more realistic description of the materials in question.

- *It is generalisable.* Normal mode QED can be extended to non-dispersive dielectrics with constant, finite permittivities [54].

9.2   Limitations

The greatest appeal of the theory, its simplicity and idealisation, naturally gives rise to intrinsic limitations:

- *It fails at short separations.* Predictions from normal mode QED with perfect conductors begin to disagree with observations from real metals as soon as the distance becomes comparable to the plasma wavelength of the metal in question or the transition wavelength of any present atom or molecule. The situation is even worse for dielectrics.

- *It ignores dispersion and material resonances.* Unlike perfect conductors, real materials exhibit an electric response which is dispersive, i.e. frequency-dependent. This is particularly relevant near material resonances where the permittivity can vary strongly with frequency. As a result, effects like the electron magnetic moment enhancement can exceed the value predicted from the perfect-conductor model [55]. Near-resonant interactions of excited atoms with such materials can greatly exceed the perfect-conductor prediction and even differ in sign [56]. Another neglected feature of real materials which is particularly relevant for metals is spatial dispersion: a non-local response.

- *It neglects absorption.* Unlike perfect conductors, real materials can absorb energy. Mathematically, this property is encoded in the imaginary part of their permittivity. The failure to account for absorption is particularly problematic when studying dissipative effects like spontaneous decay and quantum friction.

- *It only applies to equilibrium scenarios.* Approaches based on energy calculations are limited to systems in well-defined energy eigenstates. In particular, they are unable to address the dynamics which is to be expected in non-equilibrium configurations.

In the chapters that follow, we will encounter more sophisticated models of the interaction of light with matter that can overcome these limitations. We will begin in the next chapter by developing a microscopic perspective on the Casimir effect.

§10. PROBLEMS

Problem 1.1
Repulsive Casimir force

(i) Write down all the normal modes of Boyer's modified cuboid cavity consisting of five perfectly conducting plates and one infinitely permeable plate. Show that these modes satisfy the wave equation (2.9), the transversality condition (2.10) and the boundary conditions at each of the cavity walls.

(ii) Evaluate the vacuum energy inside this cavity by summing the energies of all normal modes with eigenfrequencies (7.20). Use cut-off regularisation to obtain the finite Casimir energy (7.21).

(iii) To illustrate the sign of the Casimir force, plot the contributions to the Casimir energy from modes with $n_z = 0, 1 \ldots 20$ for finite plate separation vs the corresponding result for infinite separation. Create a similar plot for Casimir's configuration of two perfectly conducting plates.

Problem 1.2
Thermal Casimir force

(i) Starting from the expression (7.23), calculate the total thermal Casimir energy between two perfectly conducting plates by using the normal modes of a cuboid cavity, performing the continuum limit in the directions parallel to the plates and subtracting the self-energies of the individual plates.

(ii) Evaluate the resulting thermal Casimir energy numerically at room temperature $T = 300\,\mathrm{K}$. Display your result normalised with respect to the zero-temperature force in a distance range $L = (1 \ldots 200)\,\mu\mathrm{m}$.

(iii) Using the high-T approximation (7.28) to the thermal photon number together with the Euler–Maclaurin formula (7.11), show that the thermal Casimir force for large temperatures is given by (7.30).

ACKNOWLEDGEMENTS

I would like to thank Claudia Eberlein, Carsten Henkel, Astrid Lambrecht and Kim Milton for stimulating discussions. Support by the DFG (grant BU 1803/3-1) is gratefully acknowledged.

§11. BIBLIOGRAPHY

[1] European Physical Society, "The United Nations proclaims an International Year of Light in 2015", press release, 20th December 2013, `http://www.eps.org/resource/resmgr/events/EPS_IYL2015Adopted.pdf`.

[2] I. Newton, "A letter of Mr. Isaac Newton, professor of the Mathematicks in the University of Cambridge; containing his new theory about light and colors", *Phil. Trans. R. Soc. Lond.* **6**(69-80), 3075 (1671).

[3] I. Newton, *Opticks: Or, a Treatise of the Reflections, Refractions, Inflections and Colours of Light* (Royal Society, London, 1704).

[4] C. Huygens, *Traité de la Lumière* (Pieter van der Aa, Leiden, 1690).

[5] T. Young, "The Bakerian Lecture: On the theory of light and colours", *Phil. Trans. R. Soc. Lond.* **92**, 12 (1802).

[6] A. Einstein, "Über eine die Erzeugung und Verwandlung des Lichts betreffenden heuristischen Gesichtspunkt", *Ann. Phys. (Leipzig)* **17**(6), 132 (1905).

[7] W. Rueckner and J. Peidle, "Young's double-slit experiment with single photons and quantum eraser", *Am. J. Phys.* **81**(12), 951 (2013).

[8] J. C. Maxwell, "A dynamical theory of the electromagnetic field", *Phil. Trans. R. Soc. Lond.* **155**, 459 (1865).

[9] Lord Rayleigh, "LIII. Remarks upon the law of complete radiation", *Phil. Mag.* **49**(301), 539 (1900).

[10] W. Heisenberg, "Über den anschaulichen Inhalt der quantentheoretischen Kinematik und Mechanik", *Z. Phys.* **43**(3-4), 172 (1927).

[11] J. D. Jackson, *Classical Electrodynamics* (Wiley, New York, 1998).

[12] V. A. Fock, "Konfigurationsraum und zweite Quantelung", *Z. Phys.* **75**(9-10), 622 (1932).

[13] L. Boltzmann, "Studien über das Gleichgewicht der lebendigen Kraft zwischen bewegten materiellen Punkten", *Wien. Ber.* **58**, 517 (1868).

[14] S. N. Bose, "Plancks Gesetz und Lichtquantenhypothese", *Z. Phys.* **26**(1) 178 (1924).

[15] A. Einstein, "Quantentheorie des einatomigen idealen Gases", *Sitzungsber. Preuss. Akad. Wiss.* 245 (1924).

[16] J. H. Jeans, "XI. On the partition of energy between matter and Æther", *Phil. Mag.* **10**(55), 91 (1905).

[17] W. Wien, "Ueber die Energieverteilung im Emissionsspectrum eines schwarzen Körpers", *Ann. Phys. (Leipzig)* **294**(8), 662 (1896).

[18] M. Planck, "Zur Theorie des Gesetzes der Energieverteilung im Normalspektrum", *Verh. Dt. Phys. Ges.* **2**, 202 (1900).

[19] M. Planck, "Über die Begründung des Gesetzes der schwarzen Strahlung", *Ann. Phys. (Leipzig)* **342**(4), 642 (1912).

[20] W. E. Lamb and R. C. Retherford, "Fine structure of the hydrogen atom by a microwave method", *Phys. Rev.* **72**(3), 241 (1947).

[21] E. A. Power and S. Zienau, "Coulomb gauge in non-relativistic quantum electrodynamics and the shape of spectral lines", *Philos. Trans. R. Soc. Lond. Ser. A* **251**(999), 427 (1959).

[22] R. G. Woolley, "Molecular quantum electrodynamics", *Proc. R. Soc. Lond. Ser. A* **321**(1547), 557 (1971).

[23] W. Thomas, "Über die Zahl der Dispersionselektronen, die einem stationären Zustande zugeordnet sind (Vorläufige Mitteilung)", *Naturwissenschaften* **13**(28), 627 (1925).

[24] W. Kuhn, "Über die Gesamtstärke der von einem Zustande ausgehenden Absorptionslinien", *Z. Phys.* **33**(1), 408 (1925).

[25] F. Reiche and W. Thomas, "Über die Zahl der Dispersionselektronen, die einem stationären Zustande zugeordnet sind", *Z. Phys.* **34**(1), 408 (1925).

[26] H. A. Bethe, "The electromagnetic shift of energy level", *Phys. Rev.* **72**(4), 339 (1947).

[27] A. Einstein, "Zur Quantentheorie der Strahlung", *Phys. Z.* **18**, 121 (1917).

[28] A. A. Markov, "Extension of the law of large numbers to dependent quantities [Russian]", *Izv. Fiz.-Matem. Obsch. Kazan Univ.* **15**, 135 (1906).

[29] P. W. Milonni, J. R. Ackerhalt and W. A. Smith, "Interpretation of radiative corrections in spontaneus emission," *Phys. Rev. Lett.* **31**(15), 958 (1973).

[30] J. Schwinger, "On quantum-electrodynamics and the magnetic moment of the electron", *Phys. Rev.* **73**(4), 416 (1948).

[31] P. Kusch and H. M. Foley, "The magnetic moment of the electron", *Phys. Rev.* **74**(3), 250 (1948).

[32] M. Kreuzer, "Mass an magnetic moment of localised electrons near conductors", *J. Phys. A: Math. Gen.* **21**(15), 3285 (1988).

[33] D. P. Craig and T. Thirunamachandran, *Molecular Quantum Electrodynamics*, (Dover, New York, 1998).

[34] J. D. van der Waals, "Over de Continuiteit van den Gas- en Vloeistoftoestand", Ph.D. thesis, Universiteit Leiden (1873).

[35] H. B. G. Casimir and D. Polder, "The influence of retardation on the London–van der Waals forces", *Phys. Rev.* **73**(4), 360 (1948).

[36] C. I. Sukenik, M. G. Boshier, D. Cho, V. Sandoghdar and E. A. Hinds, "Measurement of the Casimir–Polder force", *Phys. Rev. Lett.* **70**(5), 560 (1993).

[37] D. J. Heinzen and M. S. Feld, "Vacuum radiative level shift and sponteneous-emission linewidth of an atom in an optical resonator", *Phys. Rev. Lett.* **59**(2), 2623 (1987).

[38] E. M. Purcell, "Spontaneous emission probabilities at radio frequencies", *Phys. Rev.* **69**(11–12), 681 (1946).

[39] P. Goy, J. M. Raimond, M. Gross and S. Haroche, "Observation of cavity-enhanced single-atom spontaneous emission", *Phys. Rev. Lett.* **50**(24) 1903 (1983).

[40] R. G. Hulet, E. S. Hilfer and D. Kleppner, "Inhibited Spontaneous Emission by a Rydberg Atom", *Phys. Rev. Lett.* **55**(20), 2137 (1985).

[41] G. Barton, N. S. J. Fawcett, "Quantum electromagnetics of an electron near mirrors", *Phys. Rep.* **170**(1), 1 (1988).

[42] H. B. G. Casimir, "On the attraction of two perfectly conducting plates", *Proc. K. Ned. Akad. Wet.* **51**, 793 (1948).

[43] W. M. R. Simpson, S. A. R. Horsley and U. Leonhardt, "Divergence of Casimir stress in inhomogeneous media", *Phys. Rev. A* **87**(1), 043806 (2013).

[44] M. P. Hertzberg, R. L. Jaffe, M. Kardar and and A. Scardicchio, "Attractive Casimir forces in a closed geometry", *Phys. Rev. Lett.* **95**(25), 250402 (2005).

[45] V. Hushwater, "Repulsive Casimir force as a result of vacuum radiation pressure", *Am. J. Phys.* **65**(5), 381 (1997).

[46] T. H. Boyer, "Van der Waals forces and zero-point energy for dielectric and permeable materials", *Phys. Rev. A* **9**(5), 2078 (1974).

[47] C. Genet, A. Lambrecht and S. Reynaud, "Casimir force and the quantum theory of lossy optical cavities", *Phys. Rev. A* **67**(4), 043811 (2003).

[48] J. Mehra, "Temperature correction to the Casimir effect", *Physica* **37**(1), 145 (1967).

[49] A. Einstein, "Die Feldgleichungen der Gravitation", *Sitzungsber. Preuss. Akad. Wiss.* 844 (1915).

[50] S. Weinberg, "The cosmological constant problem", *Rev. Mod. Phys.* **61**(1), 1 (1989).

[51] S. A. Fulling, K. A. Milton, P. Parashar, A. Romeo, K. V. Shajesh and J. Wagner, "How does Casimir energy fall?", *Phys. Rev. D* **76**(2), 025004 (2007).

[52] J. Schwinger, L. L. DeRaad Jr. and K. A. Milton, "Casimir effect in dielectrics", *Ann. Phys.* **115**(1), 1 (1978).

[53] P. W. Milonni, *The Quantum Vacuum* (Academic Press, New York, 1994).

[54] R. J. Glauber and M. Lewenstein, "Quantum Optics of Dielectric Media", *Phys. Rev. A* **43**(1), 467 (1991).

[55] R. Bennett and C. Eberlein, "Anomalous magnetic moment of an electron near a dispersive surface", *Phys. Rev. A* **88**(1), 012107 (2013).

[56] H. Failache, S. Saltiel, M. Fichet, D. Bloch and M. Ducloy, "Resonant van der Waals repulsion between excited Cs atoms and sapphire surface", *Phys. Rev. Lett.* **83**(26), 5467 (1999).

[57] K. A. Milton, *The Casimir Effect: Physical Manifestations of Zero-Point Energy* (World Scientific, Singapore, 2001).

[58] J. Schwinger, "On gauge invariance and vacuum polarization", *Phys. Rev.* **82**(5), 664 (1951).

Chapter 2

Van der Waals and Casimir–Polder dispersion forces

EPHRAIM SHAHMOON

> The whole is more than the sum of its parts
>
> – Aristotle, Metaphysica

> It has been said: The whole is more than the sum of its parts. It is more correct to say that the whole is something else than the sum of its parts...
>
> – Kurt Koffka

§12. INTRODUCTION

In Chapter 1, the Casimir force was described as resulting from the change in the electromagnetic field's vacuum (zero-point) energy due to the presence of two mirrors: the mirrors impose boundary conditions on the electromagnetic vacuum modes, giving rise to a discrete set of confined modes $\{n\}$ with frequencies $\omega_n(L)$ that depend on the distance L between them. Then, summing over the zero point energies of the modes $\sum_n \frac{1}{2}\hbar\omega_n(L)$ (and subtracting those of the continuum of modes for $L \to \infty$), we obtain the attractive potential between the mirrors as a function of L, $U \propto -1/L^3$, resulting in an attractive force $F \propto 1/L^4$. Such a description treats the degrees of freedom of the electromagnetic field — namely, its modes — but does not take into account those of matter, i.e. the particles that the mirrors are comprised of. Instead, the mirrors are treated just by considering the boundary conditions that they impose on the field degrees of freedom.

Nevertheless, at the end of the day, the Casimir effect is that of an interaction (typically attraction) between two chunks of matter (the mirrors). These two chunks of matter are made from small neutral particles that, in some way, must attract each other in order to give rise to the overall attraction between the mirrors. This gives a different perspective of the Casimir effect by associating it with some collective form

(a)   (b)

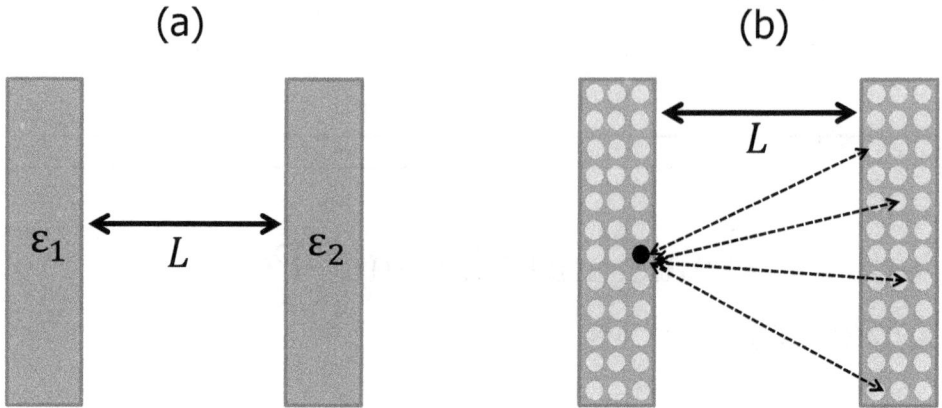

Figure 2.1: (a) The setup of the Casimir effect: two mirrors, e.g. made from dielectrics ε_1 and ε_2 (or perfect conductors as in Chapter 1) are at a distance L apart. Their vacuum-induced interaction energy scales as $-1/L^3$. (b) A microscopic description of the Casimir effect: each mirror is comprised of many particles (bright spheres). The interactions between each mirror particle (dark sphere) and all particles of the other mirror underlie the overall Casimir force.

of a much more basic interaction between pairs of neutral particles (Fig. 2.1). In fact, mirrors can be characterised by their bulk dielectric properties which are, in turn, derived from the polarisabilities of their constituent particles, such as molecules or atoms. Hence, it is the interaction between these neutral particles, the well-known van der Waals interaction, that makes the most basic building block of the Casimir effect.

Historically, the story goes in exactly the opposite direction: the Casimir effect, and its description mentioned above, were given by Casimir as an insightful example of a force driven by vacuum energy, motivated by an early work that he was involved in – the analysis of the van der Waals force in relation to the stability of colloidal systems*.

Going back to basic statistical physics, the emergence of the solid and liquid phases of matter relies on the existence of interactions between its constituent particles, namely, on the fact that matter is not an ideal gas. In 1873 van der Waals suggested a model for a pairwise inter-molecular or atomic potential consisting of two parts: long-range attraction and hard-core repulsion. This model succeeds in reproducing the gas-liquid-solid phase diagrams of matter and appears in the well-known van der Waals gas equation of state [1]. The question is then, what is the

---

*The story goes, that as a chief scientist of Philipps Laboratories, Casimir was investigating the stability of liquid paint and found out that, if the description of the van der Waals interaction known at the time was universally true, paint would get stuck in the paint bucket in contradiction to common experience. This observation apparently led to his later work with Polder [4] which showed how the van der Waals force is in fact altered by electromagnetic retardation effects, as discussed below.

origin of the long-range attractive force suggested by van der Waals?

Keesom calculated the interaction potential between two molecules with a perma-nent dipole moment. At a finite temperature T, the molecules are randomly rotating and there exists a non-vanishing average potential scaling as $U \propto -T/r^6$, where r is the distance between the molecules [2]. A similar result may be obtained when the dipole of one molecule is induced by that of the other molecule, whose dipole is in turn induced by thermal fluctuations, rather than being permanent. Namely, classical physics considerations showed that an attractive $1/r^6$ potential may exist between any pair of particles, so long as they are polarisable – in other words, that their dipoles can be induced by an electric field. The potential is driven by thermal fluctuations, and hence grows linearly with T. This result, whilst providing a phys-ical mechanism for an attractive force between a large class of neutral particles, was nevertheless inconsistent with experimental evidence, which showed that at very low temperatures the force seemed to diminish very slowly with temperature and even reach a constant value, rather than decreasing linearly with temperature T [2,3].

This calls quantum physics into play, and in 1930 London [2] used lowest order quantum mechanical perturbation theory, combined with electrostatic considera-tions, to obtain an attraction varying as $1/r^6$ between polarisable molecules at zero temperature. London interpreted this interaction, now widely known as the dis-persion force, as originating from the zero-point motion of the molecular degrees of freedom, which is a strictly quantum mechanical effect. London's remarkable result did not end the story however, since experimental results in the late 1940s led to a suggestion made by Overbeek, that at long distances r the intermolecular potential decreases more rapidly with r than the $1/r^6$ scaling, possibly due to retardation [4]. Indeed, using an analysis based on quantum electrodynamics, Casimir and Polder (1948) [4] were able to obtain an interaction potential at zero temperature that falls off as $1/r^7$ at large distances where electrodynamic wave retardation effects become significant, and reduces to that of London at shorter distances. More in-terestingly, their treatment had to take into account the degrees of freedom of the electromagnetic field modes and their vacuum fluctuations at zero temperature.

This suggested the interpretation that the van der Waals force may be viewed as being driven by the quantum, rather than thermal, fluctuations of the electromag-netic field. To put it in other words, this means that the van der Waals force, or any force between chunks of matter made from particles which interact by means of it, may be attributed to the zero-point (or vacuum) energy of the electromagnetic field. In fact, this is exactly the point highlighted by Casimir's treatment of the attraction between two mirrors [5]. Indeed, vacuum-induced forces between parti-cles underlie vacuum forces between bulk media; however, the actual mathematical relation between the two is not trivial due to the non-additivity of these interactions, as discussed in Section 16 below.

In the same paper [4] Casimir and Polder considered another related problem – that of the attraction between an atom and a perfectly-conducting mirror. This problem forms the basis of interaction effects between particles and surfaces [6,7], like the stiction of atoms and molecules to surfaces or the deflection of atomic beams. Whereas previous analyses suggested an atom-surface interaction scaling as $1/r^3$, where r is the distance of the atom from the surface [8, 9], the quantum electrodynamic treatment by Casimir and Polder showed that at large r, retardation effects lead to a $1/r^4$ scaling. Their interpretation and analysis again showed that it is the zero-point energy of the electromagnetic field and the atom that is the source of this atom-surface potential.

§13. Van der Waals interaction

Let us begin with the physical picture of the van der Waals interaction. We consider two polarisable dipoles 1 and 2 (Fig. 2.2). These dipoles can be thought of as being comprised of electrons randomly moving around the position of a positive charge like an ion. The random fluctuations of the positions of the electrons are assumed to have no preferred direction, as in the case of the spherical symmetric ground state of an electron in a hydrogen-like atom, for example. While on average the net polarisation vanishes, at any given moment there is an electron at some distance r_e from the ion and there exists a finite dipole moment er_e, e being the electron charge. Hence the fluctuations in the electronic motion give rise to fluctuations in the polarisation of the atom. In turn, such a polarised dipole, say dipole 1, creates an electric field $E(r)$ at a distance r and induces the polarisation of dipole 2, $d = \alpha E(r)$, α being its *polarisability*, i.e. its linear response to an applied electric field. The electrostatic dipole field scales as $1/r^3$, and recalling the energy of an induced dipole [10]*,

$$U = -\frac{1}{2}dE = -\frac{1}{2}\alpha E^2, \tag{13.1}$$

we obtain the scaling $U \propto -1/r^6$ for the energy of dipole 2, interpreted as the van der Waals interaction energy.

What can be the source of the isotropic fluctuations in the electron motion? Temperature is one option, as well as some random external field. However, here we are interested in the case of zero temperature and no external fields, so that the system as a whole, including all atoms and electromagnetic field modes, is in its ground state. One way to look at this situation was already mentioned here: considering the symmetric ground-state electronic wave function of a hydrogen-like atom, there exist quantum fluctuations in the electronic position, the so called zero-point motion of the electron. A different point of view ignores the inherent atomic ground-state fluctuations but considers the ground state of the electromagnetic field – the vacuum. For each field mode, this is the ground state of a quantum harmonic oscillator and hence it exhibits fluctuations in the amplitude of the field around a zero mean – these are the vacuum, or zero-point, fluctuations. These field fluctuations can then induce fluctuations in dipole 1 (or 2) and drive the van der Waals interaction.

---

*The energy of an existing dipole d in the presence of an electric field E is given by $-dE$, as in the electrodynamic Hamiltonian of Eqs. (5.6) and (13.67). However, when the dipole is *induced* by the field E, the energy required to polarise it is given by $-(1/2)\alpha E^2$ [10], in analogy to the charging energy of a capacitor with capacitance C that is connected to a voltage source V, $(1/2)CV^2$ (the factor 1/2 appears as a result of an integration [10]).

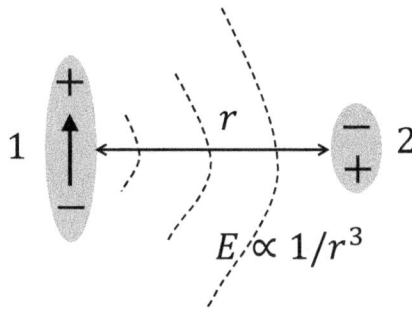

Figure 2.2: Quasistatic description of the van der Waals force. A dipole moment is created in particle 1 via its internal random dipole fluctuations. In turn, this dipole moment creates a dipole field, scaling as $E \sim 1/r^3$ in the quasi-electrostatic regime. The field arrives at particle 2, at a distance r and with polarisability α, inducing its dipole moment. The resulting dipolar energy of particle 2, $U = -(1/2)\alpha E^2 \sim -1/r^6$, is interpreted as the van der Waals interaction energy.

Whatever point of view we take, whether these are the quantum fluctuations of matter (dipoles), of the field, or even both that drive everything, the source of the van der Waals force are the quantum fluctuations of a system in its ground state. Any fluctuations require an energy source; here it is that contained in the vacuum or in the ground state of matter*. Classically, such sources of energy do not exist.

The above discussion provides a quick and intuitive explanation for the $1/r^6$ scaling of the van der Waals interaction at short distances, where the electrostatic dipole field $\propto 1/r^3$ can be considered. This quasistatic approach is made more quantitative in Section 13.1, whereas the full electrodynamic treatment, valid at all distances, is presented in Section 13.2.

13.1   Quasistatic analysis

Following an approach similar to that of London [2], we consider two identical polarisable dipoles 1 and 2 in their ground states $|g_{1,2}\rangle$, e.g. a pair of atoms (or molecules). For simplicity of illustration, out of the discrete set of excited states of each atom, we consider only the first excited state $|e_{1,2}\rangle$ that is dipole-coupled to the ground state. The dipole operator of atom $\nu = 1, 2$ is given by

$$\hat{\boldsymbol{d}}_\nu = \boldsymbol{d}_\nu |e_\nu\rangle\langle g_\nu| + \boldsymbol{d}_\nu^* |g_\nu\rangle\langle e_\nu| \tag{13.2}$$

with $\boldsymbol{d}_\nu = \langle e_\nu | \hat{\boldsymbol{d}}_\nu | g_\nu \rangle$ being the dipole matrix element

$$\boldsymbol{d}_\nu = d \left(\sin\theta_\nu \cos\phi_\nu \boldsymbol{e}_x + \sin\theta_\nu \sin\phi_\nu \boldsymbol{e}_y + \cos\theta_\nu \boldsymbol{e}_z \right), \tag{13.3}$$

---

*Viewing the energy of the vacuum (ground)-state as the energy source responsible for the necessary quantum (vacuum) fluctuations may suggest some intuition for its existence. It does not however mean that this energy source can be depleted, which would violate energy conservation (see discussion on virtual photons in Section 13.2.2).

where θ_ν and ϕ_ν are the orientation angles of dipole ν in spherical coordinates, $d = |\boldsymbol{d}_\nu|$ for both atoms, and $\boldsymbol{e}_i$ is a unit vector in direction $i = x, y, z$.

Quantum mechanically, the dipoles may fluctuate around their zero mean even when they are in their ground states, hence they may interact. The interaction Hamiltonian between the two dipoles is taken to be the electrostatic dipole-dipole interaction [11], with the classical dipole moments replaced by the dipole operators,

$$\hat{H}_I = \frac{1}{4\pi\varepsilon_0 r^3}\left[\hat{\boldsymbol{d}}_1 \cdot \hat{\boldsymbol{d}}_2 - 3(\hat{\boldsymbol{d}}_1 \cdot \boldsymbol{e}_r)(\hat{\boldsymbol{d}}_2 \cdot \boldsymbol{e}_r)\right]. \tag{13.4}$$

Here $\boldsymbol{r} = r\boldsymbol{e}_r$ is the distance between the dipoles, $r = |\boldsymbol{r}|$ being its magnitude. We can now use second order perturbation theory, Eq. (5.13) in Chapter 1, to find the energy correction of the ground state of the system $|g_1 g_2\rangle$ due to the interaction $\hat{H}_I$, interpreted as the van der Waals energy U,

$$U = -\sum_{I \neq g_1 g_2} \frac{|\langle I|\hat{H}_I|g_1 g_2\rangle|^2}{E_I - E_{g_1 g_2}}. \tag{13.5}$$

Here $|I\rangle$ are intermediate states with non-perturbed energies E_I whereas $E_{g_1 g_2}$ denotes the energy of the ground state $|g_1 g_2\rangle$. Considering the Hamiltonian $\hat{H}_I$, it operates on each atom $\nu = 1, 2$ with the dipole operator $\hat{\boldsymbol{d}}_\nu$ from Eq. (13.2), which connects the transitions $|g_\nu\rangle \leftrightarrow |e_\nu\rangle$. Then, the only possible intermediate state in (13.5) is $|I\rangle = |e_1 e_2\rangle$. Denoting the energy difference between the ground and excited levels in both identical atoms by $E = E_e - E_g$, we have $E_I - E_{g_1 g_2} = E_{e_1 e_2} - E_{g_1 g_2} = 2E_e - 2E_g = 2E$, so that

$$U = -\frac{1}{2E}\left|\langle e_1 e_2|\hat{H}_I|g_1 g_2\rangle\right|^2. \tag{13.6}$$

Choosing the z axis of the spherical coordinates to be that of $\boldsymbol{r}$, i.e. $\boldsymbol{e}_r = \boldsymbol{e}_z$, we obtain for the matrix element,

$$\langle e_1 e_2|\hat{H}_I|g_1 g_2\rangle = \frac{d^2}{4\pi\varepsilon_0 r^3}$$
$$\times (\sin\theta_1 \cos\phi_1 \sin\theta_2 \cos\phi_2 + \sin\theta_1 \sin\phi_1 \sin\theta_2 \sin\phi_2 - 2\cos\theta_1 \cos\theta_2). \tag{13.7}$$

Inserting this expression to Eq. (13.6) and assuming randomly oriented atoms, we should average the result over a uniform distribution $1/(4\pi)$ for the solid angle of each atom; namely, we integrate over the result with

$$\frac{1}{(4\pi)^2}\int_{-1}^{1} d\cos\theta_1 \int_0^{2\pi} d\phi_1 \int_{-1}^{1} d\cos\theta_2 \int_0^{2\pi} d\phi_2,$$

obtaining

$$U = -\left(\frac{1}{4\pi\varepsilon_0}\right)^2 \frac{d^4}{3E}\frac{1}{r^6}. \tag{13.8}$$

$$E_0(r_1) \qquad\qquad E_0(r_2)$$

$$\textcircled{1}\text{-}\text{-}\text{-}\text{-}\text{-}\text{-}\text{-}\text{-}\!\!\rightarrow\textcircled{2}$$

$$E_{sc}(r_2)$$

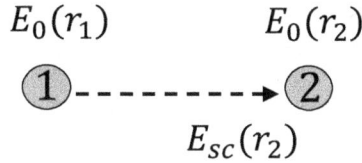

Figure 2.3: Scattering approach to the van der interaction: the vacuum field $\hat{\boldsymbol{E}}_0(\boldsymbol{r})$ exists in all space and interacts with both dipoles at their positions r_1 and r_2, hence it is also scattered by the dipoles. The scattered field from dipole 1, $\hat{\boldsymbol{E}}_{sc}(\boldsymbol{r})$, arrives at dipole 2, resulting in an interaction interpreted as the van der Waals interaction U.

Using the static polarisability of a randomly oriented two-level atom or molecule, $\alpha = (2/3)(d^2/E)$ [see Eq. (13.40) below], we finally find

$$U = -\left(\frac{1}{4\pi\varepsilon_0}\right)^2 \frac{3}{4} E\alpha^2 \frac{1}{r^6}, \tag{13.9}$$

reproducing London's result [2, 3, 12].

In analogy to our earlier discussion, we used the electrostatic dipole-dipole interaction potential that follows from the electrostatic dipole field, both scaling as $1/r^3$. Second order perturbation theory in energy then gave us the expected $1/r^6$ dependence. However, the problem of interacting fluctuating dipoles is essentially *dynamic*, whereas the electrostatic considerations taken here assume *instantaneous* responses and interactions. This implies that the above treatment is merely an effective quasi-electrostatic limit of the full problem which is valid only at short distances r. The full dynamical analysis is the topic of the next section.

13.2 Electrodynamic analysis

We present two different approaches to the calculation of the van der Waals dispersion interaction. The first approach gives a more physical picture of the problem based on a theory of scattering, whereas the second is based on systematic quantum electrodynamics perturbation theory.

13.2.1 Scattering of vacuum fluctuations

General formalism
Here we adopt an approach* where the interaction between the polarisable dipoles is treated as an electrodynamic scattering problem of oscillating dipoles driven by a fluctuating vacuum field. Within this approach, only the field is treated as a dynamical variable and hence quantised, whereas the dipoles are characterised merely by their linear response to the electric field, i.e. their polarisabilities. We consider the physical picture, mentioned above, and presented in Fig. 2.3: two polarisable

---

*In fact, we generalise treatments such as those presented in [3, 14].

dipoles, e.g. atoms, located at positions r_1 and r_2, are subject to a random, fluctuating field which exists in all space. In classical physics, this field is a random process with some probability distribution, whereas the equivalent in our case is the electric field operator $\hat{E}_0(r)$, assuming all field modes are in the vacuum state $|0\rangle$. The vacuum fluctuations at point r_1, $\hat{E}_0(r_1)$, then drive fluctuations in dipole 1, which as a result scatters a field $\hat{E}_{sc}(r)$ that reaches dipole 2 at r_2. The scattered field can in principle be the product of multiple scattering between the two dipoles. The total field that is felt by dipole 2 is then given by the sum of the scattered and incident (driving) vacuum fields,

$$\hat{E}(r_2) = \hat{E}_{sc}(r_2) + \hat{E}_0(r_2). \tag{13.10}$$

Our goal is to find the electromagnetic energy of dipole 2 due to the field $\hat{E}(r_2)$, deducing the van der Waals interaction energy from its scattering-related part.

Within the scattering approach described in the following it will be constructive to write the Hermitian electric field operator (the quantised version of the real electric field that appears in Maxwell's equations), denoted here $\hat{E}^H$, as a sum of a non-Hermitian field $\hat{E}$ and its Hermitian conjugate,

$$\hat{E}^H = \hat{E} + \hat{E}^\dagger, \quad \hat{E} = \sum_k \hat{E}_k, \tag{13.11}$$

where the non-Hermitian field is expanded in terms of normal mode field operators with mode index k. The vacuum fluctuations that drive the scattered field and hence the interaction, $\hat{E}_0^H = \hat{E}_0 + \hat{E}_0^\dagger$, are given by the non-Hermitian quantum field

$$\hat{E}_0(r) = \sum_k \hat{E}_{0,k}(r), \quad \hat{E}_{0,k}(r) = E_k(r)\hat{a}_k = i\sqrt{\frac{\hbar\omega_k}{2\varepsilon_0}}u_k(r)\hat{a}_k. \tag{13.12}$$

Here $\hat{a}_k$, $E_k(r)$ and ω_k are the dimensionless field operator, electric field mode function and frequency, respectively, of the mode k, where the electric field mode $E_k(r)$ is the product of the electromagnetic field per single photon in mode k, $\sqrt{\hbar\omega_k/(2\varepsilon_0)}$, and the dimensionless, mathematically normalised, spatial field mode function $u_k(r)$, from Eq. (2.8) in Chapter 1 [also Eq. (13.28) below]. For example, in free space $u_k(r)$ and ω_k become [Eqs. (2.16) and (2.13), respectively, Chapter 1],

$$u_{k\lambda}(r) = \frac{1}{\sqrt{V}}e^{ik\cdot r}e_{k\lambda}, \quad \omega_{k\lambda} = c|k|, \tag{13.13}$$

where $V = L^3$ is the quantisation volume, $k = n(2\pi)/L$ and $e_{k\lambda}$ is the polarisation unit vector. The average energy of dipole 2 can be expanded in its contributions from different frequencies, each such contribution being given by Eq. (13.1)

$$\begin{aligned}
U &= -\frac{1}{2}\sum_\omega \alpha_2(\omega)\langle \hat{E}_\omega^H(r_2) \cdot \hat{E}_\omega^H(r_2)\rangle \\
&= -\frac{1}{2}\sum_\omega \alpha_2(\omega)\langle \hat{E}_{0,\omega}^H(r_2) \cdot \hat{E}_{0,\omega}^H(r_2) + \hat{E}_{sc,\omega}^H(r_2) \cdot \hat{E}_{sc,\omega}^H(r_2) \\
&\quad + \hat{E}_{sc,\omega}^H(r_2) \cdot \hat{E}_{0,\omega}^H(r_2) + \hat{E}_{0,\omega}^H(r_2) \cdot \hat{E}_{sc,\omega}^H(r_2)\rangle, \tag{13.14}
\end{aligned}$$

where

$$\hat{\boldsymbol{E}}_{0,\omega}^H(\boldsymbol{r}) = \sum_{k;\omega_k=\omega} \hat{\boldsymbol{E}}_{0,k}^H(\boldsymbol{r}), \tag{13.15}$$

and the sum $\sum_{k;\omega_k=\omega}$ is performed over all modes k with a frequency ω_k equal to ω. Here $\alpha_2(\omega)$ is the *dynamical polarisability** of dipole 2 and we have used Eq. (13.10) for the field at $\boldsymbol{r}_2$. The averaging $\langle...\rangle$ is performed quantum mechanically, and working with the Heisenberg picture, we use the vacuum state for averaging, implying $\langle \hat{O} \rangle = \langle 0|\hat{O}|0\rangle$ for any operator $\hat{O}$ throughout our derivation. Looking at the second line of Eq. (13.14), the first term,

$$U_0 = -\frac{1}{2} \sum_{\omega} \alpha_2(\omega) \langle \hat{\boldsymbol{E}}_{0,\omega}^H(\boldsymbol{r}_2) \cdot \hat{\boldsymbol{E}}_{0,\omega}^H(\boldsymbol{r}_2) \rangle, \tag{13.16}$$

describes the energy correction of dipole 2 due to its interaction with the vacuum. U_0 reflects the cost in energy of dressing the dipole with vacuum fluctuations, hence it is related to the Lamb shift and it may formally diverge (see Chapter 1). However, this term is not related to the *interaction* between the dipoles. Therefore, in analogy to the Casimir effect and Lamb shifts discussed in Chapter 1, we may *renormalise* the interaction energy U by subtracting U_0 from it. Moreover, assuming the point-like dipoles to be weak scatterers, we can ignore the second term in the second line, $\propto \langle \hat{\boldsymbol{E}}_{sc,\omega}^H(\boldsymbol{r}_2) \cdot \hat{\boldsymbol{E}}_{sc,\omega}^H(\boldsymbol{r}_2) \rangle$ since it is of second order in the scattered field. This leaves us with the interaction energy,

$$U = -\frac{1}{2} \sum_{\omega} \alpha_2(\omega) \langle \hat{\boldsymbol{E}}_{sc,\omega}^H(\boldsymbol{r}_2) \cdot \hat{\boldsymbol{E}}_{0,\omega}^H(\boldsymbol{r}_2) + \hat{\boldsymbol{E}}_{0,\omega}^H(\boldsymbol{r}_2) \cdot \hat{\boldsymbol{E}}_{sc,\omega}^H(\boldsymbol{r}_2) \rangle. \tag{13.17}$$

The only thing that is missing now in order to evaluate the energy U is the scattered field operator: the driving quantum (vacuum) field is known [Eqs. (13.12) and (13.15)] and the averaging is performed with respect to the vacuum state $|0\rangle$.

Since in the Heisenberg picture the electromagnetic field operators satisfy the Maxwell equations, we can use methods known from classical electrodynamics in order to obtain the scattered field operator. Beginning with Maxwell's curl equations for the non-Hermitian electric and magnetic fields $\hat{\boldsymbol{E}}_\omega$ and $\hat{\boldsymbol{B}}_\omega$, respectively, in the presence of electric polarisation density $\hat{\boldsymbol{P}}_\omega$, all oscillating at frequency ω,

$$\boldsymbol{\nabla} \times \frac{1}{\mu_0} \hat{\boldsymbol{B}}_\omega = -i\omega \left(\varepsilon_0 \hat{\boldsymbol{E}}_\omega + \hat{\boldsymbol{P}}_\omega \right),$$
$$\boldsymbol{\nabla} \times \hat{\boldsymbol{E}}_\omega = i\omega \hat{\boldsymbol{B}}_\omega, \tag{13.18}$$

we insert the latter into the former and using $\mu_0\varepsilon_0 = 1/c^2$, we obtain the wave equation for the non-Hermitian electric field,

$$\boldsymbol{\nabla} \times \boldsymbol{\nabla} \times \hat{\boldsymbol{E}}_\omega(\boldsymbol{r}) - \frac{\omega^2}{c^2} \hat{\boldsymbol{E}}_\omega(\boldsymbol{r}) = \omega^2 \mu_0 \hat{\boldsymbol{P}}_\omega(\boldsymbol{r}). \tag{13.19}$$

---

*The dynamical polarisability $\alpha(\omega)$ is the linear response of the electric polarisation of a particle to an applied electric field oscillating at frequency ω, see Chapter 3 and Problem 2.2 in this chapter.

We can define the so-called *dyadic Green function* of the above wave equation via,

$$\boldsymbol{\nabla} \times \boldsymbol{\nabla} \times \boldsymbol{G}_j(\omega, \boldsymbol{r}, \boldsymbol{r}') - \frac{\omega^2}{c^2} \boldsymbol{G}_j(\omega, \boldsymbol{r}, \boldsymbol{r}') = \delta(\boldsymbol{r} - \boldsymbol{r}')\boldsymbol{e}_j,$$
$$G_{ij}(\omega, \boldsymbol{r}, \boldsymbol{r}') \equiv \boldsymbol{e}_i \cdot \boldsymbol{G}_j(\omega, \boldsymbol{r}, \boldsymbol{r}'). \qquad (13.20)$$

Hence, the dyadic Green function $\boldsymbol{G}_j(\omega, \boldsymbol{r}, \boldsymbol{r}')$ represents the electric field response to a point dipole source located at position $\boldsymbol{r}'$, oscillating at frequency ω and polarised in the j direction ($j = x, y, z$). Its matrix form $G_{ij}(\omega, \boldsymbol{r}, \boldsymbol{r}')$ then describes the i component ($i = x, y, z$) of the electric field's response to a dipole polarised in the j direction*. In general, the two interacting dipoles can be embedded in different geometries, e.g. free space, waveguide or a cavity, each geometry imposing different boundary conditions on Eqs. (13.19) and (13.20) and hence resulting in a different Green function. At this point we wish to keep the discussion general and not specify the geometry, so we do not specify the exact form of the dyadic Green function yet.

Using the Green function we can formally solve the wave equation (13.19) for the i component of the electric field $\hat{E}^i_\omega(\boldsymbol{r}) = \boldsymbol{e}_i \cdot \hat{\boldsymbol{E}}_\omega(\boldsymbol{r})$,

$$\hat{E}^i_\omega(\boldsymbol{r}) = \hat{E}^i_{0,\omega}(\boldsymbol{r}) + \omega^2 \mu_0 \sum_j \int_V d\boldsymbol{r}' G_{ij}(\omega, \boldsymbol{r}, \boldsymbol{r}') \hat{P}^j_\omega(\boldsymbol{r}'), \qquad (13.21)$$

with $\hat{E}^i_{0,\omega}(\boldsymbol{r}) = \boldsymbol{e}_i \cdot \hat{\boldsymbol{E}}_{0,\omega}(\boldsymbol{r})$ and $\hat{P}^i_\omega(\boldsymbol{r}) = \boldsymbol{e}_i \cdot \hat{\boldsymbol{P}}_\omega(\boldsymbol{r})$. Since polarisation exists only at the positions of the point-like polarised dipoles, and assuming a linear response of the dipoles to the field characterised by polarisabilities $\alpha_{1,2}(\omega)$, we have

$$\hat{\boldsymbol{P}}_\omega(\boldsymbol{r}) = \alpha_1(\omega)\hat{\boldsymbol{E}}_\omega(\boldsymbol{r}_1)\delta(\boldsymbol{r} - \boldsymbol{r}_1) + \alpha_2(\omega)\hat{\boldsymbol{E}}_\omega(\boldsymbol{r}_2)\delta(\boldsymbol{r} - \boldsymbol{r}_2), \qquad (13.22)$$

so that

$$\begin{aligned} \hat{E}^i_\omega(\boldsymbol{r}) &= \hat{E}^i_{0,\omega}(\boldsymbol{r}) + \hat{E}^i_{sc,\omega}(\boldsymbol{r}), \\ \hat{E}^i_{sc,\omega}(\boldsymbol{r}) &= \omega^2 \mu_0 \sum_j \left[\alpha_1(\omega)G_{ij}(\omega, \boldsymbol{r}, \boldsymbol{r}_1)\hat{E}^j_\omega(\boldsymbol{r}_1) + \alpha_2(\omega)G_{ij}(\omega, \boldsymbol{r}, \boldsymbol{r}_2)\hat{E}^j_\omega(\boldsymbol{r}_2) \right]. \end{aligned}$$
$$(13.23)$$

Here we have already separated the field into its components resulting from the original driving vacuum fluctuations and their scattered field, as in Eq. (13.10). Eq. (13.23) presents a formal solution for the scattered field as a function of itself: the right hand side of the equation, resulting from the convolution of Eq. (13.21), includes the total field, and hence its scattered component, at positions $\boldsymbol{r}_1$ and $\boldsymbol{r}_2$ as a source term, reminiscent of the Lippmann-Schwinger equation of quantum mechanical scattering theory. As we are interested only in the scattered field at the position of dipole 2, and considering again the weak scattering regime, we can take the lowest order scattering approximation (Born approximation): we neglect

---

*Note that the dyadic Green function is often presented in tensor form as in Chapters 3 and 4, see e.g. Eq. (21.20) in Chapter 3

multiple scattering so that the scattered field at dipole 2 results only from the original vacuum fluctuations scattered by dipole 1. This amounts to considering only the r_1 source term in the right hand side of Eq. (13.23) and approximating $\hat{E}^j_\omega(r_1) \approx \hat{E}^j_{0,\omega}(r_1)$, obtaining

$$\hat{E}^i_{sc,\omega}(r_2) \approx \omega^2 \mu_0 \sum_j \alpha_1(\omega) G_{ij}(\omega, r_2, r_1) \hat{E}^j_{0,\omega}(r_1). \tag{13.24}$$

Inserting the scattered field into the interaction energy from Eq. (13.17), we find

$$U = -\frac{1}{2}\mu_0 \sum_\omega \omega^2 \alpha_1(\omega)\alpha_2(\omega) \sum_{ij} \left[G_{ij}(\omega, r_2, r_1)\langle \hat{E}^j_{0,\omega}(r_1)\hat{E}^{i\dagger}_{0,\omega}(r_2)\rangle + \text{c.c.} \right], \tag{13.25}$$

where c.c. stands for complex conjugate. Here the normal ordered and phase dependent operator products of the type, $\hat{E}^{i\dagger}_{0,\omega}(r_1)\hat{E}^j_{0,\omega}(r_2)$ and $\hat{E}^i_{0,\omega}(r_1)\hat{E}^j_{0,\omega}(r_2)$, respectively, that are contained in the product of the Hermitian field operators of Eq. (13.17), were dropped since the averaging is performed with respect to the vacuum state, for which $\langle \hat{a}^\dagger_k \hat{a}_{k'}\rangle, \langle \hat{a}_k \hat{a}_{k'}\rangle = 0$. Eq. (13.25) reflects the interesting fact that the interaction energy between the dipoles depends on the product of the Green function in the considered geometry with the spatial correlation function of the driving field fluctuations. This point is relevant also to dipolar interactions induced by fields other than vacuum fluctuations, as further discussed in Section 15 below and Problem 2.5.

Using $\hat{E}^j_{0,\omega}(r)$ from Eqs. (13.15) and (13.12), we can find the vacuum field's spatial correlation,

$$\langle \hat{E}^j_{0,\omega}(r_1)\hat{E}^{i\dagger}_{0,\omega}(r_2)\rangle = \sum_{k;\omega_k=\omega}\sum_{k';\omega_{k'}=\omega} \langle \hat{E}^j_{0,k}(r_1)\hat{E}^{i\dagger}_{0,k'}(r_2)\rangle$$

$$= \sum_{k;\omega_k=\omega}\sum_{k';\omega_{k'}=\omega} \frac{\hbar\omega}{2\varepsilon_0} u^j_k(r_1)u^{i*}_{k'}(r_2)\langle \hat{a}_k \hat{a}^\dagger_{k'}\rangle$$

$$= \sum_{k;\omega_k=\omega} \frac{\hbar\omega}{2\varepsilon_0} u^j_k(r_1)u^{i*}_k(r_2), \tag{13.26}$$

with $u^i_k(r) = e_i \cdot u_k(r)$ and where $\langle 0|\hat{a}_k \hat{a}^\dagger_{k'}|0\rangle = \delta_{kk'}$ was used. Inserting this correlation into Eq. (13.25), using $\sum_\omega \sum_{k;\omega_k=\omega} = \sum_k$, and multiplying by $1 = \int_0^\infty d\omega \delta(\omega - \omega_k)$ ($\omega_k > 0$ for any mode k), we have

$$U = -\frac{\hbar}{4\varepsilon_0^2 c^2} \int_0^\infty d\omega \omega^3 \alpha_1(\omega)\alpha_2(\omega) \sum_{ij} G_{ij}(\omega, r_2, r_1) \sum_k \delta(\omega - \omega_k)u^j_k(r_1)u^{i*}_k(r_2) + \text{c.c.}, \tag{13.27}$$

where we used $\mu_0\varepsilon_0 = 1/c^2$. Then, considering that the spatial modes $u_k(r)$ are the eigenfunctions of the electric-field wave equation (13.19) [see Eq. (2.8), Chapter 1],

$$\nabla \times \nabla \times u_k(r) - \frac{\omega_k^2}{c^2}u_k(r) = 0, \tag{13.28}$$

they can be related to the dyadic Green function [11]

$$\sum_k \delta(\omega - \omega_k) u_k^j(\mathbf{r}') u_k^{i*}(\mathbf{r}) = \frac{2\omega}{\pi c^2} \text{Im} \{ G_{ij}(\omega, \mathbf{r}, \mathbf{r}') \}. \tag{13.29}$$

Exercise 13.1
Prove the general relation, Eq. (13.29): (i) Write the Green function as an expansion in the modes u_k, $G_j(\omega, \mathbf{r}, \mathbf{r}') = \sum_k b_k(\omega, \mathbf{r}') u_k(\mathbf{r})$. Find the expansion coefficients $b_k(\omega, \mathbf{r}')$. (ii) Take the imaginary part of $G_{ij} = \mathbf{e}_i \cdot \mathbf{G}_j$. *Hint:* you may first prove the identity (valid under integration and proven by contour integration),

$$\lim_{\eta \to 0^+} \frac{1}{(x + i\eta)^2 - x_0^2} = \text{P} \frac{1}{x^2 - x_0^2} - i \frac{\pi}{2x_0} [\delta(x - x_0) - \delta(x + x_0)],$$

where P denotes the principal value under integration.

Then U can be written completely in terms of the Green function $G_{ij}(\omega, \mathbf{r})$ as

$$U = -\frac{\hbar}{\pi \varepsilon_0^2 c^4} \int_0^\infty d\omega \, \omega^4 \alpha_1(\omega) \alpha_2(\omega) \sum_{ij} \text{Re} \{ G_{ij}(\omega, \mathbf{r}_2, \mathbf{r}_1) \} \text{Im} \{ G_{ij}(\omega, \mathbf{r}_2, \mathbf{r}_1) \}.$$

$$(13.30)$$

The expression (13.30) describes the van der Waals interaction energy U in an arbitrary embedding geometry in terms of the Green function of electromagnetic field propagation in that geometry, $G_{ij}(\omega, \mathbf{r}, \mathbf{r}')$, as formulated in classical electrodynamics. U has the form of a summation, or integration, over fluctuation-induced forces at all frequencies. The integrand contains a spectral overlap between the linear responses of the two dipoles $\alpha_{1,2}(\omega)$, a power-law ω^4 and the product of the real and the imaginary parts of the Green function. The real part originates in the propagation of the electromagnetic field that is scattered from one dipole to another, whereas the imaginary part results from the spatial correlation of the field fluctuations that drive the interaction (see also the discussion on the fluctuation-dissipation theorem in Chapter 3). In the van der Waals case at zero temperature the source of fluctuations is quantum mechanical, hence the prefactor $\hbar$ appears in U.

Therefore, while the van der Waals effect is driven by *quantum* fluctuations, Eq. (13.30) reveals that it is essentially a problem of scattering in *classical* electrodynamics. Namely, the calculation of U requires the knowledge of the classically calculated Green function of the electromagnetic field in the relevant embedding geometry, along with the polarisabilities of the dipoles, for all frequencies ω.

Free space: general expression
Let us apply the general formula for the interaction energy U, Eq. (13.30), in order to reproduce the Casimir and Polder result for the van der Waals energy. The dyadic Green function in free space is [11]

$$G_{ij}(k, \mathbf{r}) = \frac{e^{ikr}}{4\pi r} \left[\left(1 + \frac{ikr - 1}{k^2 r^2} \right) \delta_{ij} + \left(-1 + \frac{3 - 3ikr}{k^2 r^2} \right) \frac{r^i r^j}{r^2} \right], \tag{13.31}$$

where

$$k = \omega/c, \quad \mathbf{r} = \mathbf{r}_2 - \mathbf{r}_1, \quad r = |\mathbf{r}|, \quad r^i = \mathbf{e}_i \cdot \mathbf{r}. \tag{13.32}$$

Exercise 13.2
Verify that the free-space Green function G_{ij} from Eq. (13.31) satisfies Eq. (13.20).

We thus find,

$$
\begin{aligned}
\operatorname{Re}\{G_{ij}(k,\boldsymbol{r})\} &= \frac{1}{4\pi r}\left[\left(\cos(kr) - \frac{\sin(kr)}{kr} - \frac{\cos(kr)}{(kr)^2}\right)\delta_{ij}\right.\\
&\quad\left. + \left(-\cos(kr) + \frac{3\sin(kr)}{kr} + \frac{3\cos(kr)}{(kr)^2}\right)\frac{r^i r^j}{r^2}\right],\\
\operatorname{Im}\{G_{ij}(k,\boldsymbol{r})\} &= \frac{1}{4\pi r}\left[\left(\sin(kr) + \frac{\cos(kr)}{kr} - \frac{\sin(kr)}{(kr)^2}\right)\delta_{ij}\right.\\
&\quad\left. + \left(-\sin(kr) - \frac{3\cos(kr)}{kr} + \frac{3\sin(kr)}{(kr)^2}\right)\frac{r^i r^j}{r^2}\right]. \quad (13.33)
\end{aligned}
$$

Using

$$
\sum_{ij}\delta_{ij}\delta_{ij} = 3, \quad \sum_{ij}\frac{r^i r^j r^i r^j}{r^4} = 1, \quad \sum_{ij}\delta_{ij}\frac{r^i r^j}{r^2} = 1, \quad (13.34)
$$

with the expressions from (13.33), we get

$$
\begin{aligned}
&\sum_{ij}\operatorname{Re}\{G_{ij}(k,\boldsymbol{r})\}\operatorname{Im}\{G_{ij}(k,\boldsymbol{r})\}\\
&= \left(\frac{1}{4\pi r}\right)^2\left[\sin(2kr) + 2\frac{\cos(2kr)}{kr} - 5\frac{\sin(2kr)}{(kr)^2} - 6\frac{\cos(2kr)}{(kr)^3} + 3\frac{\sin(2kr)}{(kr)^4}\right].
\end{aligned}
$$
$$(13.35)$$

Then, by inserting the above into Eq. (13.30) we obtain the van der Waals interaction energy

$$
U = -\frac{\hbar}{16\pi^3\varepsilon_0^2 c^6}\int_0^\infty d\omega\,\omega^6\alpha_1(\omega)\alpha_2(\omega)F(r\omega/c), \quad (13.36)
$$

where

$$
F(x) = \frac{\sin(2x)}{x^2} + 2\frac{\cos(2x)}{x^3} - 5\frac{\sin(2x)}{x^4} - 6\frac{\cos(2x)}{x^5} + 3\frac{\sin(2x)}{x^6}. \quad (13.37)
$$

Eqs. (13.36) and (13.37) provide a general result for the van der Waals interaction at all ranges r. In the following we concentrate on the two limits of short and long distances r between the dipoles.

Free Space: London-van der Waals quasistatic regime
Considering distances r much shorter than any other relevant length-scale of the problem, we can keep only the dominant $1/r^6$ contribution in $F(r\omega/c)$, obtaining

$$
U \approx -\frac{3\hbar}{16\pi^3\varepsilon_0^2 r^6}\int_0^\infty d\omega\,\alpha_1(\omega)\alpha_2(\omega)\sin(2r\omega/c), \quad (13.38)
$$

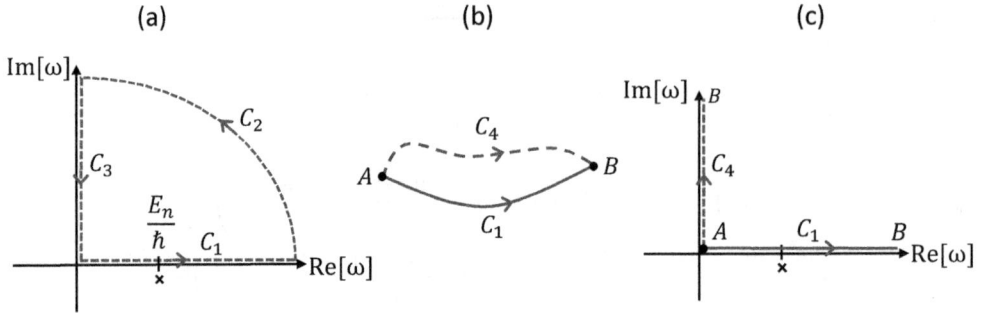

(a)   (b)   (c)

Figure 2.4: Wick rotation. This mathematical procedure is very common in Casimir physics for the evaluation of integrals over all positive frequencies (C_1) by transforming them into integrals on the imaginary frequency axis (C_3). Two equivalent approaches: (a) integration on a closed contour, see text below Eq. (13.41); (b) a deformation of any integration contour C_1 between two points A and B on the complex plane into a contour C_4 between the same points, and without crossing any poles (i.e. there are no poles enclosed in between C_1 and C_4), does not change the result of the integration. Likewise, an integral on the positive real axis, C_1 in (c), can be seen as a contour connecting the points $A = 0$ and $B = \infty$ and can be rotated to an integral on the imaginary axis, contour C_4: since ∞ is a unique point on the complex plane, B is also equal to $i\infty$, and since $\alpha(\omega)$ does not have any poles in the upper half plane (see text), the contours C_1 and C_4 do not enclose any poles, so that $I_1 = I_4 = -I_3$.

or

$$U \approx \mathrm{Im}\left\{I_1\right\}, \quad I_1 = -\frac{3\hbar}{16\pi^3\varepsilon_0^2 r^6} \int_0^\infty d\omega\, \alpha_1(\omega)\alpha_2(\omega)e^{i2r\omega/c}. \tag{13.39}$$

In order to make sense of this approximation, we note that all relevant length-scales in the integrand of Eq. (13.36) are contained in the dipolar polarisabilities $\alpha_{1,2}(\omega)$. For concreteness we consider the expression for the ground-state polarisability of an isotropic, randomly oriented system with a discrete spectrum of excited electronic states that are dipole-coupled to the ground state, as in the case of e.g. an atom or molecule [12] (see Problem 2.2),

$$\alpha(\omega) = \frac{2}{3}\sum_n \frac{E_n|\mathbf{d}_n|^2}{E_n^2 - \hbar^2\omega^2}. \tag{13.40}$$

Here n is an index running over the excited states with energies E_n above the ground-state energy and dipolar matrix elements $\mathbf{d}_n$ to the ground states. The relevant length-scales are then set by the corresponding wavelengths of the transitions $\lambda_n = 2\pi\hbar c/E_n$, so that the short-range approximation of Eqs. (13.38) and (13.39) typically holds for

$$r \ll \lambda_n = \frac{2\pi\hbar c}{E_n} \quad \forall n. \tag{13.41}$$

A common way to evaluate the integral I_1 in Eq. (13.39) is to move to a complex ω-plane as depicted in Fig. 2.4a: we consider the contour C that is comprised of three parts: C_1 running on the real axis from $\omega = 0$ to $\omega = \infty$, C_2 forming a quarter of a circle at absolute value $|\omega| \to \infty$ and C_3 on the imaginary axis from $i\infty$ to $i0$. Denoting the total integral by I and its components corresponding to C_i ($i = 1, 2, 3$) with I_i, we recognise that I_1 is indeed that from Eq. (13.39). Assuming the functions $\alpha_{1,2}(\omega)$ decay with $|\omega|$, $\alpha_{1,2}(\omega \to \infty) \to 0$, we can neglect I_2. This assumption is typically justified since in any case the real physics for sufficiently high frequencies goes beyond that of the dipolar description of the particles (see Section 15). Moreover, we assume that the poles of $\alpha_{1,2}(\omega)$ in the right half plane, $\text{Re}\,\{\omega\} > 0$, are found below the real axis (Fig. 2.4a).* For the polarisability model of Eq. (13.40) this amounts to slightly shifting the poles E_n below the real axis, $E_n \to E_n - i\eta$ with $\eta \to 0^+$. Physically, this means we add a width, or dissipation η, to the atomic transitions, later taken to zero. The mathematical consequence is that the contour C does not encircle any poles so that by the residue theorem we have

$$0 = I = I_1 + I_3 \quad \Rightarrow \quad U = -\text{Im}\,\{I_3\}. \tag{13.42}$$

The result of this procedure, known as the *Wick rotation*, is an integral for U over imaginary frequencies $\omega = iu$,

$$U \approx -\frac{3\hbar}{16\pi^3\varepsilon_0^2 r^6} \int_0^\infty du\, \alpha_1(iu)\alpha_2(iu) e^{-(2r/c)u}. \tag{13.43}$$

Here it is assumed that the product $\alpha_1(iu)\alpha_2(iu)$ is real, which is justified by the form of the polarisability in Eq. (13.40).† An alternative and equivalent description of the wick rotation is presented in Fig. 2.4, (b) and (c) (see caption).

By reconsidering the short-distance limit of (13.41), we further approximate $e^{-(2r/c)u} \approx 1$, obtaining the van der Waals energy as

$$U \approx -\frac{3\hbar}{16\pi^3\varepsilon_0^2 r^6} \int_0^\infty du\, \alpha_1(iu)\alpha_2(iu). \tag{13.44}$$

This result recovers the $1/r^6$ dependence we have previously found using electrostatic considerations. The reason is that by neglecting the exponential factor in the integrand of Eq. (13.43) within the short-range limit, we have eliminated all wave effects from the original electrodynamic problem and obtained its quasistatic limit.

For the case of the polarisability of Eq. (13.40), assuming identical atoms $\alpha_1 = \alpha_2 = \alpha$, we get

$$U \approx -\frac{1}{12\pi^3\hbar\varepsilon_0^2 r^6} \sum_n \sum_m |d_n|^2 |d_m|^2 \frac{r}{c} \int_0^\infty dx \frac{x_m}{x_m^2 + x^2} \frac{x_n}{x_n^2 + x^2}, \quad x_{n,m} \equiv \frac{E_{n,m} r}{\hbar c}, \tag{13.45}$$

with the integration variable $x = ur/c$. Using

$$\int_0^\infty dx \frac{a}{a^2 + x^2} \frac{b}{b^2 + x^2} = \frac{\pi}{2(a + b)} \tag{13.46}$$

---

*$\alpha(\omega)$ is a linear response function, hence is expected to be analytic in the upper half plane [1]. Therefore, if this function has any poles, they should be below the real axis.

†In fact, any linear response function is expected to be real on the imaginary-frequency axis [1].

we finally find

$$U \approx -\frac{1}{24\pi^2 \varepsilon_0^2 r^6} \sum_n \sum_m \frac{|\mathbf{d}_n|^2 |\mathbf{d}_m|^2}{E_n + E_m}, \tag{13.47}$$

which is London's result [2]. The result from the previous section, Eq. (13.8) is recovered by considering the contribution of a single dominant transition with energy $E = E_n = E_m$ and dipole element $d = |\mathbf{d}_n| = |\mathbf{d}_m|$.

Free Space: Casimir and Polder's retarded regime

Consider now the limit of large inter-dipolar distance r

$$r \gg \lambda_n = \frac{2\pi\hbar c}{E_n} \quad \forall n. \tag{13.48}$$

This is the so-called *retarded* regime, at which the separation between the dipoles is much larger than any of the typical dipolar wavelengths, so that wave effects of the mediating virtual photons become significant.

Examining the integral for the van der Waals energy in Eq. (13.36), the oscillatory function $F(r\omega/c)$ effectively cuts the integral around the inverse of its oscillation rate $2c/r$. Therefore, the major contribution to this integral originates from the frequency range

$$\omega \ll 2c/r \ll E_n/\hbar \quad \forall n, \tag{13.49}$$

for which $\alpha(\omega) \approx \alpha(0)$ [see Eq. (13.40) for $\alpha(\omega)$]. This means that in the retarded regime we can replace $\alpha_{1,2}(\omega)$ by $\alpha_{1,2}(0)$ in the energy expression Eq. (13.36), obtaining

$$U \approx -\frac{\hbar \alpha_1(0) \alpha_2(0)}{16\pi^3 \varepsilon_0^2 c^6} \int_0^\infty d\omega\, \omega^6 F(r\omega/c). \tag{13.50}$$

Moving to the dimensionless integration variable $x = \omega r/c$ this becomes

$$U \approx -\frac{\hbar c \alpha_1(0) \alpha_2(0)}{16\pi^3 \varepsilon_0^2 r^7} I \propto -\frac{1}{r^7}, \tag{13.51}$$

with I a dimensionless number given by the integral

$$I = \int_0^\infty dx\, x^6 F(x). \tag{13.52}$$

Equation (13.51) already provides us with an interesting feature of the van der Waals energy in the retarded regime, namely, its $1/r^7$ scaling with distance r, in contrast to the more familiar quasistatic scaling $1/r^6$. This result was first obtained by Casimir and Polder in 1948, where they examined modifications to the London quasistatic result due to retardation [4].

In order to finish the job, we need to evaluate the integral I. This integral diverges, however it can be handled by introducing a cut-off: our theory is a non-relativistic and low-energy theory that considers electrons bound to atoms and treats atoms as dipoles (see Section 15). Hence, physically, there exists an upper frequency cut-off for its validity, ω_{co}, which in turn implies a dimensionless cut-off $b = \omega_{co}r/c$ in the integral I. Mathematically, we first use an exponential cut-off: instead of replacing the upper limit of the integration by b, we smoothly cut the integral by multiplying the integrand with an exponential function $e^{-x/b}$, thus removing the divergence and *regularising* the integral. We can then take the limit $b \to \infty$ and obtain a result independent of the cut-off b,

$$I = \lim_{b \to \infty} \int_0^\infty dx x^6 F(x) e^{-x/b} = \lim_{b \to \infty} \frac{16b^6(11 - 56b^2 + 368b^4)}{(1 + 4b^2)^5} = \frac{23}{4}. \tag{13.53}$$

Finally, inserting this result for I into Eq. (13.51), we obtain Casimir and Polder's result [4] for the retarded van der Waals energy,

$$U \approx -\frac{23\hbar c}{64\pi^3 \varepsilon_0^2} \frac{\alpha_1(0)\alpha_2(0)}{r^7}. \tag{13.54}$$

Confined geometry: one dimension

Let us now consider a simple example of an embedding geometry other than free-space, namely, that of one dimension (1d). The field is confined to propagate only along the z-axis and it is polarised along the x and y transverse plane,

$$\hat{\boldsymbol{E}} = \hat{E}^x \boldsymbol{e}_x + \hat{E}^y \boldsymbol{e}_y. \tag{13.55}$$

This seemingly pedagogical problem is in fact realised in electric transmission-line structures [15, 16], for which

$$\boldsymbol{\nabla} \cdot \hat{\boldsymbol{E}} = \partial_x \hat{E}^x + \partial_y \hat{E}^y = 0, \tag{13.56}$$

typical of a 1d system along the z axis. Then, using the identity $\boldsymbol{\nabla} \times \boldsymbol{\nabla} \times = -\boldsymbol{\nabla}^2 + \boldsymbol{\nabla}\boldsymbol{\nabla}\cdot$ in the wave equation (13.19) together with Eqs. (13.20) and (13.56), we find that the dyadic Green function is a scalar

$$G_{ij} = g\delta_{ij}, \tag{13.57}$$

δ_{ij} being the Kronecker delta. Moreover, for 1d propagation the Green function depends spatially only on z such that Eq. (13.20) finally becomes

$$(\partial_z^2 + k^2)g(k, z - z') = -\frac{1}{A}\delta(z - z'), \tag{13.58}$$

with $k = \omega/c$ as before. A is the effective area of the field modes that propagate in 1d. In waveguides or transmission lines A is typically proportional to the area where the electric field is confined in the transverse direction. Solving this equation for $g(k, z - z')$ we find

$$g(k, z - z') = \frac{i}{2Ak} e^{ik|z-z'|}, \tag{13.59}$$

yielding

$$\mathrm{Re}\{g(k, z - z')\} = -\frac{1}{2Ak}\sin(k|z - z'|),$$

$$\mathrm{Im}\{g(k, z - z')\} = \frac{1}{2Ak}\cos(k|z - z'|). \tag{13.60}$$

Exercise 13.3

Find the scalar Green function $g(k, z - z')$ in 1d, Eq. (13.59). *Hint*: take the Fourier transform of Eq. (13.58) and then perform the inverse Fourier transform by contour integration.

Inserting this Green function into our general van der Waals formula, Eq. (13.30) we obtain

$$U = \frac{\hbar c}{4\pi\varepsilon_0^2 A^2} \int_0^\infty dk\, \alpha_1(ck)\alpha_2(ck)k^2 \sin(2k|z_2 - z_1|). \tag{13.61}$$

Performing a Wick rotation, as we did for the quasistatic regime in free-space, we finally obtain the 1d van der Waals interaction in a converging integral form

$$U = -\frac{\hbar c}{4\pi\varepsilon_0^2 A^2} \int_0^\infty du\, \alpha_1(icu)\alpha_2(icu)u^2 e^{-2u|z_2 - z_1|}. \tag{13.62}$$

This integral can be performed analytically using the polarisability of Eq. (13.40) yielding a result expressed with special functions [16]. The behaviour of U in the retarded regime,

$$|z_1 - z_2| \gg \lambda_n = \frac{2\pi\hbar c}{E_n} \quad \forall n, \tag{13.63}$$

is simple to obtain. As in free space, we replace $\alpha_{1,2}(icu)$ by $\alpha_{1,2}(0)$, perform the integration and get

$$U = -\frac{\hbar c}{16\pi\varepsilon_0^2} \frac{\alpha_1(0)\alpha_2(0)}{A^2} \frac{1}{|z_1 - z_2|^3}. \tag{13.64}$$

This result reveals a $1/r^3$ scaling with distance $r = |z_1 - z_2|$, which decays much more slowly than the $1/r^7$ scaling in free space [Eq. (13.54)]. This means that van der Waals interaction carried by vacuum fluctuations that propagate in 1d has a much longer range than its 3d, free-space counterpart. Intuitively, vacuum fluctuations can be "transmitted" through 1d-like electrical wires just as electronic signals are, establishing a long-range van der Waals interaction.

Furthermore, comparing the coefficients of the 1d (U_{1d}) and free space (U_{fs}) results, Eqs. (13.64) and (13.54) respectively, we find that their ratio scales as

$$\frac{U_{1d}}{U_{fs}} \propto \frac{A^2}{r^4}. \tag{13.65}$$

Roughly speaking, the 1d interaction becomes dominant as the confining length scale $\sqrt{A}$ becomes much smaller than the typical inter-dipolar distance r. This matches the intuition that 1d behaviour is a characteristic of systems that are tightly confined in the transverse dimensions.

13.2.2 Quantum electrodynamics perturbation theory

In Chapter 1, Section 5, several vacuum-induced effects on atoms were surveyed, including a brief discussion of the van der Waals interaction. The framework used for the analysis was time-independent perturbation theory of quantum electrodynamics. This non-relativistic variant of quantum electrodynamics provides a useful technical and conceptual tool in atomic, molecular and optical physics [12, 17, 18]. From the technical side it provides a systematic perturbative approach towards various dispersive, vacuum-induced effects, by associating them with different orders of energy corrections for atomic systems due to the atom-vacuum interaction: for example, the Lamb shift is related to a second order energy correction of a single atom, whereas the van der Waals interaction is described by a fourth order correction of a two-atom system. Moreover, the geometry-dependence of the electric field modes enters here via the mode functions $\boldsymbol{u}_k(\boldsymbol{r})$ rather than via the Green function $G_{ij}(\omega, \boldsymbol{r}, \boldsymbol{r}')$. This can become very useful in certain waveguide or cavity geometries where the normal modes are known analytically and are more convenient to work with (see also Section 14.2 below).

From a more conceptual point of view the perturbative technique and its associated diagrams, as presented in Chapter 1, put forward the concept of virtual photons. The scattering approach highlights the characterisation of the vacuum as a noisy field, the quantum essence of which enters via its correlations taken as those of quantum zero-point fluctuations. Here instead, the noisy character is not explicitly used, however quantum fluctuations enter via the possibility for intermediate, non-energy conserving and short lived processes, where virtual photons are created and destroyed as per Heisenberg's time-energy uncertainty relation. As we shall see below, this allows in some limiting cases to identify the dominant intermediate processes and to consider only the diagrams associated with them in the calculation.

Another principal difference is that the quantum electrodynamics approach treats both the dipoles and the field as dynamical variables, hence both are quantised and appear in the Hamiltonian. This is in contrast to the scattering approach used above, where the dipoles were effectively traced out and merely characterised by their response to the field, $\alpha(\omega)$.

We begin our discussion of the van der Waals interaction by considering two identical polarisable dipoles 1 and 2, both possessing a ground level $|g\rangle$ and a discrete set of excited levels $\{|n\rangle\}$ with energies E_n above that of $|g\rangle$. The two dipoles and the quantum field, with Hamiltonians $\hat{H}_A$ and $\hat{H}_F$, respectively,

$$\hat{H}_A = \sum_{\nu=1}^{2} \sum_{n_\nu} E_{n_\nu} |n_\nu\rangle\langle n_\nu|, \quad \hat{H}_F = \sum_k \hbar\omega_k \hat{a}_k^\dagger \hat{a}_k, \tag{13.66}$$

are coupled via the Hamiltonian $\hat{H}_{AF}$ [Eq. (5.6), Chapter 1],

$$\hat{H}_{AF} = -\sum_{\nu=1}^{2} \hat{\boldsymbol{d}}_\nu \cdot \hat{\boldsymbol{E}}(\boldsymbol{r}_\nu) = -\hbar \sum_{\nu=1}^{2} \sum_{n_\nu} \sum_k \left(|n_\nu\rangle\langle g_\nu| + |g_\nu\rangle\langle n_\nu|\right) \left(ig_{kn_\nu}\hat{a}_k - ig_{kn_\nu}^*\hat{a}_k^\dagger\right). \tag{13.67}$$

Here

$$\hat{\boldsymbol{E}}(\boldsymbol{r}) = \sum_k i\sqrt{\frac{\hbar\omega_k}{2\varepsilon_0}} \left[\boldsymbol{u}_k(\boldsymbol{r})\hat{a}_k - \boldsymbol{u}_k^*(\boldsymbol{r})\hat{a}_k^\dagger\right] \tag{13.68}$$

is the quantised Hermitian electric field, used in the previous Section as the driving vacuum fluctuations from Eq. (13.12), and

$$g_{kn_\nu} = \sqrt{\frac{\omega_k}{2\varepsilon_0\hbar}} \boldsymbol{d}_{n_\nu} \cdot \boldsymbol{u}_k(\boldsymbol{r}_\nu) \qquad (13.69)$$

is the dipolar coupling between the dipole $\nu = 1, 2$ and the k field mode, for the $|g\rangle$ to $|n\rangle$ transition with dipole matrix element $\boldsymbol{d}_n = \langle n|\hat{\boldsymbol{d}}|g\rangle$ taken to be real.

The van der Waals energy is obtained by perturbation theory as the fourth-order correction to the energy of the global unperturbed ground state $|G\rangle = |g_1, g_2, 0\rangle$, where both atoms are in their ground state $|g\rangle$ and all field modes k are in the vacuum $|0\rangle$ [12],

$$U = -\sum_{I_1, I_2, I_3} \frac{\langle G|\hat{H}_{AF}|I_3\rangle\langle I_3|\hat{H}_{AF}|I_2\rangle\langle I_2|\hat{H}_{AF}|I_1\rangle\langle I_1|\hat{H}_{AF}|G\rangle}{(E_{I_1} - E_G)(E_{I_2} - E_G)(E_{I_3} - E_G)}. \qquad (13.70)$$

Here $|I_j\rangle$ are intermediate (virtual) states and E_m is the energy of the state $|m\rangle$, in analogy to the general expression for the second order energy correction of Eq. (5.13) (Chapter 1). The summation over all possible sets of virtual processes can be organised by grouping them in diagrams, as in the example of Fig. 1.7 (Chapter 1). In total, there are 12 kinds of diagrams [12], 6 of them presented in Fig. 2.5, where the other six are obtained by replacing the indices n_1 and n_2, i.e. changing the roles of dipoles 1 and 2. For example, diagram 1 presents the process with intermediate states,

$$|I_1\rangle = |n_1, g_2, 1_k\rangle, \quad |I_2\rangle = |g_1, g_2, 1_k 1_{k'}\rangle, \quad |I_3\rangle = |g_1, n_2, 1_{k'}\rangle, \qquad (13.71)$$

where $|1_k\rangle = \hat{a}_k^\dagger|0\rangle$. The physical picture provided by the diagram is that of a virtual photon in mode k that is emitted from dipole 1 while it goes to the excited state $|n_1\rangle$ in vertex I, followed by this dipole's transition back to the ground state $|g_1\rangle$ while emitting a virtual photon of mode k' (vertex II). Then, the k virtual photon is absorbed by dipole 2 which goes to the excited state $|n_2\rangle$ in vertex III. Finally dipole 2 comes back to its ground state $|g_2\rangle$ by absorbing a virtual photon k' (vertex IV). The transitions in some of the vertices, I and IV in particular, may seem strange at first sight, however they are fully supported by the interaction Hamiltonian $\hat{H}_{AF}$, which contains matrix elements for all transitions mentioned above. We shall get back to the physical meaning of these processes in our discussion on virtual photons below. The energies, with respect to that of $|G\rangle$, which correspond to the above states, are given by

$$E_{I_1} - E_G = \hbar\omega_k + E_{n_1}, \quad E_{I_2} - E_G = \hbar\omega_k + \hbar\omega_{k'}, \quad E_{I_3} - E_G = \hbar\omega_{k'} + E_{n_2}. \quad (13.72)$$

Upon inserting the intermediate states and their energies from Eqs. (13.71) and (13.72) into Eq. (13.70) for the van der Waals energy, we obtain the contribution of diagram 1 to the interaction energy U,

$$U_1 = -\hbar^4 \sum_{n_1, n_2} \sum_{k, k'} \frac{g_{kn_1}^* g_{k'n_1}^* g_{kn_2} g_{k'n_2}}{(E_{n_1} + \hbar\omega_k)(\hbar\omega_k + \hbar\omega_{k'})(E_{n_2} + \hbar\omega_{k'})}. \qquad (13.73)$$

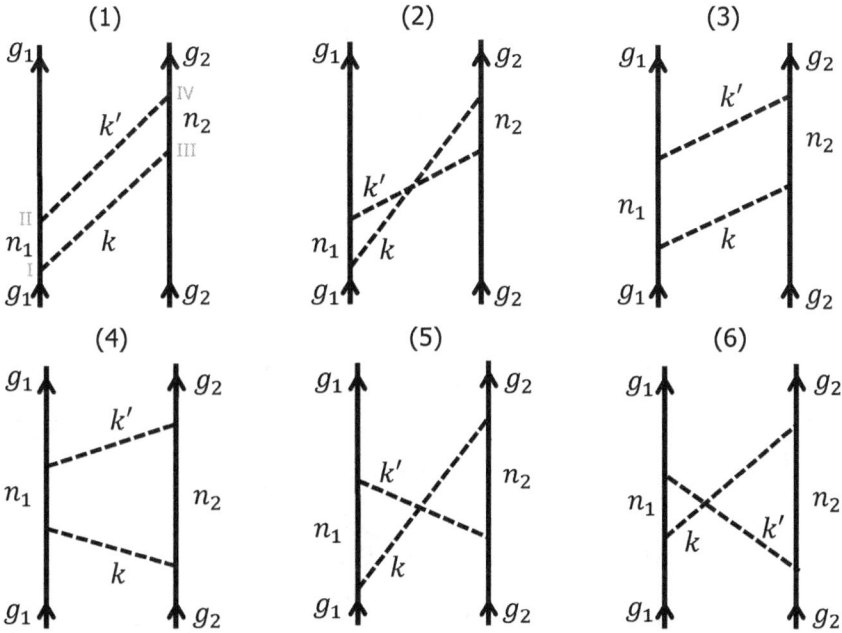

Figure 2.5: Quantum electrodynamics perturbation-theory approach to van der Waals interaction. Six of 12 possible time-ordered diagrams representing virtual processes that contribute to the energy correction of the state $|G\rangle = |g_1, g_2, 0\rangle$, Eq. (13.70). The other six diagrams are obtained by replacing the indices n_1 and n_2. Here dipolar states are represented by full lines whereas (virtual) photon propagation is represented by dashed lines.

The summations over the dipolar excited states $|n_{1,2}\rangle$ and the photon modes k, k' are due the fact that diagram 1 represents a family of virtual processes where any pair of excited dipolar states or photon modes can participate. Hence, the sum over intermediate states $|I_{1,2,3}\rangle$ includes all possible sets of states of the form shown in Eq. (13.71). Likewise, the energy contribution of the diagram discussed in Chapter 1 (Fig. 1.7 therein), denoted in Fig. 2.5 as diagram 3, becomes

$$U_3 = -\hbar^4 \sum_{n_1,n_2} \sum_{k,k'} \frac{g^*_{kn_1} g_{kn_2} g^*_{k'n_1} g_{k'n_2}}{(E_{n_1} + \hbar\omega_k)(E_{n_1} + E_{n_2})(E_{n_2} + \hbar\omega_{k'})}. \tag{13.74}$$

To conclude, the total energy U, summed over all diagrams $q = 1, 2, ..., 12$ is given by

$$U = \sum_{q=1}^{12} U_q = -\hbar^4 \sum_{n_1,n_2} \sum_{k,k'} \sum_{q=1}^{12} \frac{N_q}{D_q}. \tag{13.75}$$

Here N_q is the numerator of the energy contribution of diagram q,

$$N_q = \begin{cases} g^*_{kn_1} g^*_{k'n_1} g_{kn_2} g_{k'n_2}, & \text{for } q = 1, 2, 3; \\ g_{kn_1} g^*_{k'n_1} g^*_{kn_2} g_{k'n_2}, & \text{for } q = 4, 6; \\ g_{k'n_1} g^*_{kn_1} g^*_{k'n_2} g_{kn_2}, & \text{for } q = 5; \end{cases} \tag{13.76}$$

where for $q \in [6, 12]$ the indices n_1 and n_2 have to be replaced. The denominators D_q are given by [12]

$$
\begin{aligned}
D_1 &= (E_{n_1} + \hbar\omega_k)(\hbar\omega_k + \hbar\omega_{k'})(E_{n_2} + \hbar\omega_{k'}), \\
D_2 &= (E_{n_1} + \hbar\omega_k)(\hbar\omega_k + \hbar\omega_{k'})(E_{n_2} + \hbar\omega_k), \\
D_3 &= (E_{n_1} + \hbar\omega_k)(E_{n_1} + E_{n_2})(E_{n_2} + \hbar\omega_{k'}), \\
D_4 &= (E_{n_2} + \hbar\omega_k)(E_{n_1} + E_{n_2})(E_{n_2} + \hbar\omega_{k'}), \\
D_5 &= (E_{n_1} + \hbar\omega_k)(E_{n_1} + E_{n_2} + \hbar\omega_k + \hbar\omega_{k'})(E_{n_2} + \hbar\omega_k), \\
D_6 &= (E_{n_2} + \hbar\omega_{k'})(E_{n_1} + E_{n_2} + \hbar\omega_k + \hbar\omega_{k'})(E_{n_2} + \hbar\omega_k), \\
D_7 &= (E_{n_2} + \hbar\omega_k)(\hbar\omega_k + \hbar\omega_{k'})(E_{n_1} + \hbar\omega_{k'}), \\
D_8 &= (E_{n_2} + \hbar\omega_k)(\hbar\omega_k + \hbar\omega_{k'})(E_{n_1} + \hbar\omega_k), \\
D_9 &= (E_{n_2} + \hbar\omega_k)(E_{n_1} + E_{n_2})(E_{n_1} + \hbar\omega_{k'}), \\
D_{10} &= (E_{n_2} + \hbar\omega_k)(E_{n_1} + E_{n_2})(E_{n_1} + \hbar\omega_k), \\
D_{11} &= (E_{n_2} + \hbar\omega_k)(E_{n_1} + E_{n_2} + \hbar\omega_k + \hbar\omega_{k'})(E_{n_1} + \hbar\omega_k), \\
D_{12} &= (E_{n_1} + \hbar\omega_{k'})(E_{n_1} + E_{n_2} + \hbar\omega_k + \hbar\omega_{k'})(E_{n_1} + \hbar\omega_k).
\end{aligned} \tag{13.77}
$$

Eqs. (13.75)-(13.77) provide a general recipe for the calculation of the van der Waals interaction between a pair of dipoles embedded in an arbitrary geometry: the embedding geometry imposes boundary conditions on Eq. (13.28) for the electric field modes, and hence determines the spatial functions $u_k(r)$ and frequencies ω_k of the modes. These in turn determine g_{kn_ν} via Eq. (13.69), and therefore the expression for U.

Free space

Using the modes from Eq. (13.13), the summation over the general mode index k is replaced by an integration over wave vectors and a summation over polarisations, $\frac{V}{(2\pi)^3}\int d\mathbf{k}\sum_\lambda$,

$$U = -\sum_{n_1,n_2}\int\frac{d\mathbf{k}}{(2\pi)^3}\int\frac{d\mathbf{k'}}{(2\pi)^3}\sum_{\lambda,\lambda'}\sum_{q=1}^{12}\frac{\hbar\omega_k}{2\varepsilon_0}\frac{\hbar\omega_{k'}}{2\varepsilon_0}$$

$$\times(\mathbf{e}_{k\lambda}^*\cdot\mathbf{d}_{n_1})(\mathbf{e}_{k'\lambda'}^*\cdot\mathbf{d}_{n_1})(\mathbf{e}_{k\lambda}\cdot\mathbf{d}_{n_2})(\mathbf{e}_{k'\lambda'}\cdot\mathbf{d}_{n_2})\frac{e^{i(\mathbf{k}+\mathbf{k'})\cdot\mathbf{r}}}{D_q}, \qquad (13.78)$$

with $\mathbf{r} = \mathbf{r}_1 - \mathbf{r}_2$ and where the numerators of all diagrams are equivalent. In Ref. [12], it is shown how by first summing the contributions of all diagrams, i.e. performing the sum $\sum_q$ in Eq. (13.78), and only then performing the integrations, the Casimir and Polder result [4] is obtained,

$$U = -\frac{1}{36\pi^3\varepsilon_0^2\hbar c r^2}\sum_{n_1,n_2}|\mathbf{d}_{n_1}|^2|\mathbf{d}_{n_2}|^2\int_0^\infty du\frac{k_{n_1}k_{n_2}u^4e^{-2ur}}{(k_{n_1}^2+u^2)(k_{n_2}^2+u^2)}$$

$$\times\left[1+\frac{2}{ur}+\frac{5}{(ur)^2}+\frac{6}{(ur)^3}+\frac{3}{(ur)^4}\right], \qquad (13.79)$$

with $k_n = E_n/(\hbar c)$. This expression for the van der Waals energy is valid at all inter-dipolar distances r and is equivalent to Eq. (13.36) found above. Below we focus only on the two limiting cases of the retarded and nonretarded regimes.

Exercise 13.4
Use the Wick rotation technique and the polarisability from Eq. (13.40) in order to put the general result obtained in Section 13.2.1, Eq. (13.36), in the same form obtained by Casimir and Polder, Eq. (13.79).

Virtual photons revisited

Getting back to diagram 1 for example, we note that the interaction in the first vertex, representing the transition $|G\rangle \to |I_1\rangle$, does not conserve energy: it describes the *emission* of a photon from dipole 1, initially in the *ground* state, after which the dipole jumps up to the *excited* state $|n_1\rangle$. Namely, beginning with the ground state $|G\rangle$ the system moves to the excited state $|I_1\rangle$ gaining an energy of $E_{I_1} - E_G = \hbar\omega_k + E_{n_1}$ that comes essentially from nowhere. This transition is enabled by the *non-resonant* term $\hat{a}_k^\dagger|n_1\rangle\langle g_1|$ in the Hamiltonian which is typically neglected in resonant quantum-optics calculations [18]. Non-resonant, non-energy conserving processes are in fact allowed in quantum mechanics for short enough times. This follows from the Heisenberg uncertainty relation, typical of Fourier transforms, and in particular its energy-time variant,

$$\Delta E\Delta t \le \frac{\hbar}{2}. \qquad (13.80)$$

Therefore, processes where an energy ΔE is not conserved can last for durations of the order $\Delta t \leq \hbar/\Delta E$. This means that the missing energy ΔE has to be restored at a time $\Delta t \leq \hbar/\Delta E$. Indeed, the final state in the diagram is the initial state $|G\rangle$, so that all photons created from the vacuum in intermediate states are finally given back to the vacuum. This is also why these photons are called *virtual photons*: they can never be picked up and *recorded* by exciting and inducing a signal current in a detector system, since otherwise these photons would never be annihilated back to the vacuum, consequently violating energy conservation for infinitely-long times.

The fourth vertex also represents a strictly non-energy conserving interaction since it describes the *annihilation* of a photon, more precisely a virtual photon, accompanied by a *decrease* in the energy of the dipole, generated by the *non-resonant* Hamiltonian term $\hat{a}_k |g_2\rangle\langle n_2|$. Even the interactions of vertices II and III do not in general conserve energy, since $\hbar\omega_{k'}$ and $\hbar\omega_k$ are not in general equal to E_{n_1} and E_{n_2}, respectively.

The important thing to remember with virtual processes, however, is that energy has to be conserved in the *overall* process, rather than on a single vertex. Energy conservation is satisfied provided that the energy of the initial state and the final state of the complete diagram are equal. In the van der Waals interaction, in all diagrams, both initial and final states are identical, $|G\rangle$, such that energy conservation of the complete process is trivially satisfied.

Free space: Casimir and Polder's retarded regime

The expression for the van der waals energy in free space, Eq. (13.78), contains integrations over $\boldsymbol{k}$ and $\boldsymbol{k'}$ with an oscillatory integrand $e^{i(\boldsymbol{k}+\boldsymbol{k'})\cdot\boldsymbol{r}}$. Then, the integrals are effectively cut around $|\boldsymbol{k}| \sim 1/r$, with $r = |\boldsymbol{r}|$. Consider the limit of large inter-dipolar distances r with respect to all relevant transition wavelength, inequality (13.48). Then, as argued previously, the major contribution to the integral originates from frequencies

$$\omega_k = c|\boldsymbol{k}| \ll c/r \ll E_n/\hbar \quad \forall n. \tag{13.81}$$

Using the concept of virtual photons, we can also find a more physical explanation for the dominance of intermediate processes involving low frequencies as per inequality (13.81). Since the dipoles are at a distance r apart and the interaction-mediating virtual photons travel at speed c, their required lifetime Δt is at least r/c. Assuming also that r is larger than all transition wavelengths $r \gg \lambda_n$, as in (13.48), we thus have

$$2\pi\hbar/E_n = \lambda_n/c \ll r/c \leq \Delta t \quad \forall n. \tag{13.82}$$

On the other hand, the creation of a virtual photon at frequency ω requires an energy mismatch of at least $\hbar\omega$. Time-energy uncertainty, inequality (13.80), then states that the lifetime of this virtual photon is bounded by

$$\Delta t \leq \frac{1}{2\omega}. \tag{13.83}$$

Combining these lower and upper bounds on Δt, we have

$$\omega \ll E_n/\hbar \quad \forall n, \tag{13.84}$$

which gives the range of virtual-photon frequencies that contribute significantly to the interaction, as in the inequality (13.81).

Since the diagrams which contain the smallest denominators are expected to be dominant, inequality (13.81) implies that it is enough to consider contributions such as diagram 1, with a factor $(\hbar\omega_k + \hbar\omega_{k'})$ in their denominator and neglect all other diagrams. Since there are 4 such diagrams [see Eq. (13.77)], we can approximate U by 4 times the contribution of diagram 1, yielding

$$
U = -4 \sum_{n_1,n_2} \int \frac{d\mathbf{k}}{(2\pi)^3} \int \frac{d\mathbf{k'}}{(2\pi)^3} \sum_{\lambda,\lambda'} \frac{\hbar c|\mathbf{k}|}{2\varepsilon_0} \frac{\hbar c|\mathbf{k'}|}{2\varepsilon_0}
$$

$$
\times \sum_i e_{\mathbf{k}\lambda}^{i*} d_{n_1}^i \sum_j e_{\mathbf{k'}\lambda'}^{j*} d_{n_1}^j \sum_l e_{\mathbf{k}\lambda}^l d_{n_2}^l \sum_p e_{\mathbf{k'}\lambda'}^p d_{n_2}^p \frac{e^{i(\mathbf{k}+\mathbf{k'})\cdot\mathbf{r}}}{E_{n_1} E_{n_2} (\hbar c|\mathbf{k}| + \hbar c|\mathbf{k'}|)}.
$$

$$(13.85)$$

Here the denominator D_1 was further approximated using inequality (13.81), and $e_{\mathbf{k}\lambda}^i = \mathbf{e}_i \cdot \mathbf{e}_{\mathbf{k}\lambda}$, $d_n^i = \mathbf{e}_i \cdot \mathbf{d}_n$ are projections of the polarisation vector and dipole matrix element, respectively. Using the sum over polarisations [12],

$$
\sum_\lambda e_{\mathbf{k}\lambda}^l e_{\mathbf{k}\lambda}^{i*} = \delta_{li} - e_{\mathbf{k}}^l e_{\mathbf{k}}^i,
$$

$$(13.86)$$

with $e_{\mathbf{k}}^i = \mathbf{e}_i \cdot \mathbf{e}_{\mathbf{k}}$ the $i = x, y, z$ projection of the unit vector on the $\mathbf{k}$ wave vector direction, and assuming randomly oriented dipoles,

$$
d_{n_1}^i d_{n_1}^j \to \frac{1}{3}|\mathbf{d}_{n_1}|\delta_{ij}, \quad d_{n_2}^l d_{n_2}^p \to \frac{1}{3}|\mathbf{d}_{n_2}|\delta_{lp},
$$

$$(13.87)$$

we obtain

$$
U = -\frac{\hbar c}{4\varepsilon_0^2} \sum_{n_1,n_2} \frac{4}{9} \frac{|\mathbf{d}_{n_1}|^2 |\mathbf{d}_{n_2}|^2}{E_{n_1} E_{n_2}} \int \frac{d\mathbf{k}}{(2\pi)^3} \int \frac{d\mathbf{k'}}{(2\pi)^3} [1 + (\mathbf{e}_k \cdot \mathbf{e}_{k'})^2] \frac{|\mathbf{k}||\mathbf{k'}|e^{i(\mathbf{k}+\mathbf{k'})\cdot\mathbf{r}}}{|\mathbf{k}| + |\mathbf{k'}|}.
$$

$$(13.88)$$

Finally, by performing the integrations in spherical coordinates $\mathbf{k} = (k, \Omega)$, and making use of the expression for the polarisability in Eq. (13.40), the Casimir-Polder retarded result, Eq. (13.54), is reached [12].

Free Space: London-van der Waals quasistatic regime

The result for the short-range quasistatic regime can be obtained in free space by considering the opposite limit than that of inequality (13.81), namely, that of high frequencies. The idea is that short distances require only short virtual-photon lifetimes so that high-frequency virtual photons can become significant. Then, diagrams like diagram 3 have the smallest denominator and are hence considered as dominant, so that only these 4 diagrams [see Eq. (13.77)] are considered. We further approximate their denominators by $D_3 \approx \hbar\omega_k (E_{n_1} + E_{n_2})\hbar\omega_{k'}$, and using methods similar to those shown above for the retarded regime, London's result, Eq. (13.47), is obtained [12].

Exercise 13.5
Which diagrams out of the 12 contribute to the van der Waals interaction energy in the retarded regime? Which contribute in the quasistatic regime?

Confined geometry: one dimension and waveguides

Finally, we consider transverse electric field modes of plane waves that propagate along the z axis, characterised by the wavenumber $k \in (-\infty, \infty)$ and a polarisation index $j = x, y$,

$$u_{kj}(z) = \frac{1}{\sqrt{AL}} e^{ikz} e_j, \tag{13.89}$$

A being the effective area of the mode and L the quantisation length along z. Sums over these modes are then converted to integrals via

$$\sum_j \sum_k \rightarrow \sum_j \frac{L}{2\pi} \int_{-\infty}^{\infty} dk. \tag{13.90}$$

Using these modes in the general formula (13.75) and performing the resulting integrals over k and k' the van der Waals interaction in 1d is obtained [16] (Problem 2.4).

The above perturbative quantum-electrodynamics formalism is particularly useful for confining geometries that are more commonly characterised by their mode functions u_k. For a waveguide or fibre where the propagation is along z, the mode function of a guided mode typically has the form

$$u_{\beta m}(r) = E_{\beta m}(x, y) \frac{1}{\sqrt{L}} e^{i\beta z}. \tag{13.91}$$

Here β is the wavenumber along the propagation axis and m is the index of the transverse mode. Typically m includes three indices: one for the polarisation and two for the transverse xy dependence; e.g. TE_{01} and TM_{11} for the lowest order transverse-electric (TE) and transverse magnetic (TM) modes of a hollow metallic waveguide, where TE and TM denote the polarisation and 01 and 11 denote the xy spatial profile, or HE_{11} for the fundamental hybrid (HE) mode of an optical fibre [24]. A crucial point for the calculation of vacuum energies is that the transverse profile $E_{\beta m}(x, y)$ may depend on β and hence on frequency via the dispersion relation of the guided mode $w(\beta)$. This w or β dependence must be taken into account when integrating over *all* frequency- or wavenumber-range contained in the vacuum and relevant for the physical situation. This is in contrast to the 1d case from Eq. (13.89), relevant to transmission lines [16], where the transverse profile $e_j/\sqrt{A}$ does not dependent on frequency. For example, in a closed structure such as a metallic waveguide, a van der Waals calculation based on the above quantum electrodynamics formalism reveals the possibility for an interaction potential that is exponentially decaying with the inter-dipolar distance [19].

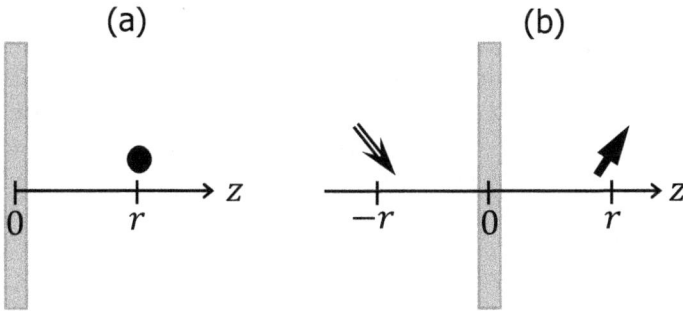

Figure 2.6: Interaction between a mirror and a polarisable particle. (a) The particle is placed at a distance r from the surface of an infinite mirror that extends over the entire xy plane at $z = 0$. (b) Quasistatic analysis by the method of images: the particle possesses a dipole moment (full arrow at $z = r$) due to its internal random dipole fluctuations. An image dipole is effectively created due to the mirror (open arrow at $z = -r$). The electrostatic dipole-dipole interaction between the dipole and its image is then interpreted as the dipole-mirror potential.

§14. CASIMIR-POLDER ATOM-SURFACE INTERACTION

In Section 13 we considered two polarisable particles that scatter vacuum fluctuations (or virtual photons) onto each other, resulting in an interaction between them. Imagine now we replace one of the particles with a mirror (Fig. 2.6a), where the term mirror is used here to describe any bulk or composite system that can reflect electromagnetic fields. Since the spatial dependence of the scattered field off an extended, large surface is very different from that scattered by a particle, we expect the interaction energy to scale very differently than the case of two particles. Moreover, the mirror, being comprised of many atoms, is a much stronger scatterer. Overall, we expect a modified and stronger interaction between a particle and a mirror compared to the van der Waals interaction between two particles.

Such dispersion forces exerted on atoms by surfaces, widely known as the *Casimir-Polder interaction*, have been studied for dielectric and metallic surfaces, in thermal equilibrium and in vacuum, as well as out of equilibrium [6–9, 25, 26]. In the following, we restrict our discussion to that of a polarisable particle in front of a perfectly-conducting mirror, both in vacuum. This simple case nevertheless demonstrates the main features and concepts of vacuum-induced interactions.

14.1 Quasistatic analysis

We consider a polarisable particle placed at $z = r$ in front of an infinite mirror made from a perfect conductor at $z = 0$ (Fig. 2.6a). Assuming random fluctuations in the electron motion inside the particle, as discussed in Section 13, the particle is temporarily polarised and has a dipole moment $\boldsymbol{d}$. For distances r shorter than any dipolar wavelength [inequality (13.41)], we can use the image method known from electrostatics, as shown in Fig. 2.6b: in order to account for the effect of the conductor on the region $z > 0$, we need to satisfy the boundary conditions imposed on the electric field, i.e. that its component parallel to the mirror vanishes on the $z = 0$ plane. This can be accomplished by introducing a fictitious image dipole at $z = -r$. Then, the potential energy of the particle in the presence of the surface is given by the dipole-dipole interaction between the dipole and its image. Since this interaction scales as $1/r^3$ [Eq. (13.4)], we obtain a longer-range interaction compared to the van der Waals $1/r^6$ counterpart, as expected.

More quantitatively, we note that the z component of the image dipole d_I is identical to that of the original dipole d, whereas their xy components are opposite (Fig. 2.6b),

$$d_I = (d \cdot e_z)e_z - \sum_{i=x,y}(d \cdot e_i)e_i = 2(d \cdot e_z)e_z - d. \tag{14.1}$$

The energy of the original dipole due to the mirror is then obtained by half the electrostatic interaction energy between the dipole and its image at a distance $2r$ along z,

$$U = \frac{1}{2}\frac{1}{4\pi\varepsilon_0(2r)^3}[d \cdot d_I - 3(d \cdot e_z)(d_I \cdot e_z)] = -\frac{1}{4\pi\varepsilon_0}\frac{1}{16r^3}[d \cdot d + (d \cdot e_z)(d \cdot e_z)]. \tag{14.2}$$

Since d originates from quantum fluctuations it is random and we need to average U with respect to the quantum state of the dipole. In analogy with the van der Waals case, we consider a dipole in its ground state $|g\rangle$ with excited states $\{|n\rangle\}$. Its dipole operator is given by the generalisation of that of a two-level atom, Eq. (13.2),

$$\hat{d} = \sum_n (|n\rangle\langle g| + |g\rangle\langle n|)\, d_n, \tag{14.3}$$

where d_n is the matrix element of the $|g\rangle \to |n\rangle$ transition assumed to be real. Replacing d in Eq. (14.2) with the operator $\hat{d}$ and averaging with $|g\rangle$, we find

$$U = -\frac{1}{4\pi\varepsilon_0}\frac{1}{16r^3}\sum_n |d_n|^2 \left(1 + \cos^2\theta_n\right). \tag{14.4}$$

Here θ_n is the angle between the matrix element d_n and the z axis, $\cos\theta_n = d_n \cdot e_z$. Finally, assuming randomly oriented dipole, we average over this angle, $(1/2)\int_{-1}^{1} d(\cos\theta_n)(\cos\theta_n)^2 = 1/3$, yielding

$$U = -\frac{1}{4\pi\varepsilon_0}\frac{1}{12r^3}\sum_n |d_n|^2. \tag{14.5}$$

This result was first derived by Lennard-Jones [8] and later generalised to the cases of metallic and dielectric mirrors [9, 13].

Forces out of correlations

We note that the averaging over the dipole fluctuations in the potential from Eq.

(14.2) requires the calculation of the *correlation* between the dipoles, or polarisations, of the two interacting objects. In general, interactions driven by fluctuations are proportional to such correlations [13, 20]. In our case, the objects are the dipole and its image, such that the relevant quantities are the second moments of the dipole itself $\langle \hat{\boldsymbol{d}} \cdot \hat{\boldsymbol{d}} \rangle$ and $\langle \hat{d}^z \hat{d}^z \rangle$, reflecting the strong correlation built by the dipole and the surface at *short distances*. At longer distances, however, correlations may decrease, and so does the strength of the interaction (see Problem 2.3), as we shall see in the next section.

This physical reasoning can be also applied to the results we have obtained in Section 13 for the van der Waals interaction: the correlations between dipole fluctuations are stronger at shorter distances, therefore in the retarded regime the interaction energy falls off more rapidly with distance, as $1/r^7$ compared to $1/r^6$. The fact that spatial correlations between fluctuations at the positions of the interacting objects are at the core of vacuum-induced forces, is evident in our general formula (13.25), where correlations of the driving electromagnetic field appear.

14.2   Electrodynamic analysis

The essence of the method of images utilised in the previous section is its treatment of the electric field in the presence of a boundary condition. Namely, the bulk mirror is treated as a boundary that creates an energy potential for the dipole, which in turn is induced by quantum fluctuations. We can take the same approach also within an electrodynamic treatment of the problem: as shown in Section 5 of Chapter 1, the energy levels of a dipole are shifted due to its interaction with the vacuum field modes, the so-called Lamb shift. Since these modes can be modified due to the boundary conditions, so can the energy shifts of the levels. Then, the dependence of this modified Lamb shift on the distance between the dipole and the boundary yields the potential energy of the dipole in the presence of the mirror.

We consider a dipole in its ground state $|g\rangle$ that is placed in an infinitely large cuboid cavity made of an empty box with perfectly-conducting walls. The box extends from $x, y, z = 0$ to $x, y, z = L$ and the dipole is placed at $x = L/2, y = L/2, z = r$. We later take the limit $L \to \infty$ such that the effect of the cavity on the dipole is only due to its wall at $z = 0$ and at a distance r apart from the dipole. Therefore, the limit $L \to \infty$ reproduces the situation we discuss, of a particle in front of a perfectly-conducting mirror.

We recall the electric field modes of the cavity from Chapter 1, Eq. (2.37),

$$
\begin{aligned}
\boldsymbol{u}_{\boldsymbol{k}\lambda}(\boldsymbol{r}) \;=\; & \sqrt{\frac{8}{V}} \, [e^x_{\boldsymbol{k}\lambda} \cos(k_x x) \sin(k_y y) \sin(k_z z) \boldsymbol{e}_x \\
& + e^y_{\boldsymbol{k}\lambda} \sin(k_x x) \cos(k_y y) \sin(k_z z) \boldsymbol{e}_y \\
& + e^z_{\boldsymbol{k}\lambda} \sin(k_x x) \sin(k_y y) \cos(k_z z) \boldsymbol{e}_z] \,,
\end{aligned}
\tag{14.6}
$$

with $\boldsymbol{k} = (k_x, k_y, k_z)$ and $\lambda = 1, 2$ being the wave vector and polarisation, respectively, and where

$$
k_i = \frac{\pi}{L} n_i, \quad n_i = 0, 1, 2, ..., \quad e^i_{\boldsymbol{k}\lambda} = \boldsymbol{e}_i \cdot \boldsymbol{e}_{\boldsymbol{k}\lambda}, \quad i = x, y, z, \quad V = L^3.
\tag{14.7}
$$

Assuming the total system is in its ground state $|G\rangle = |g, 0\rangle$, where the dipole is in its ground state and the field modes are in vacuum, we use second order perturbation theory to find the energy correction of the system [Eq. (5.13), Chapter 1],

$$U = -\sum_I \frac{|\langle G|\hat{H}_{AF}|I\rangle|^2}{E_I - E_G}, \tag{14.8}$$

where $\hat{H}_{AF}$ is the dipole-field interaction Hamiltonian of Eq. (13.67) for a single dipole. This Hamiltonian allows for transitions between $|G\rangle = |g, 0\rangle$ and virtual states of the type $|I\rangle = |n, 1_{k\lambda}\rangle$, $|n\rangle$ being an excited dipolar state, for which

$$E_I - E_G = \hbar\omega_k + E_n, \quad \langle G|\hat{H}_{AF}|I\rangle = i\hbar g_{k\lambda n} = \sqrt{\frac{\omega_k}{2\varepsilon_0\hbar}} \boldsymbol{d}_n \cdot \boldsymbol{u}_{k\lambda}(\boldsymbol{r}), \tag{14.9}$$

with $\omega_k = c|\boldsymbol{k}|$. Hence, U becomes

$$U = -\sum_n \sum_k \sum_\lambda \frac{\hbar^2|g_{k\lambda n}|^2}{\hbar\omega_k + E_n}. \tag{14.10}$$

In the last equality of Eq. (14.9) we used Eq. (13.69) with the cavity modes $\boldsymbol{u}_{k\lambda}$ from Eq. (14.6). Then,

$$
\begin{aligned}
|g_{k\lambda n}|^2 = \frac{\omega_k}{2\varepsilon_0\hbar}\sqrt{\frac{8}{V}} \Big[&(e_{k\lambda}^x d_n^x)^2 \cos^2(k_x L/2) \sin^2(k_y L/2) \sin^2(k_z r) \\
&+ (e_{k\lambda}^y d_n^y)^2 \sin^2(k_x L/2) \cos^2(k_y L/2) \sin^2(k_z r) \\
&+ (e_{k\lambda}^z d_n^z)^2 \sin^2(k_x L/2) \sin^2(k_y L/2) \cos^2(k_z r) + \sum_{i \neq j} B_{ij} d_n^i d_n^j \Big],
\end{aligned}
\tag{14.11}
$$

with B_{ij} some functions of space and where the i, j summation goes over x, y, z. Assuming a randomly oriented dipole we make use of $d_n^i d_n^j \to (1/3)|\boldsymbol{d}_n|^2\delta_{ij}$ [Eq. (13.87)] in Eq. (14.11). Then, upon inserting (14.11) into the expression for the energy in (14.10), we have

$$
\begin{aligned}
U = -\sum_n \sum_\lambda \frac{V}{(2\pi)^3} &\int_{-\infty}^{\infty} dk_x \int_{-\infty}^{\infty} dk_y \int_{-\infty}^{\infty} dk_z \frac{4}{3} \frac{\hbar\omega_k}{\varepsilon_0 V} \frac{|\boldsymbol{d}_n|^2}{\hbar\omega_k + E_n} \\
\times \Big[&(e_{k\lambda}^x)^2 \cos^2(k_x L/2) \sin^2(k_y L/2) \sin^2(k_z r) \\
&+ (e_{k\lambda}^y)^2 \sin^2(k_x L/2) \cos^2(k_y L/2) \sin^2(k_z r) \\
&+ (e_{k\lambda}^z)^2 \sin^2(k_x L/2) \sin^2(k_y L/2) \cos^2(k_z r) \Big],
\end{aligned}
\tag{14.12}
$$

where we also took the continuum limit $\sum_{\boldsymbol{k}} \to \frac{V}{(2\pi)^3} \int d\boldsymbol{k}$.

Divergence and renormalisation

The integrals over dk_i ($i = x, y, z$) in Eq. (14.12) have a diverging part for all three terms in the square brackets. In order to see why, consider the integral

$$\int_{-\infty}^{\infty} dk_i \sin^2(k_i a) = \int_{-\infty}^{\infty} dk_i \frac{1}{2}\left[1 - \cos(2k_i a)\right] = \frac{1}{2}\int_{-\infty}^{\infty} dk_i - \frac{1}{2}\int_{-\infty}^{\infty} dk_i \cos(2k_i a).$$
(14.13)

The first term clearly diverges. In fact, it diverges as the cut-off on wavenumber (or frequency) that applies to the low-energy theory we use. If we multiply the integrand by $\omega_{\boldsymbol{k}}/(\omega_{\boldsymbol{k}} + E_n/\hbar)$, as in (14.12) where $\omega_{\boldsymbol{k}}$ is a monotonically increasing function of any k_i, the integral still diverges in the same manner since $\omega_{\boldsymbol{k}}/(\omega_{\boldsymbol{k}} + E_n/\hbar) \approx 1$ for large k. The integral (14.13) with $\cos^2(k_i a) = (1/2)[1+\cos(2k_i a)]$ replacing $\sin^2(k_i a)$ leads to the same divergences.

Considering the second term, we could naively write it using Dirac delta functions as $(\pi/2)[\delta(2a) + \delta(-2a)]$ which for any $a > 0$ gives zero. However, recalling the $\omega_{\boldsymbol{k}}/(\omega_{\boldsymbol{k}} + E_n/\hbar)$ factor in the integrand, the result is not immediately obvious. Nevertheless, for $a \to \infty$, as in the case of $a = L/2$, the oscillations of this integrand are infinitely fast so that the integral averages out and is negligible.

The above conclusions mean that we may replace $\cos^2(k_i L/2)$ and $\sin^2(k_i L/2)$ by $1/2$ in the integrals of Eq. (14.12), obtaining

$$U = -\frac{1}{24\pi^3\varepsilon_0}\sum_n |\boldsymbol{d}_n|^2 \int d\boldsymbol{k} \frac{|\boldsymbol{k}|}{|\boldsymbol{k}| + E_n/(\hbar c)}\sum_\lambda \left((e_{\boldsymbol{k}\lambda}^x)^2\left[\frac{1}{2} - \frac{1}{2}\cos(2k_z r)\right]\right.$$
$$\left. + (e_{\boldsymbol{k}\lambda}^y)^2\left[\frac{1}{2} - \frac{1}{2}\cos(2k_z r)\right] + (e_{\boldsymbol{k}\lambda}^z)^2\left[\frac{1}{2} + \frac{1}{2}\cos(2k_z r)\right]\right).$$
(14.14)

This leaves us with a diverging term, the $1/2$ term, in each square bracket, so that the divergence can be eliminated by dropping these $1/2$ terms. Since these terms do not depend on the distance r they do not contribute to the force exerted on the particle by the mirror, so we can ignore them. A more elegant way to drop these terms follows the reasoning of *renormalisation* used to treat the Casimir force between two conducting plates, presented in Chapter 1 (Section 7 therein): we wish to subtract from U in Eq. (14.14) a suitable diverging reference energy [3]. As explained in Chapter 1, the introduction of the components of the setup, i.e. the particle and mirror, to the vacuum environment, costs some kind of a preparation or self energy, which is that of a particle infinitely far from the mirror. This reference energy, U_0, can be evaluated by setting $r = L/2 \to \infty$ in Eq. (14.14), averaging the $\cos(2k_z r)$ to zero hence leaving only the constant and diverging $1/2$ term,

$$U_0 = -\frac{1}{24\pi^3\varepsilon_0}\sum_n |\boldsymbol{d}_n|^2 \int d\boldsymbol{k} \frac{|\boldsymbol{k}|}{|\boldsymbol{k}| + E_n/(\hbar c)}\sum_\lambda \frac{1}{2}\left[(e_{\boldsymbol{k}\lambda}^x)^2 + (e_{\boldsymbol{k}\lambda}^y)^2 + (e_{\boldsymbol{k}\lambda}^z)^2\right].$$
(14.15)

Therefore, the subtraction of the reference energy U_0 from U *exactly* cancels out all diverging $1/2$ terms in Eq. (14.14) and we are left with

$$U = \frac{1}{48\pi^3\varepsilon_0} \sum_n |d_n|^2 \int dk \frac{|\mathbf{k}|\cos(2k_z r)}{|\mathbf{k}| + E_n/(\hbar c)} \sum_\lambda \left[(e_{\mathbf{k}\lambda}^x)^2 + (e_{\mathbf{k}\lambda}^y)^2 - (e_{\mathbf{k}\lambda}^z)^2\right]. \quad (14.16)$$

General result
Using Eq. (13.86) we perform the sum over polarisations,

$$\sum_\lambda \left[(e_{\mathbf{k}\lambda}^x)^2 + (e_{\mathbf{k}\lambda}^y)^2 - (e_{\mathbf{k}\lambda}^z)^2\right] = 2(\mathbf{e}_z \cdot \mathbf{e}_{\mathbf{k}})^2, \quad (14.17)$$

with $\mathbf{e}_{\mathbf{k}}$ the unit vector in the direction of $\mathbf{k}$.

Exercise 14.1
Obtain Eq. (14.17) using Eq. (13.86).

Moving to spherical coordinates $\mathbf{k} = (k, \theta, \phi)$ we have

$$U = \frac{1}{12\pi^2\varepsilon_0} \sum_n |d_n|^2 \int_0^\infty dk \frac{k^3}{k + E_n/(\hbar c)} \int_{-1}^1 d(\cos\theta)(\cos\theta)^2 \cos[2kr(\cos\theta)]. \quad (14.18)$$

The integral over $\cos\theta$ is performed by replacing $\cos^2\theta$ in the integrand with a double derivative with respect to $2kr$, yielding

$$U = \frac{1}{6\pi^2\varepsilon_0} \sum_n |d_n|^2 \int_0^\infty dk \frac{k^3}{k + E_n/(\hbar c)} \left[\frac{\sin(2kr)}{2kr} + 2\frac{\cos(2kr)}{(2kr)^2} - 2\frac{\sin(2kr)}{(2kr)^3}\right]. \quad (14.19)$$

This provides a general result for the vacuum-induced interaction between a polarisable particle and a perfectly-conducting mirror at any distance r. In the following, we use it as a starting point for the discussion of the quasistatic and retarded limits.

Quasistatic regime
For short distances r, as in (13.41), we take only the leading $1/r^3$ term in the square brackets of Eq. (14.19), so that

$$U = -\frac{1}{24\pi^2\varepsilon_0 r^3} \sum_n |d_n|^2 \mathrm{Im} \left\{\int_0^\infty dk \frac{e^{i2kr}}{k + E_n/(\hbar c)}\right\}. \quad (14.20)$$

Performing the Wick rotation [Fig. 2.4 and Eq. (13.42)] with $k = iu$, the integral becomes

$$\int_0^\infty du e^{-2ru} \mathrm{Im} \left\{\frac{i}{iu + E_n/(\hbar c)}\right\} = \frac{E_n}{\hbar c} \int_0^\infty du e^{-2ru} \frac{1}{u^2 + [E_n/(\hbar c)]^2}. \quad (14.21)$$

Considering the inequality (13.41) we may also take $e^{-2ru} \approx 1$, neglecting all wave effects. Finally, upon performing the resulting Lorentzian integration we find

$$U = -\frac{1}{4\pi\varepsilon_0} \frac{1}{12r^3} \sum_n |d_n|^2, \quad (14.22)$$

recovering Eq. (14.5) that was obtained by an electrostatic image-method treatment.

Retarded regime

In the retarded regime (13.48), as discussed in Section 13 and inequality (13.81), the k values that mostly contribute to the integral in Eq. (14.19) are much smaller than $E_n/(\hbar c)$. Then, we can approximate $[k + E_n/(\hbar c)]^{-1} \approx \hbar c/E_n$ in U, giving

$$U = \frac{1}{6\pi^2\varepsilon_0} \sum_n |\boldsymbol{d}_n|^2 \frac{\hbar c}{16 r^4 E_n} I, \qquad (14.23)$$

with

$$I = \int_0^\infty x^3 \left[\frac{\sin x}{x} + 2\frac{\cos x}{x^2} - 2\frac{\sin x}{x^3} \right]. \qquad (14.24)$$

Performing the integral I with an exponential *regulator* $e^{-x/b}$, taking $b \to \infty$ after the integration [as in Eq. (13.53)], we obtain $I = -4$, so that

$$U = -\frac{1}{4\pi\varepsilon_0} \frac{1}{6\pi r^4} \sum_n |\boldsymbol{d}_n|^2 \frac{\hbar c}{E_n}. \qquad (14.25)$$

The potential energy of the particle due to the mirror in the retarded regime thus scales as $1/r^4$ which is one power less than the quasistatic regime $1/r^3$, reminiscent of the van der Waals case with $1/r^7$ and $1/r^6$ scalings, respectively. As discussed in Section 14.1 above, this stronger decay of the retarded interaction is due to weaker spatial correlations between fluctuations at the dipole and at the conducting plate.

Finally, we wish to compare our result with that of Casimir and Polder, where the ground state of the dipole was taken to be that of zero total angular momentum without further assuming its random orientation [4]. This amounts to taking $d_n^i d_n^j \to d_n^i d_n^j \delta_{ij}$ in Eq. (14.11) without the assumption $d_n^i d_n^i = (1/3)|\boldsymbol{d}_n|^2$ [used in Eq. (14.12) above]. Then, following a similar calculation to that presented above, we arrive at the retarded atom-mirror potential,

$$U = -\frac{1}{4\pi\varepsilon_0} \frac{1}{4\pi r^4} \sum_n |\boldsymbol{d}_n|^2 \frac{\hbar c}{E_n}, \quad |\boldsymbol{d}_n|^2 = (d_n^x)^2 + (d_n^y)^2 + (d_n^z)^2, \qquad (14.26)$$

which is the Casimir-Polder result [4].

Exercise 14.2

Obtain the retarded potential, Eq. (14.26), using the assumption $d_n^i d_n^j \to d_n^i d_n^i \delta_{ij}$ in Eq. (14.11).

---

§15. A FEW COMMENTS

Dipole approximation

Throughout our discussion, we have treated all material particles that interact with the electromagnetic vacuum field as polarisable (induced) dipoles. This is manifest either by considering only their polarisability α, defined as their linear polarisation response to an applied electric field, or by using the field-matter interaction Hamiltonian in the dipole approximation $\hat{H}_{AF}$ from Eq. (13.67) [18].

This treatment is valid for electromagnetic field wavelengths λ that are much larger than the typical size of the atom r_a, which is essentially given by the typical distance between the farthest electron of the atom and its nucleus. Then, the atom forms a dipole of size r_a much smaller than λ. This approximation is very useful in e.g. atomic physics and quantum optics, where the applied optical fields possess wavelengths of the order of a few hundreds of nanometers or a micron, whereas the atomic size is of order of Angstroms.

Here however, we discuss interaction with the vacuum, that in principle contains *all* electromagnetic field wavelengths, even those much smaller than an Angstrom. In principle, a description within the dipole approximation is not valid for such wavelengths, so that our theory has a natural cut-off at $\lambda \gtrsim r_a$. Moreover, as typical of field theories, we usually do not want the cut-off to explicitly appear in the results of our calculations.

Nevertheless, we can use this theory to calculate non-resonant vacuum-induced interactions between matter particles that do not respond to electromagnetic fields above a certain frequency (below a certain wavelength), typically smaller than c/r_a. For example, consider the integral for the quasistatic van der Waals interaction in Eq. (13.44). Since our theory is valid only up to e.g. $\omega_{co} = 0.1c/r_a$, the integral should in principle be cut at this value. However, if $\alpha(\omega)$ becomes very small already at a frequency $\omega_a \ll \omega_{co}$, then the contribution of the integral from ω_a to ∞ is negligible. This means that the integral is effectively cut at around ω_a much before ω_{co}, so that it does not matter in practice that the integration range formally crosses ω_{co} and extends up to ∞. It does however make the analytical evaluation of the integral simpler.

When the explicit expression for $\alpha(iu)$ [Eq. (13.40)] is inserted into the integral of Eq. (13.44) [see Eq. (13.45)], we can identify the frequency-range at which the integral is effectively cut: the Lorentzian functions in the integrand become negligibly small for u above the largest frequency of transition $E_n/\hbar$. Then, e.g. for a typical alkali atom $E_n/\hbar \lesssim 2\pi \times 10^{15}$Hz, which is still much smaller than $c/r_a \sim 10^{18}$Hz. This means that the somewhat artificial extension of our low-frequency theory to $\omega \to \infty$ is at least mathematically sensible.

We note that van der Waals interactions calculated within the dipole approximation hold for inter-dipolar distances r much larger than the dipole size r_a, i.e. as long as the dipoles do not spatially overlap. For example, this can become important when such interactions are considered in the context of current state-of-the-art experimental systems involving large dipolar systems such as Rydberg atoms ($r_a \sim 100$nm) [21] or superconducting quantum bits ($r_a \sim 1\mu$m-100μm) [22,23].

Dependence on polarisability
The integral for van der Waals interaction in Eq. (13.30) exhibits a spectral overlap involving $\alpha_{1,2}(\omega)$. Hence the scaling of α with ω should affect the result. An interesting qualitative (rather than just quantitative) example is that of the van der

Waals interaction between the farthest electron and the positive ion in a Rydberg atom [14, 27]. A Rydberg atom is an atom with an electron in a highly excited state, so that this electron approximately possesses a polarisability of a free charge $\alpha_e(\omega) \sim 1/\omega^2$. This leads to a van der Waals interaction between this electron and the resulting positive ion that scales in the retarded regime as $1/r^5$ instead of $1/r^7$.

Exercise 15.1

Using Eqs. (13.36) and (13.50), show that if one of the dipoles has a polarisability $\alpha = C/\omega^p$, with C some constant and p a positive integer, then the retarded van der Waals potential in free space scales as $1/r^{7-p}$ instead of as $1/r^7$.

Another point is the fact that $\alpha(\omega)$ is in general a complex function. This is the case for systems with dipolar energy-level transitions that have a width due to coupling to a reservoir or a continuum of states. Then, losses to the reservoir at rate γ add an imaginary part (width) γ to the dipolar transition energy and consequently give rise to an imaginary part in the response of the dipole to the electric field, α. For example, in dipoles made of solid-state particles, non-radiative losses, e.g. due to coupling to phonons, may exist and add an imaginary part to α in calculations of vacuum-induced forces. The interesting point is that the fluctuation-dissipation theorem, discussed later on in Chapter 3, states that losses in the response of the system imply that the system also fluctuates. Namely, the reservoir that induces losses on the system induces also its fluctuations; and at zero temperature, these fluctuations have to be of quantum nature. Therefore, *dipolar losses imply zero-point fluctuations of the dipole*. Following our intuitive discussion in the introduction of Section 13, dipole fluctuations can serve as a source of van der Waals interactions, other than the quantum fluctuations of the vacuum field. More general treatments of vacuum-induced interactions also take loss-associated fluctuations of the dipoles into account [13].

The justification to include losses and width to energy levels in the calculation of vacuum-induced forces of *isolated* dipolar systems which interact solely with the reservoir formed by the vacuum of the continuum of photon modes, is not as obvious as for solid-state particles. Recall that in the electrodynamical calculation of the van der Waals force in Section 13.2.1, we took $\alpha(\omega)$ as a *real* function, resulting from the polarisability of a system with discrete and *sharp* energy levels, as in Eq. (13.40). In fact, the energy levels of such a system do acquire width which is associated with radiation, i.e. coupling to the photon vacuum, rather than with non-radiative processes. In the calculations of Sections 13.2.2, 14.2 and e.g. Ref. [12], which are based on perturbation theory of quantum electrodynamics, the coupling to the vacuum is treated by considering *bare* atoms with sharp energy levels that interact with the vacuum via the Hamiltonian (13.67). Then, we expect that any effect on the atoms due to the vacuum, e.g. their vacuum-induced forces, but also their level widths, would emerge in the calculation directly from this Hamiltonian. Nevertheless, the results of e.g. the van der Waals interaction in Section 13.2.2 are equivalent to those of Section 13.2.1 [12], where no effects of level broadening and width are apparent. This appears to be an artifact of lowest-order perturbation theory: the quantum-electrodynamic perturbative calculation was carried out up to fourth order, which is the lowest order in which vacuum-forces emerge. However, this does not mean that there are no higher-order corrections to this vacuum force. Such higher-order corrections may give rise to terms associated with the vacuum-interaction between dipoles that are *dressed* by the vacuum; namely, dipoles with broadened levels due to coupling with the vacuum. Alternatively, using the scattering approach of Section 13.2.1 where the polarisability is put by hand into the formalism, one can imagine that each of the interacting dipoles is first individually *dressed* by the vacuum. Then, the polarisability in Eq. (13.40) has to be replaced by that of a system of energy levels with an imaginary part (broadened levels) due to radiation losses, yielding a complex polarisability.

Forces induced by radiation

Although the focus of this book is on vacuum-induced forces it is worth mentioning the generality of our scattering-approach formalism and its usefulness in treating radiation-induced interactions. Going back to Eq. (13.25) for the interaction energy U induced by a general driving field $\hat{E}_{0,\omega}^i(\boldsymbol{r})$, we restore the normal-ordered correlations $\langle \hat{E}_{0,\omega}^{i\dagger}(\boldsymbol{r}_1)\hat{E}_{0,\omega}^j(\boldsymbol{r}_2)\rangle$ that were neglected for the case of vacuum-induced forces*. Recalling also that the interaction U in Section 13.2.1 was obtained from the energy of the dipole 2 [Eq. (13.14)], we can formally symmetrise it by adding to it the interaction energy evaluated at dipole 1 (replacing the indices 1 and 2) and multiplying by $1/2^\dagger$. Then, the radiation-induced interaction potential from Eq. (13.25) assumes the more general form

$$U = -\frac{1}{2}\mu_0 \sum_\omega \omega^2 \alpha_1(\omega)\alpha_2(\omega) \sum_{ij} \Big[G_{ij}(\omega, \boldsymbol{r}_2, \boldsymbol{r}_1)\langle \hat{E}_{0,\omega}^j(\boldsymbol{r}_1)\hat{E}_{0,\omega}^{i\dagger}(\boldsymbol{r}_2)+$$
$$\hat{E}_{0,\omega}^{i\dagger}(\boldsymbol{r}_2)\hat{E}_{0,\omega}^j(\boldsymbol{r}_1)\rangle + G_{ij}(\omega, \boldsymbol{r}_1, \boldsymbol{r}_2)\langle \hat{E}_{0,\omega}^j(\boldsymbol{r}_2)\hat{E}_{0,\omega}^{i\dagger}(\boldsymbol{r}_1) + \hat{E}_{0,\omega}^{i\dagger}(\boldsymbol{r}_1)\hat{E}_{0,\omega}^j(\boldsymbol{r}_2)\rangle + \text{c.c.}\Big].$$

$$(15.1)$$

The interaction between two particles that are driven by an incident field $\boldsymbol{E}_{0,\omega}$ is then given by a spectral overlap between the responses of the dipoles $\alpha_{1,2}$, the Green function that describes the embedding geometry G_{ij}, and the spatial correlation of the driving field at the positions of the particles. At this point of the derivation the driving field and its statistics are not yet specified, giving rise to a general formalism of radiation-induced inter-particle forces. For example, Eq. (15.1) can serve as the starting point for a calculation of the van der Waals interactions at finite temperatures, where the driving field is that of thermal rather than vacuum fluctuations. Alternatively, one may consider dipolar interactions induced by an external laser light [12, 28–32] (Problem 2.5).

---

*The phase-dependent correlations $\langle \hat{E}_{0,\omega}^i(\boldsymbol{r}_1)\hat{E}_{0,\omega}^j(\boldsymbol{r}_2)\rangle$ are still typically negligible since they give fast-oscillating contributions which vanish upon performing time-averaging [33].

†The general expression we obtained for the van der Waals potential U (induced by the vacuum), Eq. (13.30), is in fact symmetric under exchange of the indices 1 and 2 (as any interaction potential should be), so that practically we did not need to use any symmetrisation.

§16. Non-additivity of dipolar interactions

Let us get back to the starting point of this chapter: since a mirror = many atoms, then the Casimir force = many van der Waals forces. However, the question of how to relate the Casimir force between macroscopic bodies to the underlying van der Waals forces between their constituent microscopic particles is nontrivial: is the vacuum-induced interaction between composite or macroscopic polarisable objects *additive*? Namely, can it be represented by the *sum of interactions between all pairs* of particles that comprise the macroscopic objects?

More concretely, we can ask whether the Casimir force between two mirrors 1 and 2, U_C, may be obtained from a pairwise summation of the van der Waals interaction between all pairs of atoms U_{ij} where $i \in$ atom 1 and $j \in$ atom 2,

$$U_C = \sum_{i \in 1} \sum_{j \in 2} U_{ij}; \qquad (16.1)$$

or alternatively, whether the atom-mirror interaction U from Section 14 can be reproduced by a summation, over all atoms of the mirror j, of their van der Waals interaction with the atom 1 that is in front of the mirror,

$$U = \sum_j U_{1j}. \qquad (16.2)$$

As shown in the following, the answer is that Casimir-type vacuum-induced interactions are in general *non-additive* [34]. Pairwise summation can correctly reproduce general features such as the scaling with distance of Casimir forces and may give good approximate results in the case of dilute composite objects, however in general they do not yield exact results.

16.1 Example: non-additivity in atom-surface interaction

Let us illustrate the problem of non-additivity using the example of the interaction between an atom (or any other polarisable particle) and a surface in the quasistatic regime, discussed in Section 14. The xy plane at $z = 0$ forms the surface of the mirror, where its bulk extends to $z > 0$ and the atom is placed at a distance r from the surface at $z = -r$ (Fig. 2.7). Consider a mirror particle at depth z and xy position $\boldsymbol{\rho}$. The quasistatic van der Waals interaction between the atom and the mirror particle is given by Eq. (13.44) as

$$V(z, \boldsymbol{\rho}) = -\frac{C}{\left(\sqrt{(z+r)^2 + |\boldsymbol{\rho}|^2}\right)^6}, \quad C = \left(\frac{1}{4\pi\varepsilon_0}\right)^2 \frac{3\hbar}{\pi} \int_0^\infty du\, \alpha(iu) \alpha_m(iu), \quad (16.3)$$

where α and α_m are the polarisabilities of the atom and the mirror particle, respectively. Assuming identical mirror particles with particle density n_m, the interaction between the atom and mirror, U from Eq. (16.2) (assuming additivity), is given

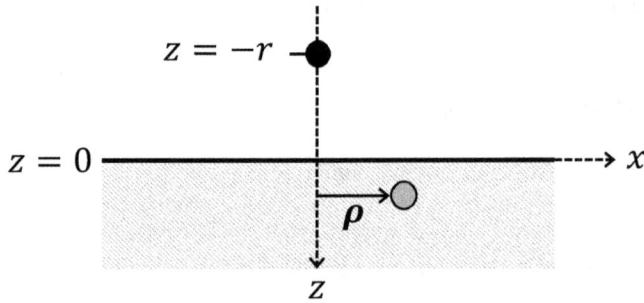

Figure 2.7: Pairwise summation approach for the atom-mirror interaction. The atom is at a distance r from the mirror surface. We focus on its van der Waals interaction with a mirror particle at $(z, \boldsymbol{\rho})$. The summation over these interactions with all mirror particles is interpreted as the vacuum-induced atom-mirror interaction.

by the summation of $V(z, \boldsymbol{\rho})$ over all mirror particles. Using cylindrical coordinates $\boldsymbol{\rho} = (\rho, \theta)$ we get

$$U = \int_0^\infty dz \int d\boldsymbol{\rho} \, n_m V(z, \boldsymbol{\rho}) = -n_m C \int_0^\infty dz \int_0^{2\pi} d\theta \int_0^\infty d\rho \rho \frac{1}{[\rho^2 + (r+z)^2]^3}.$$

(16.4)

Performing the $d\theta$ integration and changing integration variables, $\sigma = \rho^2 + (r+z)^2$, we have

$$U = -\pi n_m C \int_0^\infty dz \int_{(r+z)^2}^\infty d\sigma \frac{1}{\sigma^3} = -\frac{\pi}{2} n_m C \int_0^\infty dz \frac{1}{(r+z)^4} = -\frac{\pi}{6} \frac{n_m C}{r^3}, \quad (16.5)$$

recovering the $1/r^3$ dependence from Eq. (14.5). Inserting C from Eq. (16.3), we find

$$U = -\left(\frac{1}{4\pi\varepsilon_0}\right)^2 \frac{\hbar}{2r^3} \int_0^\infty du \, \alpha(iu) \alpha_m(iu). \quad (16.6)$$

The microscopic polarisability of the mirror-particles α_m can be related to the bulk polarisability per unit volume of the mirror, i.e. its dielectric response $\varepsilon = \varepsilon_0 \varepsilon_r$, via the well-known Clausius-Mossotti (also known as the Lorentz-Lorenz) relation [36],

$$n_m \alpha_m = 3\varepsilon_0 \frac{\varepsilon_r - 1}{\varepsilon_r + 2}. \quad (16.7)$$

Using this relation in Eq. (16.6) we finally have,

$$U = -\frac{1}{4\pi\varepsilon_0} \frac{3\hbar}{8\pi} \frac{1}{r^3} \int_0^\infty du \frac{\varepsilon_r(iu) - 1}{\varepsilon_r(iu) + 2} \alpha(iu). \quad (16.8)$$

Perfectly-conducting mirror
Taking the limit of a perfect conductor, $\varepsilon_r(iu) \to \infty$ for all imaginary frequencies*

---

*The dielectric response of a metal is discussed in Chapter 3. For example, in the Plasma model we have $\varepsilon_r(\omega) = 1 - \omega_P^2/\omega^2$, where the parameter ω_P is the so-called Plasma frequency. The

u, and using $\alpha(iu)$ from Eq. (13.40), we obtain

$$U = -\frac{1}{4\pi\varepsilon_0}\frac{3\hbar}{8\pi}\frac{1}{r^3}\int_0^\infty du\,\alpha(iu) = -\frac{1}{4\pi\varepsilon_0}\frac{1}{4\pi r^3}\sum_n |d_n|^2 \int_0^\infty du\,\frac{E_n/\hbar}{(E_n/\hbar)^2 + u^2}.$$

(16.9)

Finally, upon performing the Lorentzian integration we have

$$U = -\frac{1}{4\pi\varepsilon_0}\frac{1}{8r^3}\sum_n |d_n|^2.$$

(16.10)

This result, obtained essentially by pairwise summation of van der Waals interactions, is similar in form to that from Eq. (14.5) where additivity was not assumed (in Section 14 the effect of the mirror was fully treated via the boundary conditions it imposes on the fields). However, whereas here there is a factor $1/8$, in Eq. (14.5) there appears a factor $1/12$, making the pairwise summation result larger by a factor of $3/2$. The assumption of additivity thus reproduced the main features of the result, like the $1/r^3$ scaling with distance and the sum over the dipole matrix elements $\sum_n |d_n|^2$, however it failed in predicting the right numerical coefficients.

Dielectric mirror and dilute matter

It is constructive to compare the pairwise summation result, Eq. (16.8), to that of the general case of a dielectric mirror, where the atom-mirror potential is given by [13]

$$U = \frac{1}{4\pi\varepsilon_0}\frac{\hbar}{4\pi r^3}\int_0^\infty du\,\frac{\varepsilon_r(iu) - 1}{\varepsilon_r(iu) + 1}\alpha(iu).$$

(16.11)

Whereas the effect of the dielectric response in the former is given by the function $(3/8)[\varepsilon_r(iu) - 1]/[\varepsilon_r(iu) + 2]$, in the latter there appears a different function $(1/4)[\varepsilon_r(iu) - 1]/[\varepsilon_r(iu) + 1]$. This leads to a different numerical factor in the overall result, as demonstrated above for a perfect-conducting mirror.

However, this discrepancy disappears upon considering the case of a mirror made not from a bulk matter but rather from a dilute gas of particles. Taking the limit of particle density n_m much smaller than the inverse of the volume-scale set by the particle polarisability α_m,

$$\frac{n_m \alpha_m}{\varepsilon_0} \ll 1,$$

(16.12)

the Clausius-Mossotti relation implies $\varepsilon_r - 1 \ll 1$ so that both Eqs. (16.8) and (16.11) yield the same result,

$$U \approx -\frac{1}{4\pi\varepsilon_0}\frac{\hbar}{8\pi r^3}\int_0^\infty du[\varepsilon_r(iu) - 1]\alpha(iu).$$

(16.13)

This suggests that Casimir-like interactions are in fact approximately additive for vacuum-induced interactions involving sufficiently dilute composite systems. In the next section we discuss why this is so.

---

limit of a perfect conductor may be obtained by taking $\omega_P \to \infty$, so that in imaginary frequencies $\omega = iu$ we find $\varepsilon_r(iu) = 1 + \omega_P^2/u^2 \to \infty$ for all u.

16.2   Multiple scattering as the origin of non-additivity

The origin of the non-additivity of Casimir-like, vacuum-induced forces is most evident in our scattering theory treatment of the van der Waals interactions from
Section 13.2.1. Within this scattering approach, the dipoles interact by scattering
vacuum fluctuations on each other, where we specifically focused on the scattered
field arriving at the position of dipole 2. Our treatment assumed that the dipoles
are weak scatterers, namely, that upon being driven by an incident field $\boldsymbol{E}_{0,\omega}$, they
scatter a field $\boldsymbol{E}_{sc,\omega}$ much weaker than the incident field. The scattered field arriving at dipole 2 is thus approximated as resulting from a *single* scattering event: the
incident vacuum fluctuations are scattered by dipole 1 and arrive to dipole 2. The
ratio of the scattered to incident field then scales as [see Eq. (13.24)]

$$R_{sc} = \omega^2 \mu_0 \alpha_1(\omega) G_{ij}(\omega, \boldsymbol{r}), \tag{16.14}$$

$\boldsymbol{r}$ being the position of the scatterer (dipole 1) with respect to the place where the
field is evaluated (position of dipole 2). Hence, R_{sc} becomes the coupling parameter
of our scattering theory. This theory can then be treated perturbatively as long as
the scattering is weak,

$$R_{sc} \ll 1. \tag{16.15}$$

Considering the example of free-space in the quasistatic, short-range, regime, we
take only the dominant $1/(k^2 r^3)$ term in G_{ij} from Eq. (13.31), yielding

$$R_{sc} \sim \frac{\alpha}{\varepsilon_0 r^3} \ll 1, \tag{16.16}$$

where here we used $\omega = kc$ and $\mu_0 \varepsilon_0 = 1/c^2$. If the dipoles 1 and 2 are two out
of many identical particles that comprise a bulk or composite system with particle
density n_m, then $n_m \sim 1/r^3$ and the condition (16.12) is identical to that of (16.16).

This means that in order to be able to neglect the effect of multiple scattering,
we implicitly assumed condition (16.12) to hold, which in turn assumes that the
object whose vacuum-induced energy we wish to calculate is made of a dilute gas.
How dilute? Dilute as dictated by conditions (16.12) and (16.16); namely, that the
density of its constituent-particles is much smaller than the volume-scale set by their
polarisabilities.

Many-particles

Consider now a system of three particles: particle 1 plays the role of the atom while
the other two, 2 and 3, form the mirror in the atom-mirror configuration. We wish
to calculate the interaction energy between the mirror and the atom using scattering theory, as shown in Fig. 2.8, where each scattering event gives a factor R_{sc}:
first order perturbation theory gives a contribution of the van der Waals energies
$U_{21} \sim R_{sc}(r_{21})$ and $U_{31} \sim R_{sc}(r_{31})$ where a single scattering of the vacuum field from
the mirror-particles 2 and 3 towards the atom 1 at a distance r_{21} and r_{31}, respectively, is taken. An example for a contribution of the next order, U_{321}, includes two
consecutive scattering events; the incident vacuum fluctuations are first scattered

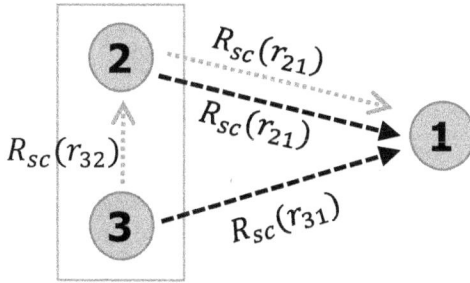

Figure 2.8: Scattering approach for a microscopic atom-mirror model: the interaction energy between a mirror comprised of particles 2 and 3, and an atom (particle 1) is analysed up to second order in scattering theory. The first order scattering processes contribute the two-body van der Waals energies U_{21} and U_{31}, that result from single scattering events represented by single dashed black arrows. They describe scattering of vacuum fluctuations off particles 2 and 3 towards particle 1, with amplitudes $U_{21} \sim R_{sc}(r_{21})$ and $U_{31} \sim R_{sc}(r_{31})$, respectively. The second order process with energy U_{321} results from two consecutive scattering events represented by dotted grey arrows: first, vacuum fluctuations are scattered off particle 3 with amplitude $R_{sc}(r_{32})$, and then they are further scattered by particle 2 towards particle 1 with amplitude $R_{sc}(r_{21})$, giving an overall amplitude $U_{321} \sim R_{sc}(r_{32})R_{sc}(r_{21})$.

by particle 3 arriving to particle 2 still *inside* the mirror, and are then scattered by dipole 2 arriving at dipole 1, such that $U_{321} \sim R_{sc}(r_{32})R_{sc}(r_{21})$. The distance between the mirror particles 2 and 3 can be defined as the length-scale associated with the density of particles in the mirror $\sim n_m^{-1/3}$, whereas their distance to the atom 1 can be taken to be of the order of the atom-mirror distance r. Then, the ratio of the second order to first order contributions scales as

$$\frac{U_{321}}{U_{21} + U_{31}} \sim \frac{R_{sc}(n_m^{-1/3})R_{sc}(r)}{R_{sc}(r) + R_{sc}(r)} \sim R_{sc}(n_m^{-1/3}) \sim \frac{\alpha_m n_m}{\varepsilon_0}, \qquad (16.17)$$

where we used Eq. (16.16) for $R_{sc}(n_m^{-1/3})$ with $n_m^{-1/3}$ as the inter-dipolar distance and where α_m is the polarisability of the mirror particles 2 and 3, reproducing the factor from condition (16.12).

Therefore, regardless of the distance r between the atom and the mirror, it is mainly the distance between the mirror particles $n_m^{-1/3}$ with respect to their polarisabilities α_m that determines the ratio between the second order and first order contributions. Then, contributions from multiple scattering events can be neglected only if the ratio of higher order to lowest order scattering, as in Eq. (16.17), is small.

The relation to the pairwise summation of van der Waals forces comes about by recalling that the van der Waals energies derived in Section 13 and used in the pairwise summation of Section 16.1 are in fact the lowest order two-body contributions U_{21} and U_{31} in the above three-body example. Hence, evaluating Casimir-like, vacuum-induced interaction energies by pairwise summation implicitly assumes that the ratio in Eq. (16.17) is small and all higher order, multiple-scattering contributions can be neglected*.

The above three-body example can of course be generalised to objects comprised of many polarisable particles [37]. For a bulk solid-state mirror the ratio (16.17) is often not small, as the typical distance between mirror-particles is of order of a few Angstroms. Then, the condition (16.12) is not satisfied and pairwise summation, despite being able to provide the correct scaling with distance, fails in giving the correct results for interactions like the atom-mirror potential and the Casimir mirror-mirror force [3].

Exercise 16.1
What is meant by the 'non-additivity' of van der Waals interactions? How many consecutive scattering events have to be taken into account in the first correction to additivity (pairwise summation)? Discuss what should be changed in the quantum electrodynamics perturbation-theory calculations from Section 13.2.2 in order to include the first correction to additivity.

The need for a macroscopic description
The method of pairwise summation can be very useful for the evaluation of the vacuum-energy of a many-particle system which is dilute in the sense of condition (16.12). However, in situations involving bulk systems this is often not the case. Therefore, although bulk matter is comprised of particles, and although the van der Waals interaction is at the heart of the Casimir effect, it is not always useful to treat vacuum-induced interactions of bulk matter from this microscopic point of view [34].

In fact, the same applies to all electrodynamic effects that are based on dipolar interactions, since they may all be described by a scattering approach similar to that used above. For example, as mentioned earlier, the origin of the dielectric response ε of a material to an applied electric field is based on the response α of the individual polarisable dipoles that comprise it. However, upon applying an external field, the dipoles scatter fields on each other, complicating the description of the overall response of the material ε in terms of the individual response α. In some specific cases this overall response can be calculated. The Clausius-Mossotti relation presented in Eq. (16.7) is one such case, that of an isotropic material. However, in general, finding ε is a difficult task. A simplification is obtained by neglecting the interaction between the dipoles, so that the overall response of the material is given by a sum over all polarisabilities of the constituent particles. Not surprisingly, this limit is reached in the case of a dilute gas, for which this approximated Clausius-Mossotti relation (where $\varepsilon_r - 1 \ll 1$) yielded the correct result, Eq. (16.13).

In any case, the conclusion is that using a microscopic description of interacting particles with polarisabilities α in order to describe the electrostatics and electro-dynamics of bulk materials is very limited and hence is typically impractical. This suggests that a macroscopic approach, where the microscopic description of parti-

---
*See also [35] or Chapter 8 of Ref. [3].

cles is *coarse-grained*, and the materials are characterised by macroscopic *response functions* such as $\varepsilon(\omega)$, is favourable*. In fact, in classical electrodynamics such a description is readily available – it is given by the Maxwell's equations in media. An equivalent macroscopic approach for *quantum* electrodynamics, especially in the context of vacuum-induced interactions, is the topic of Chapter 4.

Dependence on geometry

As a final note we wish to emphasise the dependence of non-additivity on geometry. The parameter R_{sc} from Eq. (16.14) depends on the Green function G_{ij} which depends on the geometry where the particles and bulk materials are embedded. For example, in Section 13.2 we have seen that the van der Waals interaction in 1d becomes much longer-range than its free-space counterpart. This was due to the long-range scaling with distance of G_{ij} in 1d.

Therefore, since the magnitude of the parameter R_{sc} determines the validity of the lowest-order scattering approximation, and hence the validity of pairwise summation and additivity, its modified scaling with r due to modified G_{ij} may suggest modified effects due to non-additivity in confining geometries.

§17. Problems

Problem 2.1
London-van der Waals force: generalise the quasistatic treatment from Section 13.1 to the multilevel atom case. Verify that you obtain the result in the short-range regime of the electrodynamic analysis from Section 13.2, Eq. (13.47).

Problem 2.2
Here we shall derive in a simple way the polarisability of a two-level atom in its ground state, the generalisation of which to the multilevel case is that from Eq. (13.40). Consider N photons of a specific mode with frequency and spatial function ω and u, respectively, and annihilation operator $\hat{a}$, interacting with a two-level atom with energy separation E which is in its ground state $|g\rangle$. (i) Using the field operator from Eq. (13.68) for a single mode and the dipole operator of the atom, Eq. (13.2), show that the atom-field Hamiltonian is

$$\hat{H}_I = -\hbar \left[|e\rangle\langle g| + |g\rangle\langle e| \right] \left(ig\hat{a} - ig^*\hat{a}^\dagger \right),$$

where $g = dE_0$, d being the dipole matrix element of the atom and $E_0 = u\sqrt{\omega/(2\varepsilon_0\hbar)}$ is the electromagnetic field per single photon, and where all vector properties are ignored. (ii) Use second order perturbation theory in order to find the energy correction of the ground state of the atom $+ N$ photons. What are the possible intermediate states? (iii) Recall the energy of a polarisable particle in the presence of an electromagnetic field E at a frequency ω: $U = -(1/2)\alpha(\omega)E^2$. What is the average intensity E^2 (or actually $\langle E^*E\rangle$) of N photons? Find $\alpha(\omega)$ in the limit of large photon number N. (iv) Compare your result with that of Eq. (13.40) for the two-level case: your result should be larger by a factor 3. Explain why.

Problem 2.3
Consider the quasistatic analysis of either the van der Waals (Section 2.1) or atom-surface (Section

---

*In the context of Casimir forces, some studies suggest that this macroscopic description might break down for continuously varying dielectric [38]. For further discussion of this problem, see Sections 39.2 and 39.3.

3.1) interactions. Why is this analysis approximate? For which distances between the interacting objects is it valid? Provide a physical explanation for the validity condition (13.41) using the finite speed of light and the typical oscillation frequency of the dipole.

In this quasi-electrostatic regime the interacting objects approximately follow each other *instantaneously* (they are correlated). What does this imply about their correlation and strength of interaction in the retarded regime?

Problem 2.4

Use the quantum electrodynamics perturbation theory approach to find the retarded-regime van der Waals energy in 1d, Eq. (13.64), using the 1d modes from Eq. (13.89). Include in your calculation only the 4 diagrams that are most dominant in the retarded regime. When you compare your result to that of Eq. (13.64), note that you should properly account for the random orientation of the dipoles. *Hint:* you may use the relation $\int_0^\infty d\eta \, e^{-\eta(a+b)} = 1/(a + b)$ in order to separate the integrations over k and k'.

Problem 2.5

Consider a pair of particles 1 and 2 illuminated by an external (c-number) laser field $\boldsymbol{E}_L \cos(\boldsymbol{k}_L \cdot \boldsymbol{r} - \omega_L t)$. The particles are placed at $\boldsymbol{r}_{1,2}$ and possess polarisabilities $\alpha_{1,2}(\omega)$. Find the laser-induced potential between the two particles: (i) In a general embedding geometry. (ii) In free space, for inter-particle distances in the retarded regime, $r = |\boldsymbol{r}_2 - \boldsymbol{r}_1| \gg \lambda_L = 2\pi/|\boldsymbol{k}_L|$. (iii) As in (ii) but where the constant laser envelope is now replaced by a narrow band field, i.e. $|\boldsymbol{E}_L|^2$ is replaced with $|\boldsymbol{E}_L|^2 \int d\omega S(\omega)$, where the width of the spectrum $S(\omega)$ is much smaller than ω_L. As an example, consider particles modeled as two-level atoms with an energy difference ω_a and illuminated by radiation with spectrum

$$S(\omega) = \frac{\gamma}{\gamma^2 + (\omega - \omega_L)^2}. \tag{17.1}$$

Take the parameters e.g. $\gamma/\omega_L = 0.01$, $\omega_a/\omega_L = 1.2$ and find the scaling of U as a function of r. *Hint:* you may solve integrals numerically for various values of r and plot $U(r)$. You may also try to fit $U(r)$ to an oscillating function with decreasing envelope.

ACKNOWLEDGEMENTS

I had the great pleasure to discuss some of the above topics with Carsten Henkel and Grzegorz Łach.

§18. BIBLIOGRAPHY

[1] L. D. Landau and E. M. Lifshitz, *Statistical Physics*, Part 1, 3rd ed. (Pergamon Press, Oxford, 1980).

[2] F. London, *Trans. Faraday Soc.* **33**, 8-26 (1937).

[3] P. W. Milonni, *The Quantum Vacuum: An Introduction to Quantum Electrodynamics* (Academic, 1993).

[4] H. B. G. Casimir and D. Polder, *Phys. Rev.* **73**, 360-372 (1948).

[5] H. B. G. Casimir, *Proc. K. Ned. Akad. Wet.* **51**, 793 (1948).

[6] F. Intravaia, C. Henkel and M. Antezza, in *Casimir Physics*, D. Dalvit, P. Milonni, D. Roberts and F. da Rosa (Eds.), Lecture Notes in Physics, Vol. 834, p. 345-391 (Springer Berlin Heidelberg 2011).

[7] S. Y. Buhmann and S. Scheel S, *Phys. Rev. Lett.* **100**, 253201 (2008).

[8] J. E. Lennard-Jones, *Trans. Faraday Soc.* **28**, 334 (1932).

[9] J. Bardeen, *Phys. Rev.* **58**, 727 (1940).

[10] J. D. Jackson, *Classical Electrodynamics* (Wiley, 1998).

[11] L. Novotny and B. Hecht, *Principles of Nano-Optics* (Cambridge University Press, 2006).

[12] D. P. Craig and T. Thirunamachandran, *Molecular Quantum Electrodynamics* (Academic, London, 1984).

[13] Yu. S. Barash and V. L. Ginzburg, *Sov. Phys. Usp.* **27**, 467 (1984).

[14] L. Spruch and E. Kelsey, *Phys. Rev. A* **18**, 845 (1978).

[15] D. M. Pozar, *Microwave Engineering* (John Wiley & Sons, New York, 2005).

[16] E. Shahmoon, I. Mazets and G. Kurizki, *Proc. Natl. Acad. Sci. USA* **111**, 10485 (2014).

[17] C. Cohen-Tannoudji, J. Dupont-Roc, and G. Grynberg, *Photons and Atoms: Introduction to Quantum Electrodynamics*, (John Wiley & Sons, New York, 1987).

[18] C. Cohen-Tannoudji, J. Dupont-Roc, and G. Grynberg, *Atom-Photon Interactions: Basic Processes and Applications*, (WILEY-VCH, 2004).

[19] E. Shahmoon and G. Kurizki, *Phys. Rev. A* **87**, 062105 (2013).

[20] M. Kardar and R. Golestanian, *Rev. Mod. Phys.* **71**, 1233 (1999).

[21] M. Saffman, T.G. Walker and K. Mølmer, *Rev. Mod. Phys.* **82**, 2313 (2010).

[22] J. Koch, T. M. Yu, J. Gambetta, A. A. Houck, D. I. Schuster, J. Majer, A. Blais, M. H. Devoret, S. M. Girvin, and R. J. Schoelkopf, *Phys. Rev. A* **76**, 042319 (2007).

[23] A. Blais, R. S. Huang, A. Wallraff, S. M. Girvin and R. J. Schoelkopf, *Phys. Rev. A* **69**, 062320 (2004).

[24] J. A. Kong, *Electromagnetic Wave Theory*, (John Wiley and Sons, New York, 1986).

[25] V. Sandoghdar, C. I. Sukenik, E. A. Hinds, and S. Haroche, *Phys. Rev. Lett.* **68**, 3432 (1992).

[26] C. I. Sukenik, M. G. Boshier, D. Cho, V. Sandoghdar, and E. A. Hinds, *Phys. Rev. Lett.* **70**, 560 (1993).

[27] J. F. Babb and L. Spruch, *Phys. Rev. A* **36**, 456 (1987).

[28] T. Thirunamachandran, *Mol. Phys.* **40**, 393 (1980).

[29] A. Salam, *Advances in Quantum Chemistry* **62**, 1 (2011).

[30] K. Dholakia and P. Zemánek, *Rev. Mod. Phys.* **82**, 1767 (2010).

[31] E. Shahmoon and G. Kurizki, *Phys. Rev. A* **89**, 043419 (2014).

[32] E. Shahmoon, I. Mazets and G. Kurizki, *Opt. Lett.* **39**, 3674 (2014).

[33] J. Rodríguez and D. L. Andrews, *Phys. Rev. A* **79**, 022106 (2009).

[34] I. E. Dzyaloshinskii, E. M. Lifshitz and L. P. Pitaevskii, *Advances in Physics* **10**, 165 (1961).

[35] P. W. Milonni and P. B. Lerner, *Phys. Rev. A* **46**, 1185 (1992).

[36] M. Born and E. Wolf, *Principles of Optics* (Pergamon Press, London, 1970).

[37] E. A. Power and T. Thirunamachandran, *Phys. Rev. A* **50**, 3929 (1994).

[38] W. M. R. Simpson, S. A. R. Horsley and U. Leonhardt, *Phys. Rev. A* **87**, 043806 (2013).

Chapter 3

The Casimir stress in real materials

STEFAN SCHEEL

> Ultra-modern physicists [are] ... reluctant to contemplate Nature
> unclad. Clothing she must have. At the least she must wear a
> matrix, with here and there a tensor to hold the queer garment
> together.
>
> – Sydney Evershed, *Out of the Crystal Maze: Chapters from The History of Solid State Physics (1992), 361*

Hendrik Casimir showed in his seminal work in 1948 that there exists a force between two perfectly conducting, parallel plates [1]. This is in spite of the fact that, in Casimir's thought experiment, none of the electromagnetic modes between the plates are occupied. His arguments were in fact related to the zero-point energy of the quantised electromagnetic field. Casimir, speaking at a conference in 1998 in Leipzig/Germany, told the audience about a meeting he had with Niels Bohr during which he explained to Bohr his work on van der Waals forces: 'Bohr thought this over, then mumbled something like "must have something to do with zero-point energy"' [2].

The concept of zero-point energy had been around since 1911 when Max Planck formulated his 'second quantum hypothesis'. Within this framework, the energy of the ground state of an harmonic oscillator of frequency ν had to be set to $h\nu/2$ and not zero as for a classical harmonic oscillator. Experimental evidence of such half-quantum numbers were subsequently found in vibrational spectra of molecules. With the advent of quantum mechanics in 1925/6 the existence of zero-point energy could be proven theoretically, and in 1943, Klaus Clusius published a list of experimental confirmations [3].

However, not everybody was happy with this concept. Wolfgang Pauli argued against its general validity, in particular with regard to the electromagnetic field:*

---

*'An dieser Stelle sei gleich bemerkt, daß es konsequenter ist, eine Nullpunktsenergie von $\frac{1}{2}h\nu_r$ pro Freiheitsgrad hier im Gegensatz zum materiellen Oszillator nicht einzuführen. Denn einerseits würde diese wegen der unendlichen Zahl der Freiheitsgrade zu einer unendlich großen Energie pro Volumeneinheit führen, andererseits wäre diese prinzipiell unbeobachtbar, da sie weder emittiert, absorbiert oder gestreut wird, also nicht in Wände eingeschlossen werden kann, und da sie, wie

At this point, it should be mentioned that it is more consistent not to introduce here a zero-point energy of $\frac{1}{2}h\nu$ per degree of freedom, in contrast to the material oscillator. Because, on the one hand, this would lead to an infinitely large energy per unit volume due to the infinite number of degrees of freedom; on the other hand, this infinite energy cannot be observed in principle, since it may be emitted, absorbed or diffracted — hence cannot be enclosed by walls — and since it does not create any gravitational field — as is known from experience.

Pauli rejected the existence of zero-point energy because, he supposed, it was not associated with a force field. But this is precisely what the Casimir effect is and what Hendrik Casimir showed to be true!

§19. DISPERSION AND DISSIPATION IN REAL MEDIA

Casimir's original approach to calculating the force between two plates rests on the argument that the zero-point energy between the plates is restricted with respect to the situation without these boundaries. Despite this being a valid interpretation in a variety of cases, it obscures an equally important viewpoint that we shall develop here, and which is sometimes more intuitive than the argument of mode restriction.

Let us, for the moment, concentrate on a single quantum harmonic oscillator of angular frequency $\omega = 2\pi\nu$. Its Hamiltonian is

$$\hat{H} = \hbar\omega \left(\hat{a}^\dagger \hat{a} + \frac{1}{2} \right) , \qquad (19.1)$$

where $\hat{a}$ and $\hat{a}^\dagger$ are the annihilation and creation operators of excitations of that mode. The zero-point energy $\hbar\omega/2$ in fact arises from the use of the commutation relation $[\hat{a}, \hat{a}^\dagger] = 1$ in the symmetrised form of Eq. (19.1). From the mode expansion of the electric field with spatial mode function $\boldsymbol{u}(\boldsymbol{r})$ (see also Chapter 1, §2),

$$\hat{\boldsymbol{E}}(\boldsymbol{r}) = \mathrm{i}\omega\boldsymbol{u}(\boldsymbol{r})\hat{a} + \mathrm{h.c.} \qquad (19.2)$$

it then follows that the strength of the fluctuations of the electric field in the ground state is given by

$$\langle 0|\boldsymbol{E}^2(\boldsymbol{r})|0\rangle = \omega^2|\boldsymbol{u}(\boldsymbol{r})|^2\langle 0| \left[\hat{a}, \hat{a}^\dagger \right] |0\rangle = \omega^2|\boldsymbol{u}(\boldsymbol{r})|^2 . \qquad (19.3)$$

Hence, the same commutation relation of operators that leads to the zero-point energy is responsible for the ground-state fluctuations of the electromagnetic field. It is precisely those fluctuations that we want to interpret as the source of Casimir-related forces.

Before we come to that, let us return to the setup originally envisaged by Casimir. It consisted of two perfectly conducting plates, which means that one assumes that

---

aus der Erfahrung evident ist, auch kein Gravitationsfeld erzeugt.' [4]. For more historical details, see also Ref. [5].

the plates reflect all electromagnetic waves even at the highest imaginable frequencies. This is clearly a very strong approximation. In fact, even Casimir in his original calculation had to assume a cut-off frequency beyond which the plates no longer reflect light to make the zero-point energy finite (see also Section 7, as well as Chapter 3.2.4 in Ref. [6] for details of that calculation). This means that in reality a material response that falls off at sufficiently high frequencies has to be assumed. In fact, as we shall see, any real material will have this property, and a perfectly conducting plate at all frequencies is an untenable assumption.

19.1 Dipole model of the optical response

We can construct a simple model of the material response that captures all essential properties. For that we assume a dipole consisting of two oppositely charged particles that are separated by a distance r (see Fig. 3.1). Newton's equations of motion for

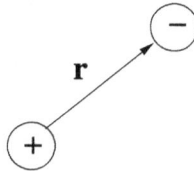

Figure 3.1: Dipole model of the linear optical response of a dielectric material.

the charged particles with charges q_i and position vectors r_i are

$$m_i \ddot{r}_i = q_i[E(r_i, t) + \dot{r}_i \times B(r_i, t)] + F_{\text{diss}} + \sum_j F_{ij}, \qquad (19.4)$$

where the forces that appear in this expression are the Lorentz force, dissipative (friction) forces F_{diss} and conservative binding forces F_{ij} between the charges. We assume now that $q_i = \pm e$ and that the applied electric field is polarised in the direction of the dipole, $E = E e_x$. The equations of motion (19.4) then reduce to those of a particle moving along a direction parallel to the dipole. We neglect the magnetic component of the Lorentz force and assume that the conservative forces are linear restoring forces. Then, for a harmonically varying field $E(t) = E_0 e^{-i\omega t}$, the equation of motion (19.4) reduces to

$$\ddot{x} + \gamma \dot{x} + \omega_0^2 x = \frac{e}{m} E_0 e^{-i\omega t}, \qquad (19.5)$$

which describes a damped harmonic oscillator where x is the displacement of the dipole from its equilibrium position, γ the friction coefficient, and ω_0 the eigenfrequency of the dipole resulting from the conservative forces. With the ansatz $x(t) = x_0 e^{-i\omega t}$ we arrive at the solution

$$(-\omega^2 - i\gamma\omega + \omega_0^2)x(t) = \frac{e}{m}E(t) \quad \rightsquigarrow \quad x(t) = \frac{\frac{e}{m}E(t)}{\omega_0^2 - \omega^2 - i\gamma\omega}. \qquad (19.6)$$

The electric dipole moment is therefore $p(t) = \alpha(\omega)E(t)$ with the (complex) polarisability

$$\alpha(\omega) = \frac{e^2}{m} \frac{1}{\omega_0^2 - \omega^2 - i\gamma\omega} \tag{19.7}$$

of a single dipole. Assuming that the medium is composed of many such dipoles with a number density n, we arrive at a linear susceptibility $\chi^{(1)}(\omega) = n\alpha(\omega)$, in which we neglect all interactions between the dipoles. Splitting the polarisability (19.7) into real and imaginary parts yields for the linear susceptibility

$$\mathrm{Re}\,\chi^{(1)}(\omega) = \frac{ne^2}{m} \frac{(\omega_0^2 - \omega^2)}{(\omega_0^2 - \omega^2)^2 + \gamma^2\omega^2}, \quad \mathrm{Im}\,\chi^{(1)}(\omega) = \frac{ne^2}{m} \frac{\gamma\omega}{(\omega_0^2 - \omega^2)^2 + \gamma^2\omega^2}. \tag{19.8}$$

Exercise 19.1

Show that the functions defined in Eq. (19.8) are the real and imaginary parts of the polarisability (19.7) times the number density n of dipoles.

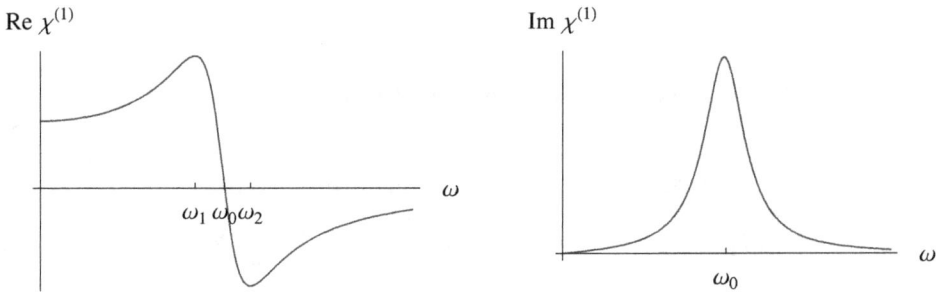

Figure 3.2: Real and imaginary parts of the linear susceptibility of an ensemble of identical dipoles with resonance frequency ω_0.

Both functions are depicted in Fig. 3.2. In the regions $\omega < \omega_1$ and $\omega > \omega_2$, an electromagnetic wave experiences a refractive index that increases with frequency (normal dispersion), whereas in the interval $\omega_1 < \omega < \omega_2$ the material shows anomalous dispersion. The imaginary part, describing absorption of the electromagnetic wave by the material, has its maximum close to the resonance frequency ω_0. Note that the response of a quantum-mechanical two-level atom to an external field shows the same functional dependence [see also Eq. (13.40)].

19.2 The Kramers–Kronig relations

Any causal linear response function, of which Eq. (19.7) is just one example, fulfils an important integral relation. In temporal space, the susceptibility is given by its (inverse) Fourier transform

$$\chi^{(1)}(t) = \frac{1}{2\pi} \int_{-\infty}^{\infty} d\omega \, e^{-i\omega t} \chi^{(1)}(\omega) \,. \tag{19.9}$$

The linear response that relates the polarisation field inside a medium to an externally applied electric field has to be causal, so that*

$$\boldsymbol{P}(\boldsymbol{r}, t) = \varepsilon_0 \int_{0}^{\infty} dt' \, \chi^{(1)}(\boldsymbol{r}, t') \boldsymbol{E}(\boldsymbol{r}, t - t') \,. \tag{19.10}$$

This in effect requires the temporal susceptibility to fulfil the relation

$$\chi^{(1)}(t) = \Theta(t) \chi^{(1)}(t) \tag{19.11}$$

where $\Theta(t)$ stands for the Heaviside step function.

Exercise 19.2
Explain why the relation (19.11) for the susceptibility in the time domain must hold.

We now Fourier transform both sides of Eq. (19.11) with the result that

$$
\begin{aligned}
\chi^{(1)}(\omega) &= \int_{-\infty}^{\infty} dt \, e^{i\omega t} \Theta(t) \chi^{(1)}(t) = \frac{1}{2\pi} \int_{-\infty}^{\infty} dt \int_{-\infty}^{\infty} d\omega' \, e^{i(\omega - \omega')t} \Theta(t) \chi^{(1)}(\omega') \\
&= \frac{1}{2\pi} \int_{-\infty}^{\infty} d\omega' \, \chi^{(1)}(\omega') \int_{0}^{\infty} dt \, e^{i(\omega - \omega')t} \\
&= \frac{1}{2\pi i} \int_{-\infty}^{\infty} d\omega' \, \chi^{(1)}(\omega') \left[i\pi\delta(\omega - \omega') - \mathcal{P} \frac{1}{\omega - \omega'} \right] \,.
\end{aligned}
\tag{19.12}
$$

Rearranging both sides of the equation, we therefore find that

$$\chi^{(1)}(\omega) = -\frac{1}{i\pi} \mathcal{P} \int_{-\infty}^{\infty} d\omega' \, \frac{\chi^{(1)}(\omega')}{\omega - \omega'} \,. \tag{19.13}$$

This expression relates the Fourier component of the susceptibility at frequency ω to an integral over all other frequency components. This is a particular example of a Kramers–Kronig relation for the linear susceptibility. Taking the real and imaginary parts of both sides of Eq. (19.13), one obtains

$$\mathrm{Re}\,\chi^{(1)}(\omega) = -\frac{1}{\pi} \mathcal{P} \int_{-\infty}^{\infty} d\omega' \, \frac{\mathrm{Im}\,\chi^{(1)}(\omega')}{\omega - \omega'} \,; \quad \mathrm{Im}\,\chi^{(1)}(\omega) = \frac{1}{\pi} \mathcal{P} \int_{-\infty}^{\infty} d\omega' \, \frac{\mathrm{Re}\,\chi^{(1)}(\omega')}{\omega - \omega'} \,. \tag{19.14}$$

---

*Here we neglect effects such as spatial dispersion and anisotropy.

Hence, real and imaginary parts of the linear susceptibility are related by this (Hilbert) integral transform. One can check that the model susceptibility (19.7) fulfils these relations and hence corresponds to a valid causal linear response.

Exercise 19.3
In Eq. (19.12), the identity

$$\int\limits_0^\infty dt\, e^{i(\omega - \omega')t} = \pi\delta(\omega - \omega') + i\mathcal{P}\frac{1}{\omega - \omega'}$$

has been used. Consider a regularisation of the integral in which the upper integration limit is replaced by a finite time T. Split the result into real and imaginary parts, sketch both functions and convince yourself that, in the limit $T \to \infty$, they converge to the expressions given above.

Kramers–Kronig relations of the type (19.13) or (19.14) are very general statements that hold for all causal response functions $\chi(t)$, because they only rely on the causality condition (19.11). In addition, it can be shown that the Fourier components $\chi(\omega)$ are analytic functions in the upper half of the complex ω-plane, with the symmetry relation $\chi(\omega) = \chi^*(-\omega^*)$ being valid. This implies that on the real frequency axis, $\mathrm{Re}\,\chi(\omega) = \mathrm{Re}\,\chi(-\omega)$ and $\mathrm{Im}\,\chi(\omega) = -\mathrm{Im}\,\chi(-\omega)$. For later use it is instructive to note that, as a consequence of these symmetry relations, on the imaginary frequency axis the Fourier components of the susceptibility are purely real.

Our simple dipole model describes the optical response of a dielectric material surprisingly well. The dipoles in question are displacements of bound electrons in the material. As there are many possible dipole transition in real materials, their optical response is then typically described by a sum of a number of individual dipole responses,

$$\varepsilon(\omega) = 1 + \sum_n \frac{\omega_{Pn}^2}{\omega_n^2 - \omega^2 - i\gamma_n\omega}, \tag{19.15}$$

where ω_{Pn} are the plasma frequencies (or oscillator strengths) of the resonances, and γ_n their widths. This is the so-called Drude–Lorentz model for dielectrics. In the limit of vanishing resonance frequencies ω_n, i.e. for unbound electrons in metals, one arrives at the Drude model for metals

$$\varepsilon(\omega) = 1 - \frac{\omega_P^2}{\omega(\omega + i\gamma)}. \tag{19.16}$$

The Drude model implies a finite static conductivity of $\sigma_0 = \omega_P^2/\gamma\varepsilon_0$. This has to be compared with the response of a free electron and ion gas with $\gamma = 0$ which, rather unrealistically, assigns an infinitely large static conductivity to it. In this plasma model, a displacement of the electrons with respect to the ions leads to an undamped oscillation of the charge cloud with a plasma frequency

$$\omega_P^2 = \frac{ne^2}{m\varepsilon_0}, \tag{19.17}$$

where n is the electron density and m their (effective) mass. Because of the assumption of vanishing dissipation, the Plasma model is equivalent to a purely real permittivity

$$\varepsilon(\omega) = 1 - \frac{\omega_P^2}{\omega^2} . \tag{19.18}$$

Such a model cannot fulfil the Kramers–Kronig relations (19.14), as they imply an imaginary part proportional to a δ-function.

Exercise 19.4
Show explicitly that the permittivity of the Drude model (19.16) fulfils the Kramers–Kronig relations. Split the function $\varepsilon(\omega)$ into its real and imaginary parts and compute one of the integrals in Eq. (19.14).

19.3 Generalised mode decomposition

The necessary existence of an imaginary part to the dielectric permittivity of a medium has far-reaching consequences. The spatial mode functions $u(r)$ that were used to construct the operator of the electric field (19.2) in free space resulted from the Helmholtz equation (see Chapter 1, §2)

$$\left[\Delta + \frac{\omega^2}{c^2}\varepsilon(\omega)\right] u(r) = 0 \tag{19.19}$$

with $\varepsilon(\omega) \equiv 1$. One can read Eq. (19.19) as an eigenvalue equation for the operator $-\Delta$ with eigenvalues $\varepsilon(\omega)\omega^2/c^2$ and eigenfunctions $u(r)$. As the Laplace operator is Hermitian, one would expect its eigenvalues to be real (which they are in free space with $\varepsilon(\omega) \equiv 1$). In this case, the eigenfunctions form a complete set of orthogonal functions that can be used to expand an arbitrary vector field. For a complex permittivity function $\varepsilon(\omega)$ this statement is no longer true. The resulting spatial mode functions are neither orthogonal nor complete as they represent damped waves, and hence cannot be used to expand vector fields.

In a quantum setting, this problem becomes even more obvious. Suppose that we are investigating a single-mode harmonic oscillator with frequency ω which we describe by its amplitude operators $\hat{a}$ and $\hat{a}^\dagger$. In a medium with absorption, we have seen that the spatial mode functions are damped, so we can expect that the temporal evolution of the expectation value of the amplitude operators is also damped,

$$\langle \hat{a}(t) \rangle = e^{-(i\omega + \Gamma/2)t} \langle \hat{a}(0) \rangle . \tag{19.20}$$

If one tried to reconstruct from Eq. (19.20) the corresponding equation of motion, one would arrive at

$$\frac{d}{dt}\langle \hat{a} \rangle = -(i\omega + \Gamma/2)\langle \hat{a} \rangle . \tag{19.21}$$

However, this equation of motion does not hold for the operator itself. Despite the fact that from

$$\dot{\hat{a}} = -(i\omega + \Gamma/2)\hat{a} \tag{19.22}$$

follows the formal solution $\hat{a}(t) = e^{-(i\omega+\Gamma/2)t}\hat{a}(0)$, the bosonic commutation relations between the amplitude operators are now invalid. If at time $t = 0$ the usual relation $[\hat{a}(0), \hat{a}^\dagger(0)] = 1$ were valid, it would not be at a later time t as $[\hat{a}(t), \hat{a}^\dagger(t)] = e^{-\Gamma t}$. Hence, the amplitude operators lose their bosonic character with time.

We now try to save the equal-time commutation relations for the amplitude operators by adding another operator $\hat{f}$ to Eq. (19.22) whose properties will be chosen such that the time evolution of the expectation values (19.20) is reproduced. For that, we set

$$\dot{\hat{a}} = -(i\omega + \Gamma/2)\hat{a} + \hat{f}(t) \tag{19.23}$$

and integrate formally with respect to time,

$$\hat{a}(t) = e^{-(i\omega+\Gamma/2)t}\hat{a}(0) + \int_0^t dt'\, e^{-(i\omega+\Gamma/2)(t-t')}\hat{f}(t'). \tag{19.24}$$

Inserting this expression into the equal-time commutation relations for the amplitude operators and using the initial condition $[\hat{a}(0), \hat{a}^\dagger(0)] = 1$, one finds that at later times

$$\left[\hat{a}(t), \hat{a}^\dagger(t)\right] = e^{-\Gamma t} + e^{-(i\omega+\Gamma/2)t}\int_0^t dt'\left[\hat{a}(0), \hat{f}^\dagger(t')\right]e^{-(-i\omega+\Gamma/2)(t-t')}$$

$$+ e^{-(-i\omega+\Gamma/2)t}\int_0^t dt'\left[\hat{f}(t'), \hat{a}^\dagger(0)\right]e^{-(i\omega+\Gamma/2)(t-t')}$$

$$+ e^{-\Gamma t}\int_0^t dt'\int_0^t dt''\left[\hat{f}(t'), \hat{f}^\dagger(t'')\right]e^{i\omega(t'-t'')+\Gamma/2(t'+t'')}. \tag{19.25}$$

If we choose the following time-dependent commutation relations,

$$\left[\hat{f}(t_1), \hat{f}^\dagger(t_2)\right]e^{i\omega(t_1-t_2)} = \Gamma\delta(t_1 - t_2), \tag{19.26a}$$

$$\left[\hat{a}(t_1), \hat{f}^\dagger(t_2)\right] = 0 \quad (t_2 > t_1), \tag{19.26b}$$

then one indeed finds that the equal-time commutation relations for the amplitude operators are $[\hat{a}(t), \hat{a}^\dagger(t)] = 1$ for all times. In order for the expectation values to have the correct time evolution (19.20), the expectation of the auxiliary operator $\hat{f}$ needs to vanish, $\langle\hat{f}(t)\rangle = 0$. The damping constant Γ is obtained from the commutation relation (19.26a) as

$$\Gamma = \int_0^\infty dt\, e^{i\omega t}\left[\hat{f}(t), \hat{f}^\dagger(0)\right]. \tag{19.27}$$

The operator $\hat{f}(t)$ is a particular example of a Langevin operator that plays an important role in the theory of open quantum systems [7]. Their task is essentially to ensure the correct equal-time commutation relations for some system operators in the presence of damping. Because of their vanishing expectation value they can be interpreted as noise operators whose correlation function just yields the damping constant Γ.

19.4   Open quantum systems and Langevin equations

The phenomenological introduction of the Langevin operators $\hat{f}$ is not quite satisfying as it does not reveal anything about their physical origin. At present, their introduction has been rather *ad hoc*. We will now construct a simple model that allows us to determine such operators explicitly. This model consists of a quantum system (or 'system' in short) which is described by a Hamiltonian $\hat{H}_{\mathrm{sys}}$. That system shall be coupled to a quantum mechanical reservoir whose evolution is governed by a Hamiltonian $\hat{H}_{\mathrm{res}}$. The (bilinear) coupling between system and reservoir is described by the interaction Hamiltonian $\hat{H}_{\mathrm{int}}$ such that the total Hamiltonian is

$$\hat{H} = \hat{H}_{\mathrm{sys}} + \hat{H}_{\mathrm{res}} + \hat{H}_{\mathrm{int}} . \tag{19.28}$$

In our example, the amplitude operators $\hat{a}$ and $\hat{a}^\dagger$ shall play the role of the system operators. They could, for example, represent a single mode of the electromagnetic field in a high-Q cavity or model the polarisation of an atom in linear approximation. The system operators are assumed to be coupled to a reservoir that is composed of a large number of harmonic oscillators, say, electromagnetic field modes, with amplitude operators $\hat{b}_i$ and $\hat{b}_i^\dagger$.* In this case, the reservoir and interaction Hamiltonians are

$$\hat{H}_{\mathrm{res}} = \sum_i \hbar \omega_i \hat{b}_i^\dagger \hat{b}_i , \qquad \hat{H}_{\mathrm{int}} = \hbar \sum_i g_i \hat{a}^\dagger \hat{b}_i + \mathrm{h.c.} \tag{19.29}$$

with coupling constants g_i. In quantum optics terms, the rotating-wave approximation (RWA), in which only energy-conserving terms are retained (see, e.g., Ref. [9]), has been employed in $\hat{H}_{\mathrm{int}}$ such that off-resonant terms such as $\hat{a}\hat{b}_i$ and $\hat{a}^\dagger \hat{b}_i^\dagger$ do not contribute to the dynamics.

   The time evolution of system and reservoir are described by Heisenberg's equations of motion which read in this case

$$\dot{\hat{a}} = \frac{1}{i\hbar} \left[\hat{a}, \hat{H} \right] = -i\omega\hat{a} - i\sum_i g_i \hat{b}_i , \tag{19.30a}$$

$$\dot{\hat{b}}_i = \frac{1}{i\hbar} \left[\hat{b}_i, \hat{H} \right] = -i\omega_i \hat{b}_i - ig_i^* \hat{a} . \tag{19.30b}$$

Formal solution of Eq. (19.30b) yields

$$\hat{b}_i(t) = e^{-i\omega_i t} \hat{b}_i(0) - ig_i^* \int_0^t \mathrm{d}t'\, e^{-i\omega_i(t-t')} \hat{a}(t') . \tag{19.31}$$

---

*This is a simplified version of the Caldeira–Leggett model used to decribe quantum Brownian motion (see, e.g., Refs. [7,8]).

Inserted back into Eq. (19.30a) gives

$$\dot{\hat{a}} = -\mathrm{i}\omega\hat{a} - \sum_i |g_i|^2 \int_0^t \mathrm{d}t'\, \mathrm{e}^{-\mathrm{i}\omega_i(t-t')}\hat{a}(t') - \mathrm{i}\sum_i g_i \mathrm{e}^{-\mathrm{i}\omega_i t}\hat{b}_i(0)\,. \qquad (19.32)$$

This integral can be solved using the Markov approximation in which the fast time evolution of the system operator $\hat{a}(t')$ is split off, $\hat{a}(t') = \mathrm{e}^{-\mathrm{i}\omega t'}\tilde{\hat{a}}(t')$, and the slowly-varying amplitude operator $\tilde{\hat{a}}(t')$ is taken out of the integral at the upper integration limit. Thus, we can write for the integral

$$\int_0^t \mathrm{d}t'\, \mathrm{e}^{-\mathrm{i}\omega_i(t-t')}\hat{a}(t') = \int_0^t \mathrm{d}t'\, \mathrm{e}^{-\mathrm{i}\omega_i(t-t')}\mathrm{e}^{\mathrm{i}\omega t'}\tilde{\hat{a}}(t')$$

$$\simeq \tilde{\hat{a}}(t) \int_0^t \mathrm{d}t'\, \mathrm{e}^{-\mathrm{i}\omega_i(t-t')}\mathrm{e}^{\mathrm{i}\omega t'} = \hat{a}(t) \int_0^t \mathrm{d}t'\, \mathrm{e}^{\mathrm{i}(\omega-\omega_i)(t-t')}\,. \qquad (19.33)$$

The integrodifferential equation (19.32) therefore reduces to an ordinary differential equation in which all operators are taken at the same time t. Thus, the Markov approximation erases all memory effects about the interaction between system and reservoir at earlier times. The remaining time integral can be further simplified for long interaction times as (see also exercise 19.3)

$$\lim_{t\to\infty} \int_0^t \mathrm{d}t'\, \mathrm{e}^{\mathrm{i}(\omega-\omega_i)(t-t')} = \pi\delta(\omega - \omega_i) + \mathrm{i}\mathcal{P}\frac{1}{\omega - \omega_i}\,. \qquad (19.34)$$

Combining everything so far, Eq. (19.32) then reduces to

$$\dot{\hat{a}} = -(\mathrm{i}\omega + \mathrm{i}\delta\omega + \Gamma/2)\hat{a} + \hat{f}(t) \qquad (19.35)$$

with the notations

$$\Gamma = 2\pi \sum_i |g_i|^2 \delta(\omega - \omega_i)\,, \qquad (19.36a)$$

$$\delta\omega = \sum_i |g_i|^2 \mathcal{P}\frac{1}{\omega - \omega_i}\,, \qquad (19.36b)$$

$$\hat{f}(t) = -\mathrm{i}\sum_i g_i \mathrm{e}^{-\mathrm{i}\omega_i t}\hat{b}_i(0)\,. \qquad (19.36c)$$

Note that the differential equation (19.35) is, apart from the appearance of the frequency shift $\delta\omega$, identical to the one we defined earlier [Eq. (19.23)] by heuristically introducing the Langevin operators.

In the framework of this model we have found that the Langevin operators $\hat{f}(t)$ are given by the initial conditions of the reservoir operators $\hat{b}_i(0)$. Using their explicit representation one finds their commutation relation

$$\left[\hat{f}(t), \hat{f}^\dagger(0)\right] = \sum_i |g_i|^2 \mathrm{e}^{-\mathrm{i}\omega_i t}\,, \qquad (19.37)$$

whose correlation function, due to the relation

$$\int_0^\infty dt\, e^{i\omega t} \left[\hat{f}(t), \hat{f}^\dagger(0)\right] = \pi \sum_i |g_i|^2 \delta(\omega - \omega_i) \equiv \Gamma/2, \qquad (19.38)$$

yields the damping constant Γ in terms of the coupling constants g_i of the system-reservoir interaction. From the Markov integral arises an additional frequency shift $\delta\omega$ whose origin also lies in the system-reservoir interaction. In fact, the damping rate Γ and the frequency shift $\delta\omega$ can be regarded as the frequency components of a constant susceptibility, and are thus related by a Kramers–Kronig relation of the kind (19.14).

19.5 Linear fluctuation-dissipation theorem

The connection between dissipation or damping and fluctuations in the system is generally valid and is expressed through the fluctuation-dissipation theorem that we want to derive here in its simplest form*. The starting point is a system in its thermal equilibrium at temperature T whose density operator $\hat{\varrho}$ is described by a Gibbs state $\hat{\varrho}_0 = e^{-\beta \hat{H}_0}/Z$ with the partition function $Z = \mathrm{tr}\, e^{-\beta \hat{H}_0}$ and the inverse temperature $\beta = (k_B T)^{-1}$. If the quantum state is displaced from its equilibrium by an external perturbation, its time evolution in the interaction picture is then governed by the von Neumann equation

$$\frac{d}{dt}\hat{\varrho}_I(t) = \frac{i}{\hbar}\left[\hat{\varrho}_I(t), \hat{H}_I(t)\right]. \qquad (19.39)$$

Integrating both sides with respect to time yields the associated integral equation

$$\hat{\varrho}_I(t) = \hat{\varrho}_I(t_0) + \frac{i}{\hbar}\int_{t_0}^t ds\, \left[\hat{\varrho}_I(s), \hat{H}_I(s)\right], \qquad (19.40)$$

that can be solved by successive approximation. In the first iteration one replaces $\hat{\varrho}_I(s)$ with the initial condition $\hat{\varrho}_I(t_0) = \hat{\varrho}_0$. In the limit $t_0 \to -\infty$ this gives the linear response

$$\hat{\varrho}_I(t) = \hat{\varrho}_0 + \frac{i}{\hbar}\int_{-\infty}^t ds\, \left[\hat{\varrho}_0, \hat{H}_I(s)\right] \qquad (19.41)$$

of the state of a quantum system to an external perturbation.

Let the Hamiltonian now be of the form $\hat{H}(t) = \hat{H}_0 - \lambda(t)\hat{V}$ such that the Gibbs state $\hat{\varrho}_0$ is stationary with respect to the unperturbed Hamiltonian $\hat{H}_0$, i.e.

---

*The general form is due to Callen and Welton [10]. The earliest example of such a relation is Einstein's relation between the diffusion constant and the mobility in Brownian motion [11]. In the context of electromagnetism an early example is the Nyquist theorem [12] which relates a random external voltage to thermal current fluctuations in an electric circuit.

$[\hat{\varrho}_0, \hat{H}_0] = 0$. Then Eq. (19.41) becomes

$$\hat{\varrho}_I(t) = \hat{\varrho}_0 - \frac{i}{\hbar} \int\limits_{-\infty}^{t} ds\, \lambda(s) \left[\hat{\varrho}_0, \hat{V}_I(s)\right] \tag{19.42}$$

and, after transforming back to the Schrödinger picture,

$$\hat{\varrho}(t) = \hat{\varrho}_0 - \frac{i}{\hbar} \int\limits_{-\infty}^{t} ds\, \lambda(s) \left[\hat{\varrho}_0, \hat{V}(s-t)\right] . \tag{19.43}$$

The expectation value of some observable $\hat{O}$ is then

$$\langle \hat{O} \rangle = \mathrm{tr}[\hat{\varrho}(t)\hat{O}] = \mathrm{tr}[\hat{\varrho}_0\hat{O}] + \int\limits_{0}^{\infty} d\tau\, \lambda(t-\tau)\Gamma(\hat{O}\hat{V};\tau) \tag{19.44}$$

where we defined the linear response function

$$\Gamma(\hat{O}\hat{V};\tau) = \frac{i}{\hbar}\mathrm{tr}\left\{\hat{\varrho}_0[\hat{O}(\tau), \hat{V}]\right\} = \frac{i}{\hbar}\langle[\hat{O}(\tau), \hat{V}]\rangle . \tag{19.45}$$

Next, we consider the symmetrised correlation function of two operators $\hat{A}$ and $\hat{B}$,

$$K(\hat{A}\hat{B};\tau) = \frac{1}{2}\langle\hat{A}(t+\tau)\hat{B}(\tau) + \hat{B}(\tau)\hat{A}(t+\tau)\rangle - \langle\hat{A}\rangle\langle\hat{B}\rangle . \tag{19.46}$$

Due to the stationarity of the Gibbs state, the time arguments can be shifted so that

$$K(\hat{A}\hat{B};\tau) = \frac{1}{2}\langle\hat{A}(\tau)\hat{B} + \hat{B}\hat{A}(\tau)\rangle - \langle\hat{A}\rangle\langle\hat{B}\rangle . \tag{19.47}$$

This expression will now be compared to the linear response function $\Gamma(\hat{A}\hat{B};\tau) = \frac{i}{\hbar}\langle[\hat{A}(\tau), \hat{B}]\rangle$.

In both functions operator products appear of the form $\hat{A}(\tau)\hat{B}$ and $\hat{B}\hat{A}(\tau)$ whose expectation values in thermal equilibrium have to be computed. Note that the unitary operator $\hat{U}(t) = e^{-i\hat{H}_0 t/\hbar}$ that evolves a quantum state in time and the Gibbs state $\hat{\varrho}_0 = e^{-\beta\hat{H}_0}/Z$ are both exponential functions of the Hamiltonian $\hat{H}_0$. With $\hat{A}(t) = e^{-i\hat{H}_0 t/\hbar}\hat{A}e^{i\hat{H}_0 t/\hbar}$ it then follows that

$$\hat{A}(\tau)e^{-\beta\hat{H}_0} = e^{\beta\hat{H}_0}e^{-\beta\hat{H}_0}\hat{A}(\tau)e^{-\beta\hat{H}_0} = e^{-\beta\hat{H}_0}\hat{A}(\tau - i\hbar\beta) , \tag{19.48}$$

which leads directly to the Kubo–Martin–Schwinger relation

$$\langle\hat{B}\hat{A}(\tau)\rangle = \langle\hat{A}(\tau - i\hbar\beta)\hat{B}\rangle . \tag{19.49}$$

We now introduce the function $f(\tau) = \langle\hat{A}(\tau)\hat{B}\rangle - \langle\hat{A}\rangle\langle\hat{B}\rangle$ with which both the linear response function and the correlation function can be expressed as

$$\Gamma(\hat{A}\hat{B};\tau) = \frac{i}{\hbar}f(\tau) - \frac{i}{\hbar}f(\tau - i\hbar\beta) , \tag{19.50a}$$

$$K(\hat{A}\hat{B};\tau) = \frac{1}{2}f(\tau) + \frac{1}{2}f(\tau - i\hbar\beta) . \tag{19.50b}$$

It is already apparent from these expressions that it must be possible to relate both functions to one another. To see this more clearly, we introduce their Fourier transforms via

$$f(\tau) = \int \frac{d\omega}{2\pi} e^{-i\omega\tau} \tilde{f}(\omega), \quad f(\tau - i\hbar\beta) = \int \frac{d\omega}{2\pi} e^{-i\omega\tau} \tilde{f}(\omega) e^{-\beta\hbar\omega}, \tag{19.51}$$

with the result that

$$\Gamma(\hat{A}\hat{B};\tau) = \frac{i}{\hbar} \int \frac{d\omega}{2\pi} e^{-i\omega\tau} \tilde{f}(\omega) \left[1 - e^{-\beta\hbar\omega}\right], \tag{19.52a}$$

$$K(\hat{A}\hat{B};\tau) = \frac{1}{2} \int \frac{d\omega}{2\pi} e^{-i\omega\tau} \tilde{f}(\omega) \left[1 + e^{-\beta\hbar\omega}\right]. \tag{19.52b}$$

This in turn means that the Fourier components of the linear response function and the correlation function are related via

$$\tilde{K}(\hat{A}\hat{B};\omega) = \frac{\hbar}{2i} \coth\left(\frac{\beta\hbar\omega}{2}\right) \tilde{\Gamma}(\hat{A}\hat{B};\omega). \tag{19.53}$$

Due to the fact that $\Gamma(\hat{A}\hat{B};\tau)$ is a causal function, its half-sided Fourier transform is the generalised susceptibility $\chi(\hat{A}\hat{B};\omega)$ which is related to the Fourier transform $\tilde{\Gamma}(\hat{A}\hat{B};\omega)$ via

$$\tilde{\Gamma}(\hat{A}\hat{B};\omega) = \chi(\hat{A}\hat{B};\omega) - \chi^*(\hat{B}\hat{A};\omega) \equiv 2i \operatorname{Im}\chi(\omega) \tag{19.54}$$

which is nothing other than the imaginary part of the susceptibility. The Fourier transform of the correlation function is the spectral density $S(\hat{A}\hat{B};\omega) \equiv S(\omega)$ of the fluctuations so that Eq. (19.53) becomes the linear fluctuation-dissipation theorem

$$S(\omega) = \hbar \coth\left(\frac{\beta\hbar\omega}{2}\right) \operatorname{Im}\chi(\omega). \tag{19.55}$$

It says that the strength of the fluctuations in a system are related to the dissipative part of the susceptibility. In the high-temperature limit, $k_B T \gg \hbar\omega$, expanding the hyperbolic function yields

$$S(\omega) \simeq \frac{2k_B T}{\omega} \operatorname{Im}\chi(\omega), \tag{19.56}$$

which is independent of $\hbar$ and thus serves as the classical limit of the fluctuation-dissipation theorem.

§20. THE STRESS TENSOR IN VACUUM

Equations of motion of any physical system can be solved either by direct integration or by the use of integrals of motion. The latter are equivalent to conserved quantities such as energy or momentum which lead to simplifications of the equations of motion. In the context of electrodynamics and Maxwell's equations, we will see that the stress tensor provides a straightforward way to compute forces acting on a system of charges and currents in vacuum. This can be extended to forces on media which leads to an expression of Casimir forces via the stress tensor of fluctuating electromagnetic fields.

Maxwell's equations imply the existence of several continuity equations. For example, from the inhomogeneous equations

$$\mathbf{\nabla} \cdot \mathbf{D} = \rho, \quad \mathbf{\nabla} \times \mathbf{H} = \dot{\mathbf{D}} + \mathbf{j}, \tag{20.1}$$

it follows that the external charge density $\rho(\mathbf{r}, t)$ and current density $\mathbf{j}(\mathbf{r}, t)$ are related by

$$\dot{\rho}(\mathbf{r}, t) + \mathbf{\nabla} \cdot \mathbf{j}(\mathbf{r}, t) = 0 \tag{20.2}$$

by virtue of the operator identity $\mathbf{\nabla} \cdot \mathbf{\nabla} \times \equiv 0$. Equation (20.2) expresses charge conservation. Integrated over a volume V and using Gauss' theorem, one finds that the temporal change of the total charge inside V is equal to the total flux through its surface ∂V.

Next, we consider the power density of the work that is performed by an electric field $\mathbf{E}$ on a current density $\mathbf{j}$. By applying Ampère's law, this immediately gives

$$-\mathbf{j} \cdot \mathbf{E} = -\mathbf{E} \cdot (\mathbf{\nabla} \times \mathbf{H}) + \mathbf{E} \cdot \dot{\mathbf{D}} = \mathbf{\nabla} \cdot (\mathbf{E} \times \mathbf{H}) + \mathbf{E} \cdot \dot{\mathbf{D}} + \mathbf{H} \cdot \dot{\mathbf{B}} \tag{20.3}$$

which can be rewritten in terms of the energy flux (or Poynting vector) $\mathbf{S} = \mathbf{E} \times \mathbf{H}$ and the energy density $W = \varepsilon_0 \mathbf{E}^2 / 2 + \mathbf{B}^2 / (2\mu_0)$ as

$$\frac{\partial W}{\partial t} + \mathbf{\nabla} \cdot \mathbf{S} = -\mathbf{j} \cdot \mathbf{E}. \tag{20.4}$$

This is the continuity equation of the energy density of the electromagnetic field in vacuum. If one identifies the term $\mathbf{j} \cdot \mathbf{E}$ as the energy density that is converted from the electromagnetic field to the system of external charges and currents, i.e. into mechanical energy, then Eq. (20.4) is also the differential version of the conservation law for the total energy (density) comprised of the energy density W of the field and the mechanical energy density $\mathbf{j} \cdot \mathbf{E}$ of the external charges.

Finally, one finds a conservation law for the linear momentum which becomes an important starting point for the calculations of Casimir-type forces. Starting from the Lorentz force density

$$\mathbf{f}(\mathbf{r}, t) = \rho(\mathbf{r}, t)\mathbf{E}(\mathbf{r}, t) + \mathbf{j}(\mathbf{r}, t) \times \mathbf{B}(\mathbf{r}, t) \tag{20.5}$$

acting on an ensemble of external charges and currents, and using Maxwell's equations to eliminate the latter,

$$\rho = \varepsilon_0 \mathbf{\nabla} \cdot \mathbf{E}, \quad \mathbf{j} = \frac{1}{\mu_0} \mathbf{\nabla} \times \mathbf{B} - \varepsilon_0 \frac{\partial}{\partial t} \mathbf{E}, \tag{20.6}$$

the rate of change of mechanical momentum (density) can be cast into the form

$$\frac{d\mathbf{p}}{dt} = \varepsilon_0 \left[\mathbf{E}(\mathbf{\nabla} \cdot \mathbf{E}) + \mathbf{B} \times \frac{\partial}{\partial t}\mathbf{E} - c^2 \mathbf{B} \times (\mathbf{\nabla} \times \mathbf{B}) \right]$$

$$= \varepsilon_0 \left[\mathbf{E}(\mathbf{\nabla} \cdot \mathbf{E}) + c^2 \mathbf{B}(\mathbf{\nabla} \cdot \mathbf{B}) - \mathbf{E} \times (\mathbf{\nabla} \times \mathbf{E}) - c^2 \mathbf{B} \times (\mathbf{\nabla} \times \mathbf{B}) \right] - \varepsilon_0 \frac{\partial}{\partial t}(\mathbf{E} \times \mathbf{B}). \tag{20.7}$$

The quantity $\boldsymbol{P} = \varepsilon_0 \boldsymbol{E} \times \boldsymbol{B}$ is identified as the momentum density of the electromagnetic field in vacuo, which is just the energy current density multiplied by $1/c^2$, $\boldsymbol{P} = \boldsymbol{S}/c^2$. It should be stressed that this interpretation only holds for fields in vacuum. In order to convert the rate equation for the total momentum density $\boldsymbol{p} + \boldsymbol{P}$ into a conservation law, we define the Maxwell stress tensor

$$\mathrm{T} = \varepsilon_0 \left[\boldsymbol{E} \otimes \boldsymbol{E} + c^2 \boldsymbol{B} \otimes \boldsymbol{B} - \frac{1}{2} \left(\boldsymbol{E}^2 + c^2 \boldsymbol{B}^2 \right) \mathbb{1} \right] = \varepsilon_0 \left[\boldsymbol{E} \otimes \boldsymbol{E} + c^2 \boldsymbol{B} \otimes \boldsymbol{B} \right] - W \mathbb{1}.$$

(20.8)

This tensor is symmetric and consists of dyadic products of the electric field and the induction field. With the help of the Maxwell stress tensor we can write

$$\frac{d}{dt}(\boldsymbol{p} + \boldsymbol{P}) = \boldsymbol{\nabla} \cdot \mathrm{T}, \tag{20.9}$$

or, equivalently,

$$\boldsymbol{f} = \boldsymbol{\nabla} \cdot \mathrm{T} - \varepsilon_0 \frac{\partial}{\partial t} (\boldsymbol{E} \times \boldsymbol{B}). \tag{20.10}$$

The integral of the Lorentz force density $\boldsymbol{f}$ given by Eq. (20.10) over some volume V gives the total electromagnetic force acting on the matter inside V,

$$\boldsymbol{F}(t) = \int_V \mathrm{d}^3 r \, \boldsymbol{f}(\boldsymbol{r}, t) = \int_{\partial V} \mathrm{d}\boldsymbol{a} \cdot \mathrm{T}(\boldsymbol{r}, t) - \varepsilon_0 \frac{\partial}{\partial t} \int_V \mathrm{d}^3 r \, \boldsymbol{E}(\boldsymbol{r}, t) \times \boldsymbol{B}(\boldsymbol{r}, t). \tag{20.11}$$

In general, the resulting force is time-dependent. However, under certain stationarity conditions, the contribution of the volume integral over the electromagnetic momentum density can be neglected, and the force is given solely by a surface integral over the Maxwell stress tensor.

§21. Lifshitz theory of dispersion forces

Hendrik Casimir's original approach to computing the ground-state force between two parallel plates was based on the summation of the zero-point energy of all available electromagnetic field modes between the perfectly conducting plates. Crucially, in this model the plates do not actually 'exist' as materials, they merely provide the boundary conditions for the electromagnetic fields that lead to the restriction of the number of available modes. Evgenii Lifshitz provided an alternative derivation for the force between two plates which, in his version, were described by a complex dielectric permittivity $\varepsilon(\omega)$. The resulting interaction force agrees with Casimir's result in the limit $\varepsilon \to \infty$ and can be seen as a natural extension to realistic materials [13]. His approach was based on the theory of fluctuating electromagnetic fields that had been developed in the early 1950's and which was based on the linear fluctuation-dissipation theorem.

21.1   Rytov's theory of fluctuating electromagnetic fields

The realisation that a linear response relation is not complete unless a noise contribution is added is, in the context of electrodynamics, attributed to the Russian
radioengineer S.M. Rytov (Refs. [14, 15]; an earlier attempt in a quasi-stationary
regime is due to Leontovich [16]). In his approach to fluctuation electrodynamics
(FED) he regarded thermal fluctuations in a medium as the source of random electromagnetic fields. His idea was to introduce into Maxwell's equations a random
field similar to the random force in the theory of Brownian motion. In modern
parlance, the fluctuating field that is derived from this random field would be called
the noise polarisation field $P_N(r, \omega)$. With this, the linear relation between the frequency components of the polarisation and the electric field in an isotropic dielectric
medium is

$$P(r, \omega) = \varepsilon_0 \chi^{(1)}(r, \omega) E(r, \omega) + P_N(r, \omega), \tag{21.1}$$

which leads to Maxwell's equations of the form

$$\nabla \times E(r, \omega) = i\omega B(r, \omega), \tag{21.2a}$$

$$\nabla \times B(r, \omega) = -i\frac{\omega}{c^2}\varepsilon(r, \omega) E(r, \omega) - i\omega\mu_0 P_N(r, \omega). \tag{21.2b}$$

Here, $\varepsilon(r, \omega)$ is again the complex dielectric function that satisfies the Kramers–
Kronig relations (19.14).

    The defining characteristic of the noise polarisation $P_N(r, \omega)$ is its correlation
function in space and frequency. In a macroscopic theory, it is always assumed that
the discrete atomic positions are not resolved, and the dielectric medium occupies
the space continuously. Hence, with regard to their spatial arguments, the noise
polarisation is δ-function correlated. Statistical stationarity in time implies also a
δ-function correlation in frequency. Together with the linear fluctuation-dissipation
theorem one thus finds for the correlation function

$$\langle P_N(r, \omega) \otimes P_N^*(r', \omega') \rangle = \frac{\hbar\varepsilon_0}{\pi} \coth\left(\frac{\hbar\omega}{2k_BT}\right) \operatorname{Im} \varepsilon(r, \omega)\delta(r - r')\delta(\omega - \omega')\mathbf{1}, \tag{21.3}$$

which is valid both for classical fields (in the limit $\hbar\omega \ll k_BT$) as well as for operator-
valued fields. In fact, the operator-valued random fields later formed the basis for
the Langevin-force approach to field quantisation in absorbing media [17–20].

21.2   Lifshitz's approach to the Casimir force

The idea of Lifshitz had been to combine the stress tensor approach to electromagnetic forces with Rytov's idea of introducing fluctuating fields into Maxwell's
equations [13]. The assumptions behind the use of the Maxwell stress tensor (20.8)
to compute the Lorentz force on a medium is that the noise currents inside the
medium that drive the fluctuating fields are considered 'external', such that the
medium is surrounded by vacuum. It should be noted that the equivalence between
the Lorentz force (with respect to external charges) and the Casimir force is only

known to hold for the case of vacuum in the interspace between the interacting objects. When the objects are themselves submerged in another dielectric, the Lorentz force approach gives contradicting results when compared with other approaches such as macroscopic quantum electrodynamics [21] (see also Chapter 4).

Let us consider the situation depicted in Fig. 3.3 in which two planar halfspaces

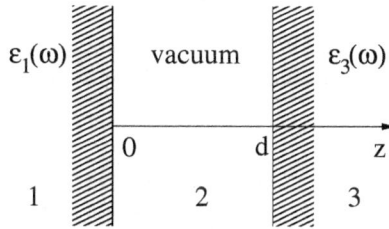

Figure 3.3:   Two parallel halfspaces filled with a dielectric medium with complex permittivities $\varepsilon_{1,3}(\omega)$, separated by a distance d.

filled with a dielectric medium with complex permittivities $\varepsilon_{1,3}(\omega)$ are separated by a distance d along the z-axis. By Eq. (20.11), the force per unit area, $f_z = F_z/L^2$, acting on the plates is then equal to the tensor component T_{zz} of the Maxwell stress tensor,

$$T_{zz} = \frac{\varepsilon_0}{2}(E_z^2 - E_\perp^2) + \frac{1}{2\mu_0}(B_z^2 - B_\perp^2),   \tag{21.4}$$

where $E_\perp$ and $B_\perp$ denote the field components in the plane parallel to the plate surfaces. As the electromagnetic fields involved here are fluctuating fields, their statistical average has to be taken. Hence, the Casimir force per unit area between the plates is given by the expectation value

$$f_z = \frac{\varepsilon_0}{2}(\langle E_z^2 \rangle - \langle E_\perp^2 \rangle) + \frac{1}{2\mu_0}(\langle B_z^2 \rangle - \langle B_\perp^2 \rangle).   \tag{21.5}$$

This expression is formally divergent as it also contains the back action of the field produced by one body on itself which does not depend on the separation d. We will see later in connection with the Green function formalism how to deal with these divergencies and simply assume that the unphysical divergent self-forces have been removed (see also Chapter 4). In the case of perfectly conducting plates, this program has been carried out using point-splitting regularisation and using an explicit image dipole construction of the Green function in Ref. [22].

In the next step it is necessary to find the solutions of Maxwell's equations in the presence of the two plates by fixing the boundary conditions on the interfaces between the halfspaces and vacuum. To this end, one expresses the fields inside the dielectric halfspaces, i.e. in regions 1 and 3, in terms of the fluctuating fields, and then uses the Maxwell boundary conditions to determine the fields in the space between the dielectric media. The correlation functions appearing in Eq. (21.5) are then given by the respective correlation functions of the fluctuating fields, i.e. the noise polarisation field, and appropriately determined reflection coefficients. We give a complete derivation in Section 21.3 and merely quote Lifshitz' result [13]:

$$f_z = -\frac{\hbar}{2\pi^2} \int_0^\infty d\omega \, \coth\left(\frac{\hbar\omega}{2k_BT}\right) \, \text{Re} \int_0^\infty dq \, qk_{3z} \sum_{\sigma=s,p} \frac{r_1^\sigma r_2^\sigma e^{2ik_{2z}d}}{1 - r_1^\sigma r_2^\sigma e^{2ik_{2z}d}}, \qquad (21.6)$$

where the integrations run over all positive frequencies ω and transverse wavenumbers q. The Fresnel reflection coefficients $r_i^\sigma(\omega)$ for s- and p-polarised waves* at the interfaces with the plates are

$$r_i^s(\omega) = \frac{k_{2z} - k_{iz}}{k_{2z} + k_{iz}}, \qquad r_i^p(\omega) = \frac{\varepsilon_i(\omega)k_{2z} - k_{iz}}{\varepsilon_i(\omega)k_{2z} + k_{iz}}, \qquad i = 1,3 \qquad (21.7)$$

and the wave vector components in z-direction in a region j are given by $k_{jz} = \sqrt{\varepsilon_j(\omega)\omega^2/c^2 - q^2}$ where we have to set $\varepsilon_2 \equiv 1$. The denominator $(1 - r_1^\sigma r_2^\sigma e^{2ik_{2z}d})^{-1}$ accounts for multiple reflections at the interfaces and is thus a result of the geometric series

$$\left(1 - r_1^\sigma r_2^\sigma e^{2ik_{2z}d}\right)^{-1} = \sum_{k=0}^\infty \left(r_1^\sigma r_2^\sigma e^{2ik_{2z}d}\right)^k. \qquad (21.8)$$

Equation (21.6) is the celebrated Lifshitz formula that generalises Casimir's result for the force per unit area between two parallel plates to dielectric media. It should be noted that the fluctuation-dissipation theorem makes its appearance in the form of the hyperbolic function of temperature, otherwise the force only depends on the reflection coefficients at the interfaces between vacuum and the dielectric media.

The integrals in Eq. (21.6) typically cannot be performed in closed form, but several useful approximations can be made. Let us for the moment discard the effects of temperature and set $\coth\frac{\hbar\omega}{2k_BT} = 1$. Due to the analyticity properties of the integrand in Eq. (21.6) with respect to the variable ω, it turns out that one can deform the integration contour in the upper complex ω-plane without encountering any poles or zeros, and thus convert the frequency integration into an integral along the imaginary axis $\omega = i\xi$ (Wick rotation, see also Chapter 2, Fig. 2.4),

$$f_z = -\frac{\hbar}{2\pi^2} \int_0^\infty d\xi \int_0^\infty dq \, q\kappa_{2z} \sum_{\sigma=s,p} \frac{r_1^\sigma(i\xi)r_2^\sigma(i\xi)e^{-2\kappa_{2z}d}}{1 - r_1^\sigma(i\xi)r_2^\sigma(i\xi)e^{-2\kappa_{2z}d}} \qquad (21.9)$$

with $\kappa_{2z} = \sqrt{\xi^2/c^2 + q^2}$. In this form, the oscillatory factors $e^{2ik_{2z}d}$ that are detrimental for numerical computations, are replaced by rapidly decreasing exponential functions $e^{-2\kappa_{2z}d}$.

---

*The polarisation indices derive from the German words *senkrecht* (meaning perpendicular) and *parallel* which refer to the direction of the electric field vector with respect to the plane of incidence on a planar surface. Some authors prefer instead the notation *transverse electric* (TE) and *transverse magnetic* (TM), respectively.

One recovers Casimir's force expression from Eq. (21.9) by formally setting $\varepsilon_1 = \varepsilon_3 \mapsto -\infty$, more precisely, by setting $r_i^s \mapsto -1$ and $r_i^p \mapsto +1$. In this way, the force becomes

$$f_z = -\frac{\hbar}{\pi^2} \int_0^\infty d\xi \int_0^\infty dq\, q\kappa_{2z} \frac{1}{e^{2\kappa_{2z}d} - 1} = -\frac{\hbar}{\pi^2} \int_0^\infty d\xi \int_{\xi/c}^\infty d\kappa_{2z} \frac{\kappa_{2z}^2}{e^{2\kappa_{2z}d} - 1}. \tag{21.10}$$

With the substitutions $x = 2\kappa_{2z}d$ and $u = 2\xi d/c$, this gives

$$f_z = -\frac{\hbar c}{16\pi^2 d^4} \int_0^\infty du \int_u^\infty dx \frac{x^2}{e^x - 1} = -\frac{\pi^2}{240} \frac{\hbar c}{d^4}. \tag{21.11}$$

Lifshitz' expression (21.6) thus coincides with Casimir's result in the limit of perfectly reflecting plates at all frequencies, as expected. The exact expression (21.6) or its counterpart on the imaginary frequency axis (21.9), respectively, can be approximately solved in two extreme distance limits.

Nonretarded limit

We begin with the regime in which the plate separation d is much smaller than all relevant resonance wavelengths of the medium, $\sqrt{|\varepsilon(0)|}d\omega_{\max}/c \ll 1$ ($\omega_{\max}$: maximum of all relevant medium resonance frequencies). Switching the integration in Eq. (21.9) from the variable q to $\kappa_{2z} = \sqrt{\xi^2/c^2 + q^2}$ with $\kappa_{2z}d\kappa_{2z} = qdq$ leads to

$$f_z = -\frac{\hbar}{2\pi^2} \int_0^\infty d\xi \int_{\xi/c}^\infty d\kappa_{2z}\, \kappa_{2z}^2 \sum_{\sigma=s,p} \frac{r_1^\sigma(i\xi)r_2^\sigma(i\xi)e^{-2\kappa_{2z}d}}{1 - r_1^\sigma(i\xi)r_2^\sigma(i\xi)e^{-2\kappa_{2z}d}}, \tag{21.12}$$

where in the reflection coefficients the wave vector components κ_{iz} have to be written as $\kappa_{iz} = \sqrt{\xi^2/c^2[\varepsilon_i(i\xi) - 1] + \kappa_{2z}^2}$. Given that the permittivities $\varepsilon_i(i\xi)$ are monotonic functions of the imaginary frequency $i\xi$, they give their largest contributions for $\xi \lesssim \omega_{\max}$. This in turn means that $d\xi/c\sqrt{\varepsilon_i(i\xi) - 1} \lesssim d\omega_{\max}/c\sqrt{|\varepsilon(0)|} \ll 1$, and the wave vector components are well approximated by $\kappa_{iz} \simeq \kappa_{2z}$. In this approximation, the reflection coefficients r_i^s for s-polarised waves vanish. After substituting $x = 2\kappa_{2z}d$ and extending the lower integration limit to 0 with little error, one finds

$$f_z \simeq -\frac{\hbar}{16\pi^2 d^3} \int_0^\infty d\xi \int_0^\infty dx\, x^2 \frac{r_1^p(i\xi)r_2^p(i\xi)e^{-x}}{1 - r_1^p(i\xi)r_2^p(i\xi)e^{-x}} = -\frac{\hbar}{8\pi^2 d^3} \int_0^\infty d\xi\, \text{Li}_3[r_1^p(i\xi)r_2^p(i\xi)]$$

$$\tag{21.13}$$

where $\text{Li}_3(x)$ is the polylogarithm function. In this nonretarded or quasistatic limit, the force between the plates, $f_z = -c_3/d^3$, scales with d^{-3}. Note that the perfect conductor limit $|\varepsilon_i| \mapsto \infty$ cannot be taken as that would violate the condition that led to our approximation. Hence, for perfectly reflecting plates there exists no nonretarded regime.

Exercise 21.1
Derive the last equality in Eq. (21.13). Use the definition

$$\mathrm{Li}_s(z) = \sum_{k=1}^{\infty} \frac{z^k}{k^s} = z + \frac{z^2}{2^s} + \frac{z^3}{3^s} + \cdots$$

of the polylogarithm function.

Retarded limit

In the opposite distance regime, the retarded limit, the plate separation is assumed to be much larger than any resonance wavelengths inside the dielectric media, $d\omega_{\min}/c \gg 1$ ($\omega_{\min}$: minimum of all relevant medium resonance frequencies). In this case, we substitute $v = c\kappa_{2z}/\xi$ and write the force per unit area (21.9) as

$$f_z = -\frac{\hbar}{2\pi^2} \int_1^{\infty} dv \int_0^{\infty} d\xi \frac{\xi^3}{c^3} v^2 \sum_{\sigma=s,p} \frac{r_1^{\sigma}(i\xi)r_2^{\sigma}(i\xi)e^{-2v\xi d/c}}{1 - r_1^{\sigma}(i\xi)r_2^{\sigma}(i\xi)e^{-2v\xi d/c}} \tag{21.14}$$

in which the wave vector components κ_{iz} are now $\kappa_{iz} = \sqrt{\varepsilon_i(i\xi) - 1 + v^2}\,\xi/c$. The frequency interval in which the integral gives the dominant contribution is set by the exponential function $e^{-2v\xi d/c}$. Hence, $2v\xi d/c \lesssim 1$ and equally, because $v \geq 1$, $2\xi d/c \lesssim 1$. Therefore, the frequency integration is restricted to values of $\xi \lesssim c/(2d)$. Together with the assumption on the plate separation it then follows that $\xi/\omega_{\min} \ll 1$ which means that all permittivities can be well approximated by their static values, $\varepsilon_i(i\xi) \simeq \varepsilon_i(0)$ [see also the discussion in Chapter 2 after Eq. (15.45)]. Performing the frequency integral yields

$$f_z \simeq -\frac{3\hbar c}{16\pi^2 d^4} \int_1^{\infty} dv \frac{1}{v^2} \sum_{\sigma=s,p} \mathrm{Li}_4\left[r_1^{\sigma}(v)r_2^{\sigma}(v)\right] \tag{21.15}$$

with

$$r_i^s(v) = \frac{v - \sqrt{\varepsilon_i(0) - 1 + v^2}}{v + \sqrt{\varepsilon_i(0) - 1 + v^2}}, \quad r_i^p(v) = \frac{\varepsilon_i(0)v - \sqrt{\varepsilon_i(0) - 1 + v^2}}{\varepsilon_i(0)v + \sqrt{\varepsilon_i(0) - 1 + v^2}}. \tag{21.16}$$

Hence, in the retarded limit, the force per unit area (21.15) on the plates, $f_z = -c_4/d^4$, scales as d^{-4} with the plate separation. This has to be compared with the nonretarded limit (21.13) with its d^{-3} scaling. Note also that Eq. (21.15) returns Casimir's result for perfectly conducting plates on account of $\mathrm{Li}_4(1) = \pi^4/90$.

This last statement requires some further elaboration. In Casimir's original calculation, it was unnecessary to distinguish between nonretarded and retarded distance regimes. The result consists of a universal distance scaling compatible with the retarded limit. Indeed, as the plates are required to be perfectly conducting at all frequencies, the quantity $\omega_{\min}$ that describes the minimal medium resonance frequency is infinitely large, so that the retardation condition $d\omega_{\min}/c \gg 1$ is fulfilled for arbitrary separations d.

As an example, Fig. 3.4 shows the Casimir force per unit area (21.9) for two identical halfspaces with permittivity*

$$\epsilon(\omega) = \frac{\Omega_L^2 - \omega^2 - i\omega\gamma_L}{\Omega_T^2 - \omega^2 - i\omega\gamma_T} \qquad (21.17)$$

with the numerical values $\Omega_L = 2.69 \times 10^{16}\,\text{rad/s}$, $\Omega_T = 1.33 \times 10^{16}\,\text{rad/s}$, $\gamma_L = 3.05 \times 10^{16}\,\text{rad/s}$, and $\gamma_T = 6.40 \times 10^{15}\,\text{rad/s}$. The solid line corresponds to the

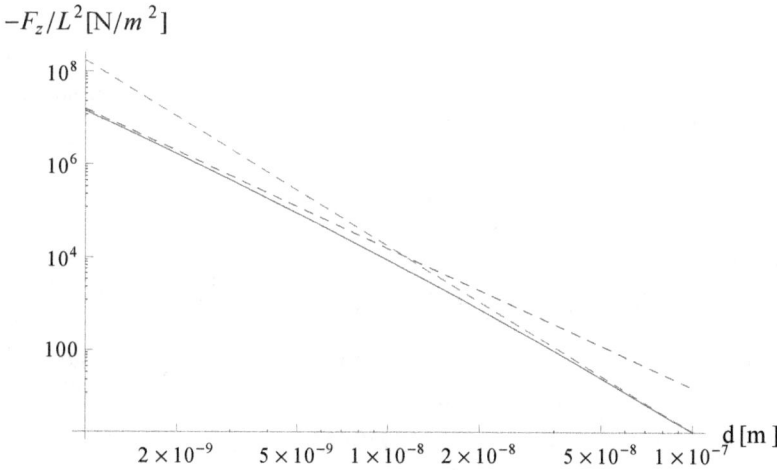

Figure 3.4: Casimir-Lifshitz force between two Si_3N_4 halfspaces. The nonretarded and retarded limits (dashed lines) are compared with the exact numerical result (solid line).

numerical integration of Eq. (21.9). The two dashed lines are the approximate results in the nonretarded [Eq. (21.13)] and retarded [Eq. (21.15)] limits with $c_3 = 1.57 \times 10^{-20}\,\text{Nm}^{-1}$ and $c_4 = 1.76 \times 10^{-28}\,\text{Nm}^{-2}$, respectively. Although both asymptotes do not agree with the exact numerical result in the intermediate region between the two extreme limits, one can nevertheless find a reasonably good approximation by writing

$$f_z = -\frac{c_4}{d^3(d+a)} \qquad (21.18)$$

with $a = c_4/c_3$ determining the length scale of the intermediate region. For the numerical example shown in Fig. 3.4, $a = 1.12 \times 10^{-8}\,\text{m}$ which is the distance at which the two asymptotes cross.

---

*This is in fact the 4-parameter semi-quantum model fit to the spectroscopic data of the permittivity for non-crystalline Si_3N_4 [23].

21.3   Green function approach

Rytov's approach to adding a fluctuating polarisation field to Maxwell's equations implies that, even without external fields, there exist electromagnetic fields driven by (zero-point or thermal) fluctuations inside a medium. Lifshitz explicitly exploited this fact by constructing the electric and magnetic field fluctuations from those of the noise polarisation field in the specific geometric configuration shown in Fig. 3.3. This approach can be generalised to arbitrary geometric arrangements of bodies, and forms the basis of field quantisation in absorbing media (see Chapter 4).

From the modified Maxwell's equations (21.2) follows the Helmholtz equation

$$\nabla \times \nabla \times \boldsymbol{E}(\boldsymbol{r},\omega) - \frac{\omega^2}{c^2}\varepsilon(\boldsymbol{r},\omega)\boldsymbol{E}(\boldsymbol{r},\omega) = \frac{\omega^2}{\varepsilon_0 c^2}\boldsymbol{P}_{\mathrm{N}}(\boldsymbol{r},\omega) \qquad (21.19)$$

for the frequency components of the electric field. This is formally solved by introducing the dyadic Green function $\mathrm{G}(\boldsymbol{r},\boldsymbol{r}',\omega)$ as the fundamental solution of the Helmholtz equation [see also Eq. (13.20)]

$$\nabla \times \nabla \times \mathrm{G}(\boldsymbol{r},\boldsymbol{r}',\omega) - \frac{\omega^2}{c^2}\varepsilon(\boldsymbol{r},\omega)\mathrm{G}(\boldsymbol{r},\boldsymbol{r}',\omega) = \delta(\boldsymbol{r}-\boldsymbol{r}')1 \qquad (21.20)$$

such that

$$\boldsymbol{E}(\boldsymbol{r},\omega) = \frac{\omega^2}{\varepsilon_0 c^2}\int \mathrm{d}^3 r'\, \mathrm{G}(\boldsymbol{r},\boldsymbol{r}',\omega)\cdot \boldsymbol{P}_{\mathrm{N}}(\boldsymbol{r}',\omega). \qquad (21.21)$$

The total field operators in the Schrödinger picture, i.e. at $t=0$, read

$$\boldsymbol{E}(\boldsymbol{r}) = \int_0^\infty \mathrm{d}\omega\, \boldsymbol{E}(\boldsymbol{r},\omega) + \mathrm{h.c.} \qquad (21.22)$$

The dyadic Green function now contains all the information about the electromagnetic properties as well as the geometry of the dielectric media under study. In view of the field quantisation in vacuo, where the electric field is expanded into a set of orthonormal modes, the field expansion (21.21) can be viewed as a generalisation of the usual mode expansion to absorbing media. The fact that Eq. (21.21) is a good starting point for macroscopic quantum electrodynamics will be shown in Chapter 4.

The field expansion via a Green function has the additional advantage that the statistical problem of determining the random field distribution is separated from solving the (classical) electromagnetic scattering problem. The dyadic Green function $\mathrm{G}(\boldsymbol{r},\boldsymbol{r}',\omega)$ has the following general properties: as a function of complex frequency ω, it is an analytic function in the upper half-plane and fulfils the Schwarz reflection principle,

$$\mathrm{G}(\boldsymbol{r},\boldsymbol{r}',\omega) = \mathrm{G}^*(\boldsymbol{r},\boldsymbol{r}',-\omega^*), \qquad (21.23)$$

which it inherits from the dielectric permittivity $\varepsilon(\boldsymbol{r},\omega)$. For a reciprocal medium the Green function also fulfils the Onsager–Lorentz reciprocity relation [24]

$$\mathrm{G}(\boldsymbol{r},\boldsymbol{r}',\omega) = \mathrm{G}^{\mathrm{T}}(\boldsymbol{r}',\boldsymbol{r},\omega) \qquad (21.24)$$

which relates the mutual interaction between two groups of sources and their reactions (or measurements). Furthermore, from the Helmholtz equation it follows that the Green function obeys the important integral relation

$$\int d^3 s \, \frac{\omega^2}{c^2} \, \mathrm{Im}\, \varepsilon(s, \omega) G(r, s, \omega) \cdot G^*(s, r', \omega) = \mathrm{Im}\, G(r, r', \omega). \qquad (21.25)$$

Exercise 21.2

Use the Helmholtz equation (21.20) to show the integral relation (21.25). *Hint*: Multiply Eq. (21.20) from the right by $G^*(r, s, \omega)$, exchange $r \leftrightarrow s$, integrate over s and subtract the complex conjugate equation.

Due to the linearity of the Helmholtz equation, the Green function can always be decomposed into a bulk contribution $G^{(0)}(r, r', \omega)$ that solves the inhomogeneous Helmholtz equation with the correct boundary condition at infinity, and a scattering contribution $G^{(S)}(r, r', \omega)$. The latter is a solution of the homogeneous Helmholtz equation and has to be introduced to satisfy the boundary conditions for the electromagnetic fields at the interfaces between different media. As the Green function represents the retarded photon propagator that describes the propagation of an electromagnetic wave of frequency ω from the excitation point r' to the observation point r, the decomposition of the Green function can be visualised as in Fig. 3.5. The bulk Green function $G^{(0)}(r, r', \omega)$ describes direct propagation from r' to r,

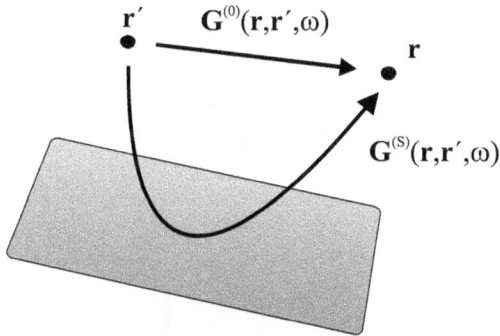

Figure 3.5: Decomposition of the Green function into a bulk part and a scattering contribution is equivalent to a decomposition into different paths that an electromagnetic wave can take between the points r' and r.

whereas the scattering part $G^{(S)}(r, r', \omega)$ describes the reflection off a surface.

The field expansion (21.21) can be viewed as a Fourier transformed linear response relation that expresses the induced electric field generated by a (noise) polarisation field, with the Green function playing the role of the linear response function. From our general considerations it then follows that the Green function obeys Kramers–Kronig-type relations similar to those fulfilled by the permittivity. As a linear response relation, Eq. (21.21) should also imply a version of the linear fluctuation-dissipation theorem. In fact, the integral relation (21.25) provides just that. The correlation function of the electric field (21.21) is

$$\langle \boldsymbol{E}(\boldsymbol{r}, \omega) \otimes \boldsymbol{E}^*(\boldsymbol{r}', \omega') \rangle$$

$$= \left(\frac{\omega^2}{\varepsilon_0 c^2} \right)^2 \int \mathrm{d}^3 s \, \mathrm{d}^3 s' \, \mathrm{G}(\boldsymbol{r}, \boldsymbol{s}, \omega) \cdot \langle \boldsymbol{P}_\mathrm{N}(\boldsymbol{s}, \omega) \otimes \boldsymbol{P}_\mathrm{N}^*(\boldsymbol{s}', \omega') \rangle \cdot \mathrm{G}^*(\boldsymbol{s}', \boldsymbol{r}', \omega'). \quad (21.26)$$

Inserting Rytov's expression for the correlation function of the noise polarisation field (21.3) and making use of the integral relation (21.25) one obtains

$$\langle \boldsymbol{E}(\boldsymbol{r}, \omega) \otimes \boldsymbol{E}^*(\boldsymbol{r}', \omega') \rangle = \frac{\hbar \omega^2}{\pi \varepsilon_0 c^2} \coth \left(\frac{\hbar \omega}{2 k_B T} \right) \operatorname{Im} \mathrm{G}(\boldsymbol{r}, \boldsymbol{r}', \omega) \delta(\omega - \omega'). \quad (21.27)$$

The same expression would have been obtainable from direct application of the fluctuation-dissipation theorem. In this way, the Casimir force can be computed directly from the expectation value of the stress tensor

$$\mathrm{T}(\boldsymbol{r}, \boldsymbol{r}') = \varepsilon_0 \langle \boldsymbol{E}(\boldsymbol{r}) \otimes \boldsymbol{E}(\boldsymbol{r}') \rangle + \frac{1}{\mu_0} \langle \boldsymbol{B}(\boldsymbol{r}) \otimes \boldsymbol{B}(\boldsymbol{r}') \rangle$$

$$- \frac{1}{2} 1 \left[\varepsilon_0 \langle \boldsymbol{E}(\boldsymbol{r}) \cdot \boldsymbol{E}(\boldsymbol{r}') \rangle + \frac{1}{\mu_0} \langle \boldsymbol{B}(\boldsymbol{r}) \cdot \boldsymbol{B}(\boldsymbol{r}') \rangle \right]. \quad (21.28)$$

in the limit $\boldsymbol{r} \to \boldsymbol{r}'$. The thermal expectation values of the electric field and the induction field are[*]

$$\langle \boldsymbol{E}(\boldsymbol{r}) \otimes \boldsymbol{E}(\boldsymbol{r}') \rangle = \frac{\hbar \mu_0}{\pi} \int_0^\infty \mathrm{d}\omega \, \omega^2 \coth \left(\frac{\hbar \omega}{2 k_B T} \right) \operatorname{Im} \mathrm{G}(\boldsymbol{r}, \boldsymbol{r}', \omega), \quad (21.29a)$$

$$\langle \boldsymbol{B}(\boldsymbol{r}) \otimes \boldsymbol{B}(\boldsymbol{r}') \rangle = \frac{\hbar \mu_0}{\pi} \int_0^\infty \mathrm{d}\omega \coth \left(\frac{\hbar \omega}{2 k_B T} \right) \boldsymbol{\nabla} \times \operatorname{Im} \mathrm{G}(\boldsymbol{r}, \boldsymbol{r}', \omega) \times \overset{\leftarrow}{\boldsymbol{\nabla}'}. \quad (21.29b)$$

As the Green function can always be decomposed as $\mathrm{G} = \mathrm{G}^{(0)} + \mathrm{G}^{(S)}$ into a sum of a bulk tensor $\mathrm{G}^{(0)}$ and a scattering part $\mathrm{G}^{(S)}$, we neglect the contribution of the bulk Green function as they represent divergent self-forces which are independent of the separation between the bodies. This finally yields the stress tensor in the form

$$\mathrm{T}(\boldsymbol{r}, \boldsymbol{r}) = \lim_{\boldsymbol{r}' \to \boldsymbol{r}} \left[\boldsymbol{\theta}(\boldsymbol{r}, \boldsymbol{r}') - \frac{1}{2} \operatorname{tr} \boldsymbol{\theta}(\boldsymbol{r}, \boldsymbol{r}') 1 \right] \quad (21.30)$$

with

$$\boldsymbol{\theta}(\boldsymbol{r}, \boldsymbol{r}') = \frac{\hbar}{\pi} \int_0^\infty \mathrm{d}\omega \coth \left(\frac{\hbar \omega}{2 k_B T} \right) \left[\frac{\omega^2}{c^2} \operatorname{Im} \mathrm{G}^{(S)}(\boldsymbol{r}, \boldsymbol{r}', \omega) \right.$$

$$\left. - \boldsymbol{\nabla} \times \operatorname{Im} \mathrm{G}^{(S)}(\boldsymbol{r}, \boldsymbol{r}', \omega) \times \overset{\leftarrow}{\boldsymbol{\nabla}'} \right].$$

---

[*]The notation $\times \overset{\leftarrow}{\boldsymbol{\nabla}'}$ refers to taking the curl operation from the right, with the differentiation acting on the second spatial argument $\boldsymbol{r}'$ and the cross product on the second tensor index of the Green function.

In the limit $T \to 0$, the hyperbolic cotangent turns to unity, and the stress tensor returns the usual zero-point result.

For computational ease it is useful to transform the frequency integration along the real axis into one along the imaginary axis, the reason being that the Green function is rapidly oscillating for real frequencies, whereas it is monotonic for imaginary frequencies. The integral conversion relies on the analyticity of the Green function and the Schwarz reflection principle. For now, we restrict ourselves to the case $T = 0$ and deal with finite temperature later. The first of the frequency integrals in Eq. (21.3) can be written as

$$\int_0^\infty d\omega\, \omega^2 \operatorname{Im} G^{(S)}(\boldsymbol{r}, \boldsymbol{r}', \omega) = \frac{1}{2i} \int_0^\infty d\omega\, \omega^2 \left[G^{(S)}(\boldsymbol{r}, \boldsymbol{r}', \omega) - G^{(S)*}(\boldsymbol{r}, \boldsymbol{r}', \omega) \right]$$

$$= \frac{1}{2i} \int_0^\infty d\omega\, \omega^2 \left[G^{(S)}(\boldsymbol{r}, \boldsymbol{r}', \omega) - G^{(S)}(\boldsymbol{r}, \boldsymbol{r}', -\omega) \right]$$

$$= \frac{1}{2i} \int_0^\infty d\omega\, \omega^2 G^{(S)}(\boldsymbol{r}, \boldsymbol{r}', \omega) - \frac{1}{2i} \int_{-\infty}^0 d\omega\, \omega^2 G^{(S)}(\boldsymbol{r}, \boldsymbol{r}', \omega). \tag{21.31}$$

Closing the integration contour in the upper half plane (see Fig. 3.6 and also Chapter 2, Fig. 2.4) and using the asymptotic behaviour $G(\boldsymbol{r}, \boldsymbol{r}', \omega) \overset{|\omega| \to \infty}{\to} -c^2/\omega^2 \delta(\boldsymbol{r} - \boldsymbol{r}')\mathbb{1}$

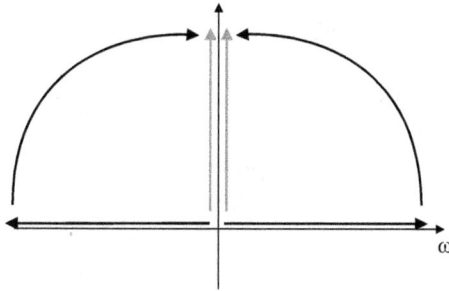

Figure 3.6: Rotation of the integration contour from the real frequency axis to the imaginary axis. The integral over a closed loop in the upper complex halfplane vanishes on account of the analyticity property of the dyadic Green function.

which follows from the Helmholtz equation (21.20) together with $\varepsilon(\boldsymbol{r}, \omega) \overset{|\omega| \to \infty}{\to} 1$, one finds that the contributions from the two large quartercircles cancel each other. The remaining integrals along the imaginary axis add up to $(\omega = i\xi)$

$$\int_0^\infty d\omega\, \omega^2 \operatorname{Im} G^{(S)}(\boldsymbol{r}, \boldsymbol{r}', \omega) = -\int_0^\infty d\xi\, \xi^2 G^{(S)}(\boldsymbol{r}, \boldsymbol{r}', i\xi). \tag{21.32}$$

An analogous relation exists for the second term in Eq. (21.3) which can now be written as

$$\boldsymbol{\theta}(\boldsymbol{r}, \boldsymbol{r}') = -\frac{\hbar}{\pi} \int\limits_0^\infty d\xi \left[\frac{\xi^2}{c^2} \mathrm{G}^{(S)}(\boldsymbol{r}, \boldsymbol{r}', \mathrm{i}\xi) + \boldsymbol{\nabla} \times \mathrm{G}^{(S)}(\boldsymbol{r}, \boldsymbol{r}', \mathrm{i}\xi) \times \overleftarrow{\boldsymbol{\nabla}}' \right] . \tag{21.33}$$

It should be noted that Eq. (21.30) is valid for an arbitrary arrangement of bodies, not just to the special case of two halfspaces. In this way, the slightly formal Green function approach provides somewhat more insight into the physics of the Casimir force which is lacking if the field expansion is used. Of course, the challenge for an actual calculation of the force between two bodies is the determination of the dyadic Green function. As this is a common problem in electromagnetic theory, a large variety of analytical and numerical methods exist.

For the situation in Fig. 3.3 of two halfspaces separated by a distance d, one commonly decomposes the scattering part of the dyadic Green function into a two-dimensional Fourier integral (Weyl expansion) as $[\boldsymbol{r} = (\boldsymbol{\varrho}, z)]$ [24, 25]

$$\mathrm{G}^{(S)}(\boldsymbol{r}, \boldsymbol{r}', \omega) = \int d^2 k_\| \, \mathrm{e}^{\mathrm{i}\boldsymbol{k}_\| \cdot (\boldsymbol{\varrho} - \boldsymbol{\varrho}')} \mathrm{G}^{(S)}(\boldsymbol{k}_\|, z, z', \omega) \tag{21.34}$$

with the Fourier components [26]

$$\mathrm{G}^{(S)}(\boldsymbol{k}_\|, z, z', \omega) = \frac{\mathrm{i}}{8\pi^2 k_z} \sum_{\sigma=s,p} \frac{\mathrm{e}^{\mathrm{i}k_z d}}{D^\sigma} \left\{ \xi^\sigma r_-^\sigma \mathrm{e}^{\mathrm{i}k_z z} \boldsymbol{e}_\sigma^+ \otimes \left(\boldsymbol{e}_\sigma^+ \mathrm{e}^{-\mathrm{i}k_z(d-z')} + r_+^\sigma \boldsymbol{e}_\sigma^- \mathrm{e}^{\mathrm{i}k_z(d-z')} \right) \right.$$
$$\left. + r_+^\sigma \mathrm{e}^{\mathrm{i}k_z(d-z)} \boldsymbol{e}_\sigma^- \otimes \left(\boldsymbol{e}_\sigma^- \mathrm{e}^{-\mathrm{i}k_z z'} + r_-^\sigma \boldsymbol{e}_\sigma^+ \mathrm{e}^{\mathrm{i}k_z z'} \right) \right\} \tag{21.35}$$

where $\boldsymbol{e}_\sigma^\pm$ are the polarisation unit vectors for $\sigma = s, p$ polarised waves travelling in the positive/negative z-direction, $r_\pm^\sigma$ are the respective reflection coefficients at the left/right plate, and $\xi^{s/p} = \mp 1$. The polarisation vectors can be given in terms of the unit vectors normal $(\boldsymbol{e}_z)$ and parallel $(\boldsymbol{e}_\| = \boldsymbol{k}_\|/k_\|)$ to the plate surfaces as

$$\boldsymbol{e}_s^\pm = \boldsymbol{e}_\| \times \boldsymbol{e}_z, \quad \boldsymbol{e}_p^\pm = \frac{1}{k} \left(k_\| \boldsymbol{e}_z \mp k_z \boldsymbol{e}_\| \right) . \tag{21.36}$$

Their normalisation and orthogonality relations follow directly from their definitions as

$$\boldsymbol{e}_s^\pm \cdot \boldsymbol{e}_s^\pm = \boldsymbol{e}_s^\pm \cdot \boldsymbol{e}_s^\mp = 1 ,$$

$$\boldsymbol{e}_p^\pm \cdot \boldsymbol{e}_p^\pm = 1, \quad \boldsymbol{e}_p^\pm \cdot \boldsymbol{e}_p^\mp = -1 + 2 \left(\frac{k_\| c}{\omega} \right)^2 ,$$

$$\boldsymbol{e}_s^\pm \cdot \boldsymbol{e}_p^\pm = \boldsymbol{e}_s^\pm \cdot \boldsymbol{e}_p^\mp = 0 . \tag{21.37}$$

Finally, the factor $D^\sigma = 1 - r_-^\sigma r_+^\sigma \mathrm{e}^{2\mathrm{i}k_z d}$ again accounts for multiple reflections between the plates.

With a view to computing the stress tensor (21.30) component T_{zz} on the plate surfaces $z = z' = 0$ and $z = z' = d$, we first note that the Green tensor turns into a diagonal tensor with respect to the unit vectors $\boldsymbol{e}_\|$, $\boldsymbol{e}_z$ and $\boldsymbol{e}_\| \times \boldsymbol{e}_z$. Next, we require the combination $T_{zz} = \theta_{zz} - \theta_\|$ and thus the Fourier components

$$
\frac{\omega^2}{c^2} \left[G_{zz}^{(S)}(\boldsymbol{k}_\|, z, z, \omega) - G_\|^{(S)}(\boldsymbol{k}_\|, z, z, \omega) \right]
$$

$$
= \frac{i}{8\pi^2 k_z} \left\{ \frac{k_z^2}{D^p} \left[2r_-^p r_+^p e^{2ik_z d} - r_-^p e^{2ik_z z} - r_+^p e^{2ik_z(d-z)} \right] \right.
$$

$$
+ \frac{k^2}{D^s} \left[2r_-^s r_+^s e^{2ik_z d} + r_-^s e^{2ik_z z} + r_+^s e^{2ik_z(d-z)} \right]
$$

$$
\left. - \frac{k_\|^2}{D^p} \left[2r_-^p r_+^p e^{2ik_z d} + r_-^p e^{2ik_z z} + r_+^p e^{2ik_z(d-z)} \right] \right\} . \tag{21.38}
$$

Finally, we need to compute the magnetic component of the stress tensor. This is easily done by observing that the curl operation under the Weyl transformation simply gives $\boldsymbol{\nabla} \times \equiv (\boldsymbol{k}_\| \pm k_z \boldsymbol{e}_z) \times$ with the help of which we find

$$
(\boldsymbol{k}_\| \pm k_z \boldsymbol{e}_z) \times \boldsymbol{e}_\sigma^\pm = k \xi^\sigma \boldsymbol{e}_{\bar{\sigma}}^\pm, \qquad \bar{s} = p, \quad \bar{p} = s. \tag{21.39}
$$

Thus, the magnetic part of the stress tensor is given again by Eq. (21.38) only with the polarisation indices $s \leftrightarrow p$ swapped. Adding all contributions, one finds for the relevant stress tensor component

$$
T_{zz}(\boldsymbol{r}, \boldsymbol{r}) = -\frac{\hbar}{4\pi^3} \int_0^\infty d\omega \; \text{Re} \int d^2 k_\| \, k_z \sum_{\sigma=s,p} \frac{r_+^\sigma r_-^\sigma e^{2ik_z d}}{D^\sigma}
$$

$$
= -\frac{\hbar}{2\pi^2} \int_0^\infty d\omega \; \text{Re} \int_0^\infty dk_\| \, k_\| k_z \sum_{\sigma=s,p} \frac{r_+^\sigma r_-^\sigma e^{2ik_z d}}{D^\sigma} . \tag{21.40}
$$

Rotating the integration contour onto the imaginary axis, we find the final result for the Casimir stress between two parallel plates at zero temperature as

$$
T_{zz}(\boldsymbol{r}, \boldsymbol{r}) = \frac{\hbar}{2\pi^2} \int_0^\infty d\xi \int_0^\infty dk_\| \, k_\| \kappa_z \sum_{\sigma=s,p} \frac{r_+^\sigma r_-^\sigma e^{-2\kappa_z d}}{D^\sigma} \tag{21.41}
$$

where $\kappa_z = \sqrt{k_\|^2 + k^2} = ik_z$.

The expression (21.41) is seen to coincide with Lifshitz' original Eq. (21.9). The Green function formalism, however, via the stress tensor approach, is generally applicable and not restricted to this specific situation. Moreover, the formulation via the stress tensor reveals some physical insights into the dependencies of the Casimir force on external parameters. In particular, it is apparent that the Casimir force is essentially a scattering phenomenon, driven by fluctuations of the quantum vacuum, and mediated by the available electromagnetic scattering modes between the involved macroscopic bodies, the latter being contained in the scattering Green tensor $\mathbf{G}^{(S)}(\boldsymbol{r}, \boldsymbol{r}', \omega)$.

In order to make contact with Casimir's original calculation via the ground-state energy, one has to note that Eq. (21.41) can be rewritten with the help of the logarithmic derivative of the multiple reflection coefficient D^σ,

$$\frac{r_+^\sigma r_-^\sigma e^{-2\kappa_z d}}{D^\sigma} = \frac{1}{2\kappa_z}\frac{\partial}{\partial d}\ln D^\sigma , \qquad (21.42)$$

as

$$T_{zz}(\boldsymbol{r},\boldsymbol{r}) = \frac{\hbar}{4\pi^2}\frac{\partial}{\partial d}\int_0^\infty d\xi \int_0^\infty dk_\parallel\, k_\parallel \sum_{\sigma=s,p}\ln D^\sigma . \qquad (21.43)$$

Recalling that $T_{zz} = -\partial\mathcal{E}/\partial d$, we can read off the expression

$$\mathcal{E} = -\frac{\hbar}{4\pi^2}\int_0^\infty d\xi \int_0^\infty dk_\parallel\, k_\parallel \sum_{\sigma=s,p}\ln D^\sigma \qquad (21.44)$$

as the ground-state energy density (i.e. energy per unit area) between the plates. Rewriting the multiple reflection coefficient further as

$$D^\sigma = 1 - r_+^\sigma r_-^\sigma e^{-2\kappa_z d} = 1 - r_+^\sigma e^{-\kappa_z d} r_-^\sigma e^{-\kappa_z d} , \qquad (21.45)$$

one obtains another useful interpretation of the scattering process as the propagation (with imaginary wavenumber κ_z) for a distance d, the reflection with coefficient r_-^σ, another propagation and reflection with r_+^σ. In this way, a generalisation to nonplanar geometries can easily be established. This is known as the scattering approach to Casimir forces which, as we have seen, is a direct consequence of the Green function formalism.

21.4   Thermal contribution to Casimir–Lifshitz forces

So far, we have concentrated on the idealised situation of zero temperature. We will see shortly under which circumstances this can be regarded as a valid approxima-tion. At finite temperature, the hyperbolic cotangent has to be reinstated into the calculation of the stress tensor [Eq. (21.27)]. In this case, however, the closure of the integration contour in the upper frequency halfplane does not result in an inte-gral along the imaginary frequency axis. This is because the hyperbolic cotangent function acquires poles on the imaginary axis $\omega = i\xi$,

$$\coth\left(\frac{\hbar\omega}{2k_B T}\right) \mapsto -i\cot\left(\frac{\hbar\xi}{2k_B T}\right) , \qquad (21.46)$$

at the points $\xi_j = j2\pi k_B T/\hbar$, the Matsubara frequencies. Thus, the integral turns into a sum over the residues at the Matsubara frequencies (see Fig. 3.7) according to the replacement rule

$$\frac{\hbar}{\pi}\int_0^\infty d\xi\, f(i\xi) \mapsto 2k_B T \sum_{j=0}^\infty \left(1 - \frac{1}{2}\delta_{j0}\right) f(i\xi_j) . \qquad (21.47)$$

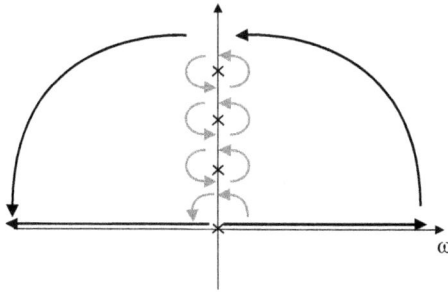

Figure 3.7: Rotation of the integration contour at finite temperature. The result is a sum over all Matsubara frequency poles on the imaginary axis.

Note that the residue at $\xi = 0$ has to be counted with only half its weight because the integration contour only performs a semi-circle around the pole.

Exercise 21.3
Derive the replacement rule (21.47) by contour integration techniques.

In the limit of vanishing temperature, the Matsubara frequencies move closer together and eventually form a continuum that retrieves the integral along the imaginary axis. The stress tensor thus takes the form

$$T_{zz}(\boldsymbol{r},\boldsymbol{r}) = \frac{k_B T}{\pi} \sum_{j=0}^{\infty} \left(1 - \frac{1}{2}\delta_{j0}\right) \int_0^{\infty} \mathrm{d}k_{\parallel}\, k_{\parallel} \kappa_z \sum_{\sigma=s,p} \frac{r_+^{\sigma}(\mathrm{i}\xi_j) r_-^{\sigma}(\mathrm{i}\xi_j) \mathrm{e}^{-2\kappa_z d}}{D^{\sigma}(\mathrm{i}\xi_j)} \qquad (21.48)$$

where all reflection coefficients have to be evaluated at the Matsubara frequencies $\xi_j{}^*$. The Matsubara sum (21.48), despite formally involving all Matsubara frequencies, can be truncated in certain situations. For example, if the temperature T is high enough (or the plate separation d large enough) such that $2\pi k_B T d/(\hbar c) \gg 1$, then only the lowest Matsubara frequency contributes to the sum as the exponential function $\mathrm{e}^{-2\kappa_z d}$ suppresses all contributions from Matsubara frequencies ξ_j with $j > 0$. In the opposite extreme case, when the distance d between the plates is very small, the poles on the imaginary axis move close together so that the zero-temperature limit is recovered.

In all the derivations in this chapter, we have explicitly assumed that the macroscopic bodies and the radiation field are in thermal equilibrium at the same temperature T. Only in this situation are we allowed to invoke the linear fluctuation-dissipation theorem in the form presented here. If one or both plates are held at a different temperature from the radiation field, the plates exchange thermal pho-

---

*As a peculiar observation one should note that any numerical evaluation of the frequency integral always involves a discretization of the integration range. This is nothing other than artificially invoking a finite temperature.

tons at different rates*. This requires a dynamical theory of Casimir forces [27, 28] which is beyond the quasistatic equilibrium theory described here. The assumption of thermal equilibrium and the use of the fluctuation-dissipation theorem has also been questioned in situations in which the macroscopic bodies are in relative motion to one another (for an example of Casimir forces between moving bodies, see Chapter 4, §26). In such cases, the Kubo–Martin–Schwinger relation (19.49) might not be applicable, and the notion of equilibrium has to be replaced by stationarity.

§22. Problems

Problem 3.1
Use the fluctuation-dissipation theorem to determine the ground-state interaction energy between a small particle, described by its polarisability $\alpha(\omega)$, and a surface.

Problem 3.2
Calculate the high-temperature limit of the thermal Casimir force (21.48) where the plate separation is larger than the thermal wavelength.

Problem 3.3
Consider a classical damped harmonic oscillator with mass m, eigenfrequency ω_0 and damping coefficient γ. Define the (dimensionless) quality factor Q of the oscillator as $Q = m\omega_0/\gamma$. Suppose the damped oscillator is driven by a fluctuating force F_{th}. Write down the equation of motion for the oscillator, calculate the spectral density $S_x(\omega)$ of the thermally driven motion and show that, at resonance, the spectral density increases linearly with the quality factor Q. How does the spectral density behave for large frequencies?

Problem 3.4
Suppose that a series of measurements provides one with the imaginary part $\operatorname{Im} G(\omega)$ of a response function. The task is to compute the corresponding real parts and hence the full complex response function $G(\omega)$ of

a) $\operatorname{Im} G(\omega) = -\dfrac{1}{1+\omega^2}$,

b) $\operatorname{Im} G(\omega) = \dfrac{2\gamma\omega}{(\omega_0^2 - \omega^2)^2 + 4\gamma^2\omega^2}$

using the Kramers–Kronig relations.

In calculations of dispersion forces it is sometimes necessary to know the values of various response functions, such as the permittivity $\varepsilon(\omega)$, at imaginary frequencies. Given the knowledge of the imaginary part $\operatorname{Im}\varepsilon(\omega)$ at all real frequencies, show that

$$\varepsilon(i\xi) = 1 + \frac{2}{\pi} \int_0^\infty d\omega' \frac{\omega'\operatorname{Im}\varepsilon(\omega')}{\omega'^2 + \xi^2}\,.$$

---

*Even in thermal equilibrium, the plates exchange thermal photons, but with equal rates of emission and absorption.

§23. Bibliography

[1] H.B.G. Casimir, "On the attraction between two perfectly conducting plates", *Proc. Kongl. Ned. Akad. v. Wetensch.* **B51**, 793 (1948).

[2] M. Bordag (ed.), *The Casimir Effect 50 Years Later* (World Scientific, Singapore, 1999).

[3] K. Clusius, *Angewandte Chemie* **56**, 241 (1943).

[4] W. Pauli, *Die allgemeinen Prinzipien der Wellenmechanik*, in H. Geiger, H. and K. Scheel (eds.), *Handbuch der Physik, Band 24/1* (Julius Springer, Berlin, 1933).

[5] H. Rechenberg, "Historical remarks on zero-point energy and the Casimir effect (1911–1998)" in Ref. [2].

[6] C. Itzykson, and J.-B. Zuber, *Quantum Field Theory* (McGraw-Hill, New York, 1980).

[7] H.-P. Breuer and F. Petruccione, *The Theory of Open Quantum Systems* (Oxford University Press, Oxford, 2002).

[8] G.W. Ford, M. Kac, and P. Mazur, "Statistical Mechanics of Assemblies of Coupled Oscillators", *J. Math. Phys.* **6**, 504 (1965).

[9] C. Cohen-Tannoudji, J. Dupont-Roc, and G. Grynberg, *Photons and atoms: Introduction to quantum electrodynamics* (Wiley, New York, 2007).

[10] H.B. Callen and T.A. Welton, "Irreversibility and generalized noise", *Phys. Rev.* **83**, 34 (1951).

[11] A. Einstein, "Über die von der molekularkinetischen Theorie der Wärme geforderte Bewegung von in ruhenden Flüssigkeiten suspendierten Teilchen", *Ann. d. Phys.* **322**, 549 (1905).

[12] H. Nyquist, "Thermal Agitation of Electric Charge in Conductors", *Phys. Rev.* **32**, 110 (1928).

[13] E.M. Lifshitz, "The theory of molecular attractive forces between solids", *Sov. Phys. JETP* **2**, 73 (1956).

[14] S.M. Rytov, *Theory of Electrical Fluctuations and Thermal Radiation* (Acad. Sci. Press, Moscow, 1953).

[15] S.M. Rytov, "Correlation theory of thermal fluctuations in an isotropic medium", *Sov. Phys. JETP* **6**, 130 (1958).

[16] M.A. Leontovich, and S.M. Rytov, "O differentsialnom zakone dlya intensivnosti elektricheskikh fluktuatsii i o vliyanii na nikh skin-effekta", *Zh. Eksp. Teor. Fiz.* **23**, 246 (1952).

[17] T. Gruner and D.-G. Welsch, "Green-function approach to the radiation-field quantization for homogeneous and inhomogeneous Kramers–Kronig dielectrics", *Phys. Rev. A* **53**, 1818 (1996).

[18] Ho Trung Dung, L. Knöll, and D.-G. Welsch, "Three-dimensional quantization of the electromagnetic field in dispersive and absorbing inhomogeneous dielectrics", *Phys. Rev. A* **57**, 3931 (1998).

[19] S. Scheel, L. Knöll, and D.-G. Welsch, "QED commutation relations for inhomogeneous Kramers–Kronig dielectrics", *Phys. Rev. A* **58**, 700 (1998).

[20] A. Tip, S. Scheel, L. Knöll, and D.-G. Welsch, "Equivalence of the Langevin and auxiliary-field quantization methods for absorbing dielectrics", *Phys. Rev. A* **63**, 043806 (2001).

[21] T.G. Philbin, "Casimir effect from macroscopic quantum electrodynamics", *New. J. Phys.* **13**, 063026 (2011).

[22] L.S. Brown and G.J. Maclay, "Vacuum stress between conducting plates: An image solution", *Phys. Rev.* **184**, 1272 (1969).

[23] E.D. Palik (ed.), *Handbook of Optical Constants of Solids II* (Academic Press, New York, 1991).

[24] W.C. Chew, *Waves and fields in inhomogeneous media* (IEEE Press, New York, 1995).

[25] S. Scheel and S.Y. Buhmann, "Macroscopic quantum electrodynamics — concepts and applications", *Acta Phys. Slovaca* **58**, 675 (2008).

[26] M.S. Tomaš, "Casimir force in absorbing multilayers", *Phys. Rev. A* **66**, 052103 (2002).

[27] M. Antezza, L.P. Pitaevskii, and S. Stringari, "Casimir-Lifshitz force out of thermal equilibrium and asymptotic nonadditivity", Phys. Rev. Lett. **95**, 113202 (2005).

[28] S.Y. Buhmann and S. Scheel, "Thermal Casimir versus Casimir-Polder Forces: Equilibrium and Nonequilibrium Forces", *Phys. Rev. Lett.* **100**, 253201 (2008).

Chapter 4

Macroscopic QED and vacuum forces

Simon Horsley

> The universe was a language with a perfectly ambiguous grammar.
> Every physical event was an utterance that could be parsed in two
> entirely different ways, one causal and the other teleological.
>
> – Ted Chiang, *Story of your life and other stories, 1998*

§24. Preliminary remarks

This chapter takes a different approach to the theory of vacuum forces, one based upon the principle of least action. As well as reclaiming and perhaps justifying some of the key equations from the previous chapters we shall find some new ones, and all within a framework familiar from basic quantum mechanics: we shall work with a Hamiltonian operator and a wavefunction to describe the field, the body, and its motion. The advantage of this formalism is that it contains no assumptions about the state of the medium or the field beyond the fact that the macroscopic Maxwell equations are valid.

In the domain of Casimir physics we are seeking to calculate tiny forces on objects that are too large for us to use a microscopic theory. Yet the force stems from an electromagnetic field with a very low amplitude, so that the account must also be quantum mechanical. As previous chapters have indicated, in this situation we might expect the electromagnetic field to obey quantised versions of the macroscopic (spatially averaged) Maxwell equations,

$$
\begin{aligned}
\boldsymbol{\nabla} \cdot \boldsymbol{D} &= \rho_f \\
\boldsymbol{\nabla} \cdot \boldsymbol{B} &= 0 \\
\boldsymbol{\nabla} \times \boldsymbol{E} &= -\frac{\partial \boldsymbol{B}}{\partial t} \\
\boldsymbol{\nabla} \times \boldsymbol{H} &= \boldsymbol{j}_f + \frac{\partial \boldsymbol{D}}{\partial t}
\end{aligned}
\qquad \longrightarrow \qquad
\begin{aligned}
\boldsymbol{\nabla} \cdot \hat{\boldsymbol{D}} &= \hat{\rho}_f \\
\boldsymbol{\nabla} \cdot \hat{\boldsymbol{B}} &= 0 \\
\boldsymbol{\nabla} \times \hat{\boldsymbol{E}} &= -\frac{\partial \hat{\boldsymbol{B}}}{\partial t} \\
\boldsymbol{\nabla} \times \hat{\boldsymbol{H}} &= \hat{\boldsymbol{j}}_f + \frac{\partial \hat{\boldsymbol{D}}}{\partial t},
\end{aligned}
\tag{24.1}
$$

where ρ_f and $\boldsymbol{j}_f$ are the free charge and current density within the medium that are not induced by the field. The set of equations on the left hand side describes the interaction of a classical electromagnetic field with a material medium (macroscopic electromagnetism), usually restricted so that the important length scales of the field are not comparable to the atomic structure of the material*. The theory implied by the right hand set of equations — often called macroscopic quantum electro-dynamics (macro–QED) — is both macroscopic and quantum mechanical, for the fields and sources in these equations are operators that represent spatial averages over complicated microscopic field and current distributions. Yet, to be a genuine piece of quantum physics it must be possible to derive these equations as operator equations of motion,

$$\frac{\partial \hat{\boldsymbol{D}}}{\partial t} = \frac{i}{\hbar}\left[\hat{H},\hat{\boldsymbol{D}}\right] \quad \frac{\partial \hat{\boldsymbol{B}}}{\partial t} = \frac{i}{\hbar}\left[\hat{H},\hat{\boldsymbol{B}}\right], \qquad (24.2)$$

otherwise we cannot be sure that (24.1) make any sense, and we might begin to doubt the theory developed in the previous chapters. This leads to the question: *can we find the Hamiltonian that has (24.1) as its equations of motion?* Because electromagnetic energy is not conserved in the presence of matter the answer to this question is not obvious, even classically. Constructing the Hamiltonian of macro–QED is the subject of the first part of this chapter[†].

The overall purpose is to show that the theory of the Casimir effect can be derived consistently from a principle of least action. This provides a coherent understanding of both the quantum theory of light in dispersive media and quantum forces due to the electromagnetic field. Indeed, the theory we develop from macro–QED goes beyond the results of Lifshitz theory developed in Chapter 3, and can be thought of as a general quantum theory of radiation pressure. To illustrate its utility, in the final part of the chapter we apply macro–QED to the problem of electromagnetic friction between closely spaced moving bodies (*quantum friction*).

The guiding principle is to describe the theory in a simple manner, which compels us to spend some time on macro–QED restricted to a single dimension. In the end the 1D results differ very little from the three dimensional results, and the generalisation rarely involves more than performing a sum over polarisations and an integration over angles.

§25. AN INTRODUCTION TO MACROSCOPIC QED

We begin this chapter with an introduction to the simplest case of macro–QED: one polarisation of the electromagnetic field propagating in a fixed direction through a uniform dispersive medium. We shall then illustrate the generalisation to three dimensions.

---

*See e.g. §103 in [1] for a discussion of macroscopic electromagnetism in cases where the scale of the field becomes comparable to the scale of the microscopic parts of the medium.

†Aspects of the theory can be found in the works of Hopfield [2], Huttner and Barnett [3], Suttorp [4], Kheirandish and Soltani [5], Scheel and Buhmann [6], and Philbin [7] (upon which this chapter is based), which is very far from a complete list.

25.1 Macroscopic QED in one dimension

Consider an electromagnetic wave propagating along the x–axis with the electric field pointing along z and the magnetic field pointing along y. This is a special case that keeps the equations simple, but to be concrete we could imagine this to be a field in a confined geometry as discussed in Chapter 2. In this case there are only

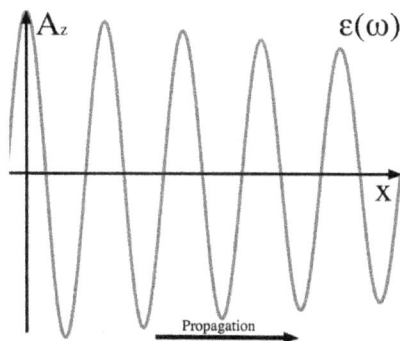

Figure 4.1: We consider electromagnetic waves polarised along the z axis with vector potential A_z, and propagating along the x axis. A material is present throughout all of space characterised by a permittivity $\varepsilon(\omega)$, the imaginary part of which determines the rate of decay of electromagnetic waves.

two non–trivial Maxwell equations,

$$\partial E_z/\partial x = \partial B_y/\partial t$$
$$\partial B_y/\partial x = \mu_0 j_z + c^{-2}\partial E_z/\partial t. \tag{25.1}$$

Rather than working in terms of the electric and magnetic fields we write the electromagnetic field in terms of the *magnetic vector potential*,

$$E_z = -\partial A_z/\partial t \qquad\qquad B_y = -\partial A_z/\partial x \tag{25.2}$$

which reduces (25.1) to a single equation: the first of (25.1) is identically fulfilled and the second is the *wave equation*

$$\left[\frac{\partial^2}{\partial x^2} - \frac{1}{c^2}\frac{\partial^2}{\partial t^2}\right] A_z(x,t) = -\mu_0 j_z(x,t). \tag{25.3}$$

It is thus evident that the electric current, $j_z(x,t)$, is the source (or sink) for the electromagnetic waves.

Exercise 25.1
Show that if the electric field only depends on x and points along z and there is no static magnetic field, then Maxwell's equations reduce to (25.1).

Now to introduce the medium. We imagine that as the wave propagates through a material it slightly displaces the charges, and in doing so induces an electric current. Such an effect can be mathematically described through writing the source on the right hand side of (25.3) as a linear function of the past behaviour of the electric field,

$$j_z(x,t) = \varepsilon_0 \int_0^\infty \chi(\tau) \frac{\partial E_z(x, t-\tau)}{\partial t} d\tau = -\varepsilon_0 \int_0^\infty \chi(\tau) \frac{\partial^2 A_z(x, t-\tau)}{\partial t^2} d\tau. \quad (25.4)$$

The function χ is the time–dependent *linear susceptibility*, and represents how the effect of the electric field persists in the material over time*. Note that we are assuming a non–conducting material (no current flows in response to a constant electric field), which is why the time derivative of E_z appears in (25.4).

A simpler understanding of the susceptibility can be found if we write Eq. (25.3) in the frequency domain. Representing the time dependence of the vector potential as a Fourier integral,

$$A_z(x,t) = \int_{-\infty}^\infty \frac{d\omega}{2\pi} \tilde{A}_z(x, \omega) e^{-i\omega t} \quad (25.5)$$

and substituting (25.4) into (25.3), gives the following equation for $\tilde{A}_z$,

$$\left[\frac{\partial^2}{\partial x^2} + \frac{\omega^2}{c^2} \varepsilon(\omega) \right] \tilde{A}_z(x, \omega) = 0, \quad (25.6)$$

where the quantity $\varepsilon(\omega)$ is the *electric permittivity*. The electric permittivity changes the wavelength within the medium from $\lambda = 2\pi c/\omega$ to $\lambda = 2\pi c/\omega\sqrt{\varepsilon}$ and is related to the time–dependent susceptibility as follows,

$$\varepsilon(\omega) = 1 + \int_0^\infty \chi(\tau) e^{i\omega\tau} d\tau. \quad (25.7)$$

The integral over τ runs from zero to infinity because we have assumed that the medium responds to the past behaviour of the field. It is only when the medium responds instantaneously to the field — i.e. $\chi(\tau) = \chi_0 \delta(\tau)$ — that $\varepsilon(\omega)$ is real and frequency independent. This is an unrealistic assumption, and in all physical cases $\varepsilon(\omega)$ is a complex function of frequency, the imaginary part quantifying the rate at which the field is absorbed into the medium. In general the real and imaginary parts of (25.7) are connected to one another through the *Kramers–Kronig* relations given in Chapter 3, Eq. (19.14)[†].

---

*Note that the convention adopted in (25.4) is such that the time dependent susceptibility has units of frequency.

†The interested reader can also find an excellent discussion of the general mathematical properties of linear susceptibilities in [1, 10].

Exercise 25.2
Show that if $\gamma > 0$ the permittivity corresponding to a single resonance at frequency ω_0

$$\varepsilon(\omega) = 1 + \frac{\omega_{P0}^2}{\omega_0^2 - \omega^2 - i\gamma\omega}$$

does not have either poles or zeros in the upper half complex frequency plane ($\mathrm{Im}[\omega] > 0$). Extend this proof to the case of 2 and then an arbitrary number of resonances (i.e. the general expression for the permittivity given in Chapter 3, Eq. (19.15)).

It can be shown (see e.g. [10]) that any permittivity satisfying the Kramers–Kronig relations which has a positive imaginary part is free from both poles and zeros in the upper half complex frequency plane.

The first task of this chapter is to find an action which yields (25.6) as an equation of motion, for a generic permittivity that satisfies the Kramers–Kronig relations. Then we can develop the theory of quantum light in media. This is not as straightforward as it sounds, because the process of dissipation means that the total energy of the field is not conserved. At the same time, the existence of a Hamiltonian (so long as it has no explicit time dependence) implies the conservation of energy. To avoid this apparent contradiction we must introduce another system to account for the energy absorbed from the field.

Mimicking a medium: finding the Lagrangian

The aim is now as follows: find a closed system made up of the electromagnetic field plus something else, where the 'something else' precisely mimics a material with permittivity $\varepsilon(\omega)$. The most general way to formulate laws of motion for a closed system is to start from the *principle of least action* [8,9], which is an approach that takes the actual laws of physics to be those that are optimal out of a set of possible alternatives. Not only does this way of formulating physical theories have a deep significance, but it can be used as the basis of both classical and quantum theories of motion.

The basic quantity of interest is the *action*, $S = \int L \, dt$, which is given by the integral over time of a Lagrangian, L. At a given time the Lagrangian depends on the instantaneous configuration of the system, and the classical equations of motion are obtained through finding the time evolution of the system that makes S take an extreme value. In macroscopic electromagnetism, both the field and the material are described as continuous functions of position. It is therefore appropriate to write the Lagrangian as the integral over space of a *Lagrangian density* $\mathcal{L}^*$,

$$L = \int \mathcal{L} \, dx. \tag{25.8}$$

Example: Consider the case of a scalar field $\phi(x, t)$. The Lagrangian density is a function of the field and its derivatives, $\mathcal{L}(\phi, \partial_x\phi, \partial_t\phi)$, and for a fixed initial and final configuration of the field,

---

*For readers unfamiliar with this object see [9, 11]

an infinitesimal change in the evolution of the field $\phi \to \phi + \delta\phi$ induces the following change in the action

$$\delta S = \int dt \int dx \left[\delta\phi \frac{\partial \mathscr{L}}{\partial \phi} + \partial_x(\delta\phi) \frac{\partial \mathscr{L}}{\partial(\partial_x \phi)} + \partial_t(\delta\phi) \frac{\partial \mathscr{L}}{\partial(\partial_t \phi)} \right]$$
$$= \int dt \int dx \left[\frac{\partial \mathscr{L}}{\partial \phi} - \frac{\partial}{\partial x}\left(\frac{\partial \mathscr{L}}{\partial(\partial_x \phi)} \right) - \frac{\partial}{\partial t}\left(\frac{\partial \mathscr{L}}{\partial(\partial_t \phi)} \right) \right] \delta\phi,$$

where the second line was obtained from an integration by parts. For the action to take an extreme value with respect to all possible evolutions of the field we must at least have $\delta S = 0$, which implies

$$\frac{\partial}{\partial x}\left(\frac{\partial \mathscr{L}}{\partial(\partial_x \phi)} \right) + \frac{\partial}{\partial t}\left(\frac{\partial \mathscr{L}}{\partial(\partial_t \phi)} \right) = \frac{\partial \mathscr{L}}{\partial \phi}. \tag{25.9}$$

Equation (25.9) is the *Euler–Lagrange* field equation, which is the classical equation of motion for a field theory. For the particular Lagrangian $\mathscr{L} = (1/2)[c^{-2}\dot{\phi}^2 - (\partial_x \phi)^2]$, (25.9) gives the wave equation without a source, $c^{-2}\ddot{\phi} - \partial_x^2 \phi = 0$.

For the case of the electromagnetic field in a dispersive medium it is useful to consider the Lagrangian density as broken up into a sum of three parts,

$$\mathscr{L} = \mathscr{L}_F + \mathscr{L}_I + \mathscr{L}_R, \tag{25.10}$$

where $\mathscr{L}_F$ is the contribution due to the electromagnetic field alone, $\mathscr{L}_R$ is the contribution from the system that mimics the response of the material (we shall call this the *reservoir* — Fig. 4.2 shows how the reservoir ought to interact with the field), and $\mathscr{L}_I$ accounts for the interaction between the two. The part due to the field is the same as in empty space,

$$\mathscr{L}_F = \frac{1}{2\mu_0} \left[\frac{1}{c^2}\left(\frac{\partial A_z(x,t)}{\partial t} \right)^2 - \left(\frac{\partial A_z(x,t)}{\partial x} \right)^2 \right], \tag{25.11}$$

which, as the reader can verify, reproduces the wave equation (without a source) when (25.9) is applied. The reservoir (which simulates the material response, and is the sink for the electromagnetic energy) is taken as a collection of simple harmonic oscillators,

$$\mathscr{L}_R = \frac{1}{2} \int_0^\infty \left[\left(\frac{\partial X_\omega(x,t)}{\partial t} \right)^2 - \omega^2 X_\omega^2(x,t) \right] d\omega. \tag{25.12}$$

After a little consideration we can see that this is a system with a tremendous number of degrees of freedom. At each point in space there is a continuum of oscillators of amplitude X_ω, each labelled with a real number ω. Specifying the instantaneous configuration of the X_ω requires us to specify a function of ω holding over the range $\omega \in [0, \infty)$, for every point in space. Meanwhile, A_z assigns just a single number to each point. Despite the apparent application of sledgehammer to nut, the inclusion of every possible natural frequency of oscillation in (25.12) is essential to reproduce the wave equation in an absorbing material.

Each reservoir oscillator is assumed to contribute an amount $\alpha(\omega)X_\omega$ to the total polarisation (dipole moment per unit volume) of the medium, where $\alpha(\omega)$ is the polarisability. Recalling the interaction energy $-\mathbf{d} \cdot \mathbf{E}$ between a dipole $\mathbf{d}$ and an electric field leads us to the following interaction Lagrangian

$$\mathscr{L}_I = -\frac{\partial A_z(x,t)}{\partial t} \int_0^\infty \alpha(\omega) X_\omega(x,t) d\omega. \tag{25.13}$$

The Lagrangian density given by the sum of (25.11–25.13) is sufficient to reproduce

Reflected

"Reservoir"   Transmitted

$X_\omega \vdots$

response mimics $\varepsilon(\omega)$

Incident

Figure 4.2: When electromagnetic radiation interacts with matter then some of it is inevitably absorbed: e.g. if a wave is incident onto an object then the sum of the reflected and transmitted intensities is less than that which is incident. To model this dissipation while keeping the system closed we introduce a reservoir of simple harmonic oscillators, each with amplitude $X_\omega(x,t)$. These mimic the interaction between the field and material, and are the sink for the electromagnetic energy.

the field equations in a medium (25.6–25.7), so long as we choose the function $\alpha(\omega)$ carefully. Applying the Euler–Lagrange equations (25.9) we find the two equations of motion

$$\left[\frac{\partial^2}{\partial x^2} - \frac{1}{c^2} \frac{\partial^2}{\partial t^2} \right] A_z = -\mu_0 \frac{\partial}{\partial t} \int_0^\infty \alpha(\omega) X_\omega d\omega \tag{25.14}$$

$$\left[\frac{\partial^2}{\partial t^2} + \omega^2 \right] X_\omega = -\alpha(\omega) \frac{\partial A_z}{\partial t}, \tag{25.15}$$

from which we can identify the relationship between the current in the medium and the configuration of the reservoir

$$j_z = \frac{\partial}{\partial t} \int_0^\infty \alpha(\omega) X_\omega d\omega. \tag{25.16}$$

To mimic an absorbing medium, the reservoir must be set up in such a way that the current (25.16) is always a sink for the field energy.

To directly compare the wave equation (25.14) with that in a material (25.6), we must eliminate the reservoir from the equation for the vector potential. This can be done through recognising that the above equation for X_ω (25.15) has a general solution in terms of two real functions $\mathcal{G}_{R,A}(t - t')$,

$$\mathcal{G}_{R,A}(t - t') = \frac{1}{\omega} \begin{cases} \Theta(t - t') \sin[\omega(t - t')] \\ \Theta(t' - t) \sin[\omega(t' - t)], \end{cases} \tag{25.17}$$

which are known respectively as the *retarded* and *advanced Green functions* of the oscillator ($\Theta(t-t')$ is a Heaviside step function). These functions satisfy a differential equation very similar to (25.15), but with a delta function on the right hand side,

$$\left[\frac{\partial^2}{\partial t^2} + \omega^2\right]\mathcal{G}_{R,A}(t - t') = \delta(t - t'),\qquad(25.18)$$

and describe the response of the oscillator to a sudden force at $t = t'$. In the retarded case the oscillator responds *after* the force has been applied, and in the advanced case the response occurs *before* the application of the force.

Exercise 25.3
Show that both of (25.17) satisfy (25.18).

Taking the retarded case $\mathcal{G}_R$ and integrating it against $\alpha(\omega)\partial A_z/\partial t$, we find the motion of the reservoir,

$$X_\omega(x,t) = -\alpha(\omega)\int_{-\infty}^{\infty}\mathcal{G}_R(t - t')\frac{\partial A_z(x,t')}{\partial t'}dt' + C_\omega(x)e^{-i\omega t} + C_\omega^*(x)e^{i\omega t}$$

$$= -\frac{\alpha(\omega)}{\omega}\int_{-\infty}^{t}\sin[\omega(t - t')]\frac{\partial A_z(x,t')}{\partial t'}dt' + C_\omega(x)e^{-i\omega t} + C_\omega^*(x)e^{i\omega t}.$$

$$(25.19)$$

The complex constant $C_\omega(x)$ multiplies a function that satisfies (25.15) with the right hand side equal to zero (the *homogeneous* solution to the equation), and is the amplitude of a current within the medium that produces radiation. In an infinite medium all radiation originates from such a current, and the state of the system can be specified entirely through the choice of $C_\omega(x)$. For the moment we'll set these to zero*, although in quantum mechanics these quantities play the role of creation and annihilation operators. Substituting (25.19) into (25.14) gives the wave equation for the vector potential, written without reference to the X_ω,

$$\left[\frac{\partial^2}{\partial x^2} - \frac{1}{c^2}\frac{\partial^2}{\partial t^2}\right]A_z = \frac{1}{c^2}\int_0^{\infty}\chi(\tau)\frac{\partial^2 A_z(x,t - \tau)}{\partial t^2}d\tau,\qquad(25.20)$$

where we identified the susceptibility from the earlier expressions (25.3–25.4), finding it equal to

$$\chi(\tau) = \frac{1}{\varepsilon_0}\int_0^{\infty}\frac{\alpha^2(\omega)}{\omega}\sin(\omega\tau)d\omega.\qquad(25.21)$$

---

*This is equivalent to imposing the initial condition that $X_\omega(x,t) = 0$ at $t = -\infty$

According to (25.7), this implies the permittivity is

$$
\begin{aligned}
\varepsilon(\omega) &= 1 + \frac{1}{\varepsilon_0} \int_0^\infty \frac{\alpha^2(\Omega)}{\Omega} \int_0^\infty \sin(\Omega\tau) e^{i\omega\tau} d\tau d\Omega \\
&= 1 + \frac{1}{\varepsilon_0} \int_0^\infty \frac{\alpha^2(\Omega)}{2\Omega} \lim_{\eta\to 0} \left[\frac{1}{\omega + \Omega + i\eta} - \frac{1}{\omega - \Omega + i\eta} \right] d\Omega \\
&= 1 + \frac{i\pi}{2\varepsilon_0\omega} \alpha^2(\omega) + \mathrm{P} \int_0^\infty \frac{\varepsilon_0^{-1}\alpha^2(\Omega)}{\Omega^2 - \omega^2} d\Omega.
\end{aligned}
\tag{25.22}
$$

Looking back to the Kramers–Kronig relations given in Chapter 3 Section 19.2, we can see that our permittivity (25.22) represents *any* causal material, so long as $\alpha(\omega)$ is given by

$$
\alpha(\omega) = \sqrt{\frac{2\omega\varepsilon_0 \mathrm{Im}[\varepsilon(\omega)]}{\pi}},
\tag{25.23}
$$

and this completes the specification of our Lagrangian. True to the Kramers–Kronig relations, the coupling $\alpha(\omega)$ between the field and the reservoir allows us the freedom to choose only the imaginary part of the permittivity.

Our first aim is thus fulfilled: we have found a closed system that has (25.6) as an equation of motion. Before going any further, let us pause for a moment to consider how this theory works. The Lagrangian (25.10) consists of the electromagnetic field coupled to an infinite number of oscillators (of every possible natural frequency) which mimic the polarisation of the material in response to the field. There are uncountably more degrees of freedom in the reservoir than the field, so that when the reservoir is initially at rest, energy flows out of the field into the medium without coming back. Of course real materials heat up and radiate the absorbed energy, but for the sake of simplicity we neglect this when we use a complex permittivity with a positive imaginary part. This idealised behaviour is mimicked by the reservoir.

Exercise 25.4
Rederive (25.22) from the equations of motion for the field and the reservoir, but this time use the advanced Green function from (25.17). What has happened to the permittivity? Can you explain this?

The Hamiltonian

The final task of this section is to find a mathematical expression for the energy of this system — the Hamiltonian, H [8] — so that we can progress on to the quantum mechanical case. For a field theory the Hamiltonian is given by the integral over a *Hamiltonian density* $H = \int \mathcal{H}\, dx$ [12], which represents the energy density of the system.

Example: In the case of a scalar field the Hamiltonian density is defined in terms of the Lagrangian density as follows

$$\mathscr{H} = \dot{\phi}\frac{\partial \mathscr{L}}{\partial \dot{\phi}} - \mathscr{L} \equiv \dot{\phi}\Pi_\phi - \mathscr{L}, \tag{25.24}$$

where a dot above a quantity denotes a time derivative, and the *canonical momentum* is given by $\Pi_\phi = \partial\mathscr{L}/\partial\dot{\phi}$ (c.f. the Hamiltonian of a point particle, $H = p\dot{x} - L$). Taking a partial derivative of $\mathscr{H}$ with respect to $\dot{\phi}$ one finds zero, which means that when we use the Hamiltonian we switch to a description in terms of the field and its canonical momentum, ceasing to use the time derivative of the field as a variable. For the particular Lagrangian density of a free scalar field, $\mathscr{L} = (1/2)[c^{-2}\dot{\phi}^2 - (\partial_x\phi)^2]$, the Hamiltonian density calculated from (25.24) is

$$\mathscr{H} = \frac{1}{2}\left[\frac{1}{c^2}\dot{\phi}^2 + \left(\frac{\partial\phi}{\partial x}\right)^2\right] = \frac{1}{2}\left[c^2\Pi_\phi^2 + \left(\frac{\partial\phi}{\partial x}\right)^2\right].$$

In our case the canonical momenta are

$$\Pi_{A_z} = \frac{\partial \mathscr{L}}{\partial \dot{A}_z} = \varepsilon_0\frac{\partial A_z}{\partial t} - \int_0^\infty \alpha(\omega)X_\omega d\omega \tag{25.25}$$

and

$$\Pi_{X_\omega} = \frac{\partial \mathscr{L}}{\partial \dot{X}_\omega} = \frac{\partial X_\omega}{\partial t}. \tag{25.26}$$

The two canonical momenta (25.25–25.26) are evidently related to the time derivatives of the field amplitudes in quite a simple way so that it is straightforward to write the Hamiltonian density in terms of these variables:

$$\mathscr{H} = \Pi_{A_z}\dot{A}_z + \int_0^\infty \Pi_{X_\omega}\dot{X}_\omega d\omega - \mathscr{L}$$

$$= \frac{1}{2}\left[\frac{1}{\varepsilon_0}\left(\Pi_{A_z} + \int_0^\infty \alpha(\omega)X_\omega d\omega\right)^2 + \frac{1}{\mu_0}\left(\frac{\partial A_z}{\partial x}\right)^2 + \int_0^\infty \left(\Pi_{X_\omega}^2 + \omega^2 X_\omega^2\right)d\omega\right]. \tag{25.27}$$

This equals the field energy plus the reservoir energy. Although we are still working with a simplified 1D theory, this Hamiltonian is of the same form as that needed to describe the full theory of macroscopic electromagnetism [7].

Exercise 25.5
Re–express the Hamiltonian $H = \int \mathscr{H}dx$ in terms of the electric and magnetic fields, and X_ω and $\dot{X}_\omega$. What interpretation can you give the Hamiltonian when it is written in this form?

The passage from classical to quantum theory

Despite the fact that quantum mechanics is a conceptual break from classical physics, the formal path for constructing a quantum field theory from a classical one is quite well established*. One straightforward route is to proceed from the expression for

---

*For example see [12].

the action and perform a path integral*, and if the reader is feeling particularly keen they might want to attempt this. However, here we take the more traditional path where the Hamiltonian is turned into an operator, and commutation relations are imposed between the fields and their canonical momenta. The quantum mechanical version of (25.27) is

$$\hat{H} = \frac{1}{2} \int dx \left[\frac{1}{\varepsilon_0} \left(\hat{\Pi}_{A_z} + \int_0^\infty \alpha(\omega) \hat{X}_\omega d\omega \right)^2 + \frac{1}{\mu_0} \left(\frac{\partial \hat{A}_z}{\partial x} \right)^2 + \int_0^\infty \left(\hat{\Pi}_{X_\omega}^2 + \omega^2 \hat{X}_\omega^2 \right) d\omega \right],$$

(25.28)

where the operators are taken to satisfy the canonical commutation relations

$$\left[\hat{A}_z(x,t), \hat{\Pi}_{A_z}(x',t) \right] = i\hbar \delta(x - x')$$

(25.29)

and

$$\left[\hat{X}_\omega(x,t), \hat{\Pi}_{X_{\omega'}}(x',t) \right] = i\hbar \delta(\omega - \omega') \delta(x - x').$$

(25.30)

The right hand sides of (25.29–25.30) are $i\hbar$ times the equivalent classical Poisson brackets[†], which is the correspondence between classical and quantum physics that was established by Dirac [14]. All the other commutation relations equal zero. Equations (25.28–25.30) are the bare bones of the quantum theory of macroscopic electromagnetism, restricted to the case of a single polarisation propagating in one direction. As described in the introduction, such a quantum theory is suitable for describing the effect of a material body (perhaps the air in this room, or a piece of metal or glass) on a very low intensity (quantum) electromagnetic field. The fact that the field amplitudes are represented by operators is significant.

In principle we could apply (25.28) to quantum mechanical problems immediately, but at the moment it is not in a very user–friendly form. For example, it would be some feat to directly determine the eigenstates of the system from (25.28). Yet as this Hamiltonian contains at most quadratic combinations of the field and reservoir operators, it is not anything more than an esoteric way of writing down the Hamiltonian of a system of many coupled simple harmonic oscillators. Therefore there is a much simpler way to write (25.28), which is to recast the system in terms of its normal modes[‡]. Our case is complicated by the fact that there are infinitely many of these coupled oscillators. Nevertheless, this transformation can be found, and one way to see what it must be is through examining the equations of motion for the operators.

---

*An extensive exposition of this technique can be found in [13].

[†]A Poisson bracket is a classical quantity that measures the degree of independence of the gradients of two quantities in phase space. For example, if two quantities A and B are functions of a scalar field ϕ and its canonical momentum Π_ϕ the Poisson bracket is defined as,

$$\{A, B\} = \int dx \left[\frac{\partial A}{\partial \phi(x)} \frac{\partial B}{\partial \Pi_\phi(x)} - \frac{\partial B}{\partial \phi(x)} \frac{\partial A}{\partial \Pi_\phi(x)} \right].$$

For a full explanation of the significance of the Poisson brackets see [9]. For the relationship between Poisson brackets and quantum mechanics see [14].

[‡]See e.g. [8].

The operator equations of motion

We shall now examine the behaviour of the quantum mechanical operators for the field and the medium, so as to better understand how to apply the theory of macro–QED. From now on — where possible — we shall work in the Heisenberg picture*, where the time dependence of the system is placed in the operators rather than the wave–function. The equations of motion for the field operators are

$$\frac{\partial \hat{A}_z}{\partial t} = \frac{i}{\hbar} \left[\hat{H}, \hat{A}_z \right] = \frac{1}{\varepsilon_0} \left(\hat{\Pi}_{A_z} + \int_0^\infty \alpha(\omega) \hat{X}_\omega d\omega \right)$$

$$\frac{\partial \hat{\Pi}_{A_z}}{\partial t} = \frac{i}{\hbar} \left[\hat{H}, \hat{\Pi}_{A_z} \right] = \frac{1}{\mu_0} \frac{\partial^2 \hat{A}_z}{\partial x^2}, \tag{25.31}$$

and those for the reservoir are

$$\frac{\partial \hat{X}_\omega}{\partial t} = \frac{i}{\hbar} \left[\hat{H}, \hat{X}_\omega \right] = \hat{\Pi}_{X_\omega}$$

$$\frac{\partial \hat{\Pi}_{X_\omega}}{\partial t} = \frac{i}{\hbar} \left[\hat{H}, \hat{\Pi}_{X_\omega} \right] = -\omega^2 \hat{X}_\omega - \frac{\alpha(\omega)}{\varepsilon_0} \left(\hat{\Pi}_{A_z} + \int_0^\infty \alpha(\omega') \hat{X}_{\omega'} d\omega' \right), \tag{25.32}$$

both of which are, after the elimination of the canonical momenta, formally identical to the classical equations of motion (25.14–25.15)

$$\left[\frac{\partial^2}{\partial x^2} - \frac{1}{c^2} \frac{\partial^2}{\partial t^2} \right] \hat{A}_z = -\mu_0 \frac{\partial}{\partial t} \int_0^\infty \alpha(\omega) \hat{X}_\omega d\omega \tag{25.33}$$

$$\left[\frac{\partial^2}{\partial t^2} + \omega^2 \right] \hat{X}_\omega = -\alpha(\omega) \frac{\partial \hat{A}_z}{\partial t}. \tag{25.34}$$

For the purposes of overall coherence it is worth noting that in this simplified case we have now found the mathematical route between the left and right hand sides of (24.1) — these are the macroscopic Maxwell equations. The operator expressions that satisfy (25.33–25.34) are simply the classical expressions, but with the unknown amplitudes — i.e. the complex quantities $C_\omega(x)$ in (25.19) — becoming operators.

As the classical motion of the reservoir (25.19) obeys an equation that is formally identical to the operator equation (25.34), the expression for the operator is therefore exactly the same,

$$\hat{X}_\omega(x, t) = -\frac{\alpha(\omega)}{\omega} \int_{-\infty}^t \sin[\omega(t - t')] \frac{\partial \hat{A}_z(x, t')}{\partial t'} dt' + N_\omega \left[\hat{C}_\omega(x) e^{-i\omega t} + \hat{C}_\omega^\dagger(x) e^{i\omega t} \right]. \tag{25.35}$$

The part of the classical motion (25.19) that was specified by the complex amplitude $C_\omega(x)$ has been replaced with the non–Hermitian operator $\hat{C}_\omega(x)$, which is the annihilation operator for excitations of current within the medium (its Hermitian adjoint is the creation operator). The real constant N_ω is for the moment undetermined, and shall be fixed by the commutation relations (25.29–25.30). To find

---

*See e.g. [14].

a representation of the vector potential operator $\hat{A}_z$ in terms of the creation and annihilation operators, (25.35) is inserted into (25.33) to give

$$\left[\frac{\partial^2 \hat{A}_z(t)}{\partial x^2} - \frac{1}{c^2}\frac{\partial^2 \hat{A}_z(t)}{\partial t^2} - \frac{1}{c^2}\int_0^\infty d\tau \chi(\tau)\frac{\partial^2 \hat{A}_z(t-\tau)}{\partial t^2}\right]$$
$$= i\mu_0 \int_0^\infty d\omega\, \alpha(\omega)N_\omega\omega\left[\hat{C}_\omega(x)e^{-i\omega t} - \hat{C}_\omega^\dagger(x)e^{i\omega t}\right] \quad (25.36)$$

which has the solution

$$\hat{A}_z(x,t) = -i\mu_0\int_0^\infty d\omega\, \omega N_\omega\alpha(\omega)\int_{-\infty}^\infty dx'g(x-x',\omega)\hat{C}_\omega(x')e^{-i\omega t} + \text{h.c.}, \quad (25.37)$$

where '+h.c.' means that the Hermitian conjugate of the expression should be added, and

$$g(x-x',\omega) = \frac{ie^{i\frac{\omega}{c}\sqrt{\varepsilon(\omega)}|x-x'|}}{2\frac{\omega}{c}\sqrt{\varepsilon(\omega)}}. \quad (25.38)$$

The above quantity is the retarded Green function for the electromagnetic field, and satisfies the wave equation in the frequency domain with a delta function on the right hand side

$$\left[\frac{\partial^2}{\partial x^2} + \frac{\omega^2}{c^2}\varepsilon(\omega)\right]g(x-x',\omega) = -\delta(x-x'). \quad (25.39)$$

Exercise 25.6

Show that (25.38) satisfies (25.39).

To (25.37) we could have added a superposition of solutions to the wave equation in the absence of a source, $\exp(\pm i\omega\sqrt{\varepsilon(\omega)}x/c - i\omega t)$. When the medium is homogeneous and absorbing, these waves grow exponentially large when x goes to either plus or minus infinity. This divergence corresponds to the fact that within an infinitely extended absorbing medium it is impossible for a monochromatic field to exist in the absence of a source (the field is being absorbed!), and therefore these waves should not be included. Yet in general, when the medium is not homogeneous there are solutions that do not diverge at infinity (e.g. waves incident from vacuum onto an absorbing material). We have two options; we can either consider all of space to be filled with an absorbing medium, and take free space as the limit $\alpha(\omega) \to 0$ at the end of every calculation; or we can include these extra solutions within the Green function to ensure that there is no energy lost from the system at infinity*. Having made this qualification, both electromagnetic field and reservoir operators can be written entirely in terms of $\hat{C}_\omega$ and $\hat{C}_\omega^\dagger$, and in the remainder of the text we shall assume that the Green function has the appropriate behaviour at infinity.

---

*A discussion of this can be found in [15].

The full expression for the $\hat{X}_\omega$ in terms of the creation and annihilation operators is

$$\hat{X}_\omega(x,t) = \lim_{\eta \to 0} \int_0^\infty d\Omega \int_\infty^\infty dx' N_\Omega \left[\frac{\mu_0 \Omega^2 \alpha(\omega)\alpha(\Omega)g(x-x',\Omega)}{(\omega+\Omega+i\eta)(\omega-\Omega-i\eta)} \right.$$

$$\left. + \delta(\Omega-\omega)\delta(x-x') \right] \hat{C}_\Omega(x')e^{-i\Omega t} + \text{h.c.} \quad (25.40)$$

Expressions for the remaining operators, $\hat{\Pi}_{X_\omega}$ and $\hat{\Pi}_{A_z}$ can be found from applying the operator equations of motion (25.31–25.32):

$$\hat{\Pi}_{A_z}(x,t) = -\int_0^\infty d\omega \, N_\omega \alpha(\omega) \int_{-\infty}^\infty dx' \left[\frac{\omega^2}{c^2}\varepsilon(\omega)g(x-x',\omega) \right.$$

$$\left. + \delta(x-x') \right] \hat{C}_\omega(x')e^{-i\omega t} + \text{h.c.} \quad (25.41)$$

and

$$\hat{\Pi}_{X_\omega}(x,t) = -\lim_{\eta \to 0} \int_0^\infty i\Omega d\Omega \int_{-\infty}^\infty dx' N_\Omega \left[\frac{\mu_0 \Omega^2 \alpha(\omega)\alpha(\Omega)g(x-x',\Omega)}{(\omega+\Omega+i\eta)(\omega-\Omega-i\eta)} \right.$$

$$\left. + \delta(\Omega-\omega)\delta(x-x') \right] \hat{C}_\Omega(x')e^{-i\Omega t} + \text{h.c.} \quad (25.42)$$

The key to understanding these expressions is that the electromagnetic field within the medium originates from a current. The $\hat{C}_\omega(x)$ and $\hat{C}_\omega^\dagger(x)$ create and annihilate the quanta of this current, and are the operators that replace the photon creation and annihilation operators of ordinary quantum electrodynamics*. The quantum theory of the electromagnetic field in a dispersive and dissipative medium is one where quanta of current are treated as the fundamental objects, and there are no photons, so to speak. Therefore when we come to discuss the Casimir effect, the description won't be anything like Casimir's original visualisation: the force will be seen to arise from the interaction of the ground state currents within the media, rather than being due to the confined electromagnetic modes between them.

Diagonalising the Hamiltonian

Having rewritten all of the field and reservoir operators in terms of $\hat{C}_\omega$ and $\hat{C}_\omega^\dagger$, we can now identify the normal modes of the system — then we might actually make use of the Hamiltonian! The process of reducing the Hamiltonian to its normal modes is often referred to as *Fano diagonalisation* due to its similarity with a procedure used by Fano in a study of the coupling of an atomic bound state to a continuum of (ionised) excited states [16].

First notice that in both field and reservoir operators, the time dependence occurs either as a factor of $\exp(-i\omega t)$, sitting next to $\hat{C}_\omega$, or as $\exp(i\omega t)$ sitting next to

---

*See e.g. [12] or [14].

$\hat{C}_\omega^\dagger$. As a consequence, taking a time derivative of any of the above operators is the same as making the substitution,

$$\hat{C}_\omega(x) \rightarrow -i\omega\hat{C}_\omega(x).$$

But from (25.31–25.32), the commutation between the Hamiltonian and $\hat{C}_\omega$ must have the same effect as a time derivative, implying,

$$\frac{i}{\hbar}\left[\hat{H}, \hat{C}_\omega\right] = -i\omega\hat{C}_\omega(x). \qquad (25.43)$$

If we knew the commutation relations between $\hat{C}_\omega(x)$ and $\hat{C}_\omega^\dagger(x)$, then we could use (25.43) to infer the expression for the Hamiltonian in terms of these operators. What we do know is that when the coupling between the field and reservoir is turned off (i.e. the function $\alpha(\omega)$ is set to zero) the reservoir reduces to a field of simple harmonic oscillators uncoupled from the field and each other, with commutation relations

$$\left[\hat{C}_\omega(x), \hat{C}_{\omega'}^\dagger(x')\right] = \delta(\omega - \omega')\delta(x - x'). \qquad (25.44)$$

We make the assumption that (25.44) also holds when the reservoir is coupled to the field — an assumption which is justified below — so that the Hamiltonian consistent with (25.43) is given by

$$\hat{H} = \frac{1}{2}\int dx \int_0^\infty d\omega\, \hbar\omega \left[\hat{C}_\omega(x)\hat{C}_\omega(x)^\dagger + \hat{C}_\omega(x)^\dagger\hat{C}_\omega(x)\right]$$

$$= \int dx \int_0^\infty d\omega\, \hbar\omega \left[\hat{C}_\omega(x)^\dagger\hat{C}_\omega(x) + \frac{1}{2}\delta(x = 0)\delta(\omega = 0)\right], \qquad (25.45)$$

which is the simplified form of the Hamiltonian (25.28) we set out to find: the system of coupled fields has been reduced to a continuum of uncoupled simple harmonic oscillators. It is a lengthy process to explicitly verify that substituting the expressions for the operators (25.40–25.42) into the Hamiltonian gives (25.45), but several authors have verified this and the interested reader should consult [3,4,7,17].

Written in these terms, the meaning of the Hamiltonian is transparent. The integrand consists of the operator $\hat{C}_\omega^\dagger(x)\hat{C}_\omega(x)$, which is analogous to the photon number operator in QED (see Chapter 1 Section 3.2) and counts how many quanta of current per unit frequency per unit volume are within the medium. In addition to this we have the term $\frac{1}{2}\delta(x = 0)\delta(\omega = 0)$ which is the ground state energy of the system. This ground state contribution can be thought of as the total energy of the system that results from the irreducible 'fluctuating' current within the reservoir (medium), and in this theory it is the equivalent of the infinite ground state energy that enters Casimir's calculation of the force between two perfect mirrors.

Exercise 25.7

In Chapter 1 the ground state energy of the electromagnetic field in empty space (in 3D) was given as Eq. (4.3)

$$\int d^3r \left\langle \frac{\varepsilon_0}{2} \hat{E}^2(r) + \frac{1}{2\mu_0} \hat{B}^2(r) \right\rangle = \int d^3r \int_0^\infty \frac{\hbar\omega^3}{2\pi^2 c^3} d\omega$$

which is infinite. However the *integrand* is finite. Meanwhile, the ground state energy of our Hamiltonian is the integral over delta functions given in (25.45), which is infinite even before we integrate over frequency and space. Why is the divergence much worse in the theory of macro–QED?

Commutation relations

The final loose end is to justify our assumption that the commutation relation between $\hat{C}_\omega$ and $\hat{C}_\omega^\dagger$ is (25.44). For this assumption to be consistent, the commutation relations (25.29–25.30) must continue to hold when the field and reservoir operators are written in terms of $\hat{C}_\omega$ and $\hat{C}_\omega^\dagger$. Taking the vector potential and its canonical momentum, and using their representation in terms of the creation and annihilation operators, (25.37) and (25.41), one obtains

$$\left[\hat{A}_z(x,t), \hat{\Pi}_{A_z}(x',t)\right] = \frac{i}{\pi c}\int_0^\infty d\omega \, |N_\omega|^2 \, \omega \sqrt{\varepsilon(\omega)} e^{i\frac{\omega}{c}\sqrt{\varepsilon(\omega)}|x-x'|} - \text{c.c.} \quad (25.46)$$

where '− c.c.' implies the subtraction of the complex conjugate, and to obtain this formula we applied the result,

$$\int_{-\infty}^\infty dx_1 e^{i\frac{\omega}{c}\left(\sqrt{\varepsilon(\omega)}|x-x_1|-\sqrt{\varepsilon^*(\omega)}|x'-x_1|\right)} = \frac{\sqrt{\varepsilon^*(\omega)} e^{i\frac{\omega}{c}\sqrt{\varepsilon(\omega)}|x-x'|}}{\frac{\omega}{c}\text{Im}[\varepsilon(\omega)]} + \text{c.c.} \quad (25.47)$$

which using the notation of (25.38) is equivalent to

$$\int_{-\infty}^\infty dx_1 \text{Im}[\varepsilon(\omega)]g(x-x_1,\omega)g^*(x'-x_1,\omega) = \frac{c^2}{\omega^2}\text{Im}[g(x-x',\omega)]. \quad (25.48)$$

In passing we note that in the above form (25.48) is a result that can be generalised to two and three dimensions, as well as to inhomogeneous media, provided that the Green function vanishes at infinity (see Chapter 3 Eq. (21.25)). To make progress we choose the constant N_ω to take the value,

$$N_\omega = \sqrt{\frac{\hbar}{2\omega}}, \quad (25.49)$$

a choice that is partly motivated by the fact that it allows us to write the right hand side of Eq. (25.46) as an integral over the entire real line,

$$\left[\hat{A}_z(x,t), \hat{\Pi}_{A_z}(x',t)\right] = \frac{i\hbar}{2\pi c}\int_{-\infty}^\infty d\omega \sqrt{\varepsilon(\omega)} e^{i\frac{\omega}{c}\sqrt{\varepsilon(\omega)}|x-x'|}. \quad (25.50)$$

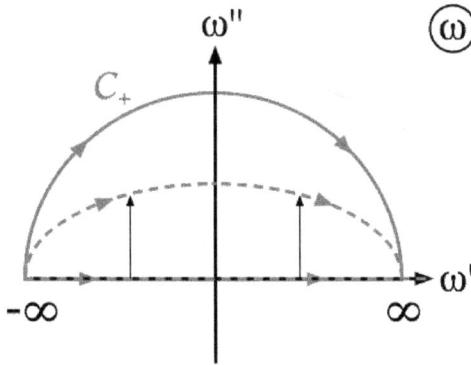

Figure 4.3: The integral along the real line given by (25.50) can be deformed into a contour integral, C_+ running along an infinite semicircle in the upper half frequency plane, $\omega = \omega' + i\omega''$, by virtue of the fact that the integrand is analytic in this region.

Assuming that $\varepsilon(\omega)$ is free from zeros and poles in the upper half frequency plane*, the integrand is analytic in this region, and the path of integration can be deformed into a semicircular contour, C_+, running from $-\infty$ to ∞ through the upper half plane†, as shown in Fig. 4.3. Having done this we can see that the commutation relation gives the desired result,

$$\left[\hat{A}_z(x,t), \hat{\Pi}_{A_z}(x',t)\right] = \frac{\hbar}{2\pi c}\int_0^\pi d\theta e^{i\theta} \lim_{|\omega|\to\infty}\left[|\omega|\sqrt{\varepsilon(\omega)}e^{i\frac{|\omega|e^{i\theta}}{c}\sqrt{\varepsilon(\omega)}|x-x'|}\right]$$

$$= \frac{i\hbar}{\pi}\delta(x-x')\int_0^\pi d\theta$$

$$= i\hbar\delta(x-x'). \tag{25.51}$$

To get to the second line of (25.51) we used (25.7) to show that $\varepsilon(|\omega| \to \infty) = 1$ in the upper half frequency plane, and applied the following representation of the delta function,

$$\delta(x) = \lim_{\lambda\to\infty}\frac{\lambda}{2}e^{-\lambda|x|}.$$

It is worth noting that (25.47) (upon which (25.51) rests) is not valid at frequencies where $\text{Im}[\varepsilon(\omega)] = 0$. The validity of this theory therefore depends upon there being *some* dissipation at all frequencies‡. The reader is left to verify that the commutation relations for the reservoir operators also do not change from (25.30). The transformed Hamiltonian (25.45) is therefore equivalent to the original expression (25.28), and we have now completed the first step towards our goal: we have shown

---

*This is true when $\varepsilon(\omega)$ has a positive imaginary part, see [10] and exercise 25.2.
†See e.g. [18].
‡With the exception of $\omega = 0$ where there can be no dissipation.

how to develop a quantum mechanical theory of light in materials that includes the effects of dispersion and dissipation*.

Exercise 25.8
Verify that when written as (25.40) and (25.42) the reservoir operators continue to satisfy (25.30).
Warning: This may take some time.
Hint: After evaluating the integrals over the delta functions, you'll still be left with integrals over frequency — combine them and use the residue theorem.

The ground state of the system

Before graduating from the safety of our 1D theory, we shall apply it to the ground state of the field and medium (the 'vacuum state'), the properties of which are responsible for Casimir forces.

Our Hamiltonian (25.45) is that of a continuum of uncoupled simple harmonic oscillators, and the ground state of this system (denoted by $|0\rangle$) can be defined via

$$\hat{C}_\omega(x)|0\rangle = 0, \qquad (25.52)$$

i.e. if we try to find a state with less current in the medium we get identically zero. Due to Heisenberg's uncertainty principle, the quantum mechanical ground state of a simple harmonic oscillator has an irreducible spread of possible positions and momenta. In this case the material is represented by a continuum of simple harmonic oscillators, the ground state of which exhibits an irreducible fluctuation in the current. The electromagnetic field (which can be thought of as originating from this current) therefore also has a spread of possible values. In this example we use the theory of Section 25.1 to compute the properties of the ground state field and current, rederiving the results of the fluctuation dissipation theorem given in Chapter 3 Section 19.5.

In accordance with their definitions in terms of the vector potential, the electric and magnetic field operators are given by

$$\hat{E}_z = -\frac{\partial \hat{A}_z}{\partial t}, \qquad \hat{B}_y = -\frac{\partial \hat{A}_z}{\partial x}. \qquad (25.53)$$

For this particular case we'll concentrate on the properties of the electric field

$$\hat{E}_z(x,t) = -\mu_0 \int_0^\infty d\omega\, \omega^2 \sqrt{\frac{\hbar}{2\omega}} \alpha(\omega) \int_{-\infty}^\infty dx'\, g(x-x',\omega)\hat{C}_\omega(x')e^{-i\omega t} + \text{h.c.} \quad (25.54)$$

the expectation value of which is zero in the ground state,

$$\langle 0|\hat{E}_z(x,t)|0\rangle = 0. \qquad (25.55)$$

---

*Further details on various aspects of this theory, along with its applications can be found in the extensive review paper of Scheel and Buhmann [6].

Meaning that although the field itself is not zero, if we did many measurements at a fixed point inside the medium we would find the average value to be zero. However if we performed two measurements of the field at separate points x_1 and x_2 multiplied the results together and then averaged we would not find zero:

$$\langle 0|\hat{E}_z(x_1,t)\hat{E}_z(x_2,t)|0\rangle = \frac{\mu_0}{c^2}\int_0^\infty d\omega\,\omega^4\frac{\hbar}{\pi}\mathrm{Im}[\varepsilon(\omega)]\int_{-\infty}^\infty dx'g(x_1-x',\omega)g^*(x_2-x',\omega)$$

$$= \frac{\hbar\mu_0}{\pi}\int_0^\infty d\omega\omega^2\mathrm{Im}[g(x_1-x_2,\omega)], \qquad (25.56)$$

where we have applied the result (25.47) to obtain the second line. The ground state of the field is thus spatially *correlated* within the medium, with the value of the field at one point in space being related to the value at another point. The integrand of (25.56) equals the result that would have been obtained from an application of the linear fluctuation–dissipation theorem, and represents the correlation of the field at a fixed frequency. As an aside, notice that (25.56) can be continuously brought towards the limit $\varepsilon(\omega) \to 1$, so that the theory also applies to empty space*.

As x_1 and x_2 approach one another, this correlation becomes a measure of the electric field intensity,

$$\lim_{x_1 \to x_2} \langle 0|\hat{E}_z(x_1)\hat{E}_z(x_2)|0\rangle = \frac{\hbar\mu_0 c}{2\pi}\int_0^\infty d\omega\,\omega\,\mathrm{Im}\left[\frac{i}{\sqrt{\varepsilon(\omega)}}\right], \qquad (25.57)$$

which diverges because there is a contribution to the field intensity at every frequency, which does not fall to zero as ω increases. This intensity has an associated field energy, which is also infinite, and is part of the divergent ground state energy in (25.45). This divergent contribution must be dealt with in calculations of the Casimir effect via a formal procedure we have called *renormalisation*, which we shall return to in Section 26.4.

Exercise 25.9
Show that the ground state expectation value of the following operator

$$\hat{S}_x(x,t) = -\frac{1}{2\mu_0}\left[\hat{E}_z(x,t)\hat{B}_y(x,t)+\hat{B}_y(x,t)\hat{E}_z(x,t)\right]$$

is zero. What is the physical interpretation for this?

As we have already established, the source of this field can be effectively thought of as a current within the medium. The operator for this current can be inferred from the right hand side of the equation of motion for $\hat{A}_z$ (25.36), and is

$$\hat{j}(x,t) = -i\int_0^\infty d\omega\,\alpha(\omega)\sqrt{\frac{\hbar\omega}{2}}\left[\hat{C}_\omega(x)e^{-i\omega t}-\hat{C}_\omega^\dagger(x)e^{i\omega t}\right].$$

---

*This limit is delicate. As a rule of thumb, never take the limit $\mathrm{Im}[\varepsilon(\omega)] \to 0$ in macro–QED until the end of a calculation.

This quantity also has zero expectation value,

$$\langle 0|\hat{j}(x,t)|0\rangle = 0,$$

but its correlation function is proportional to a delta function, meaning that it is *not* correlated in space

$$\langle 0|\hat{j}(x_1,t)\hat{j}(x_2,t)|0\rangle = \delta(x_1 - x_2)\frac{\hbar\varepsilon_0}{\pi}\int_0^\infty d\omega\,\omega^2\mathrm{Im}[\varepsilon(\omega)]. \qquad (25.58)$$

The reason for the lack of spatial correlation is that the excitations of the medium are independent from one another*. Meanwhile the electromagnetic field obeys a wave equation containing spatial derivatives and as a consequence different points are not independent, leading to the correlation (25.56).

25.2   Macroscopic QED in three dimensions

The extension of the results given in Section 25.1 to three dimensional electromagnetism is fairly straightforward, for the price of little more than an occasionally cumbersome notation. In three dimensions the Lagrangian for the field is of the same overall form as (25.11), but with the electric and magnetic fields being given in terms of both the scalar and vector potentials

$$\boldsymbol{E} = -\boldsymbol{\nabla}\varphi - \dot{\boldsymbol{A}} \qquad\qquad \boldsymbol{B} = \boldsymbol{\nabla}\times\boldsymbol{A}. \qquad (25.59)$$

For our purposes we can think of the potentials as a nice shorthand for the fields, although in general we must emphasise that in quantum mechanics the potentials are the more fundamental quantities. The three parts of the Lagrangian density (25.10) are now given by

$$\mathscr{L}_F = \frac{\varepsilon_0}{2}\left[\left(\boldsymbol{\nabla}\varphi + \dot{\boldsymbol{A}}\right)^2 - c^2(\boldsymbol{\nabla}\times\boldsymbol{A})^2\right] \qquad (25.60)$$

$$\mathscr{L}_R = \frac{1}{2}\int_0^\infty\left[\left(\frac{\partial\boldsymbol{X}_\omega}{\partial t}\right)^2 - \omega^2\boldsymbol{X}_\omega^2\right]d\omega \qquad (25.61)$$

$$\mathscr{L}_I = -(\boldsymbol{\nabla}\varphi + \dot{\boldsymbol{A}})\cdot\int_0^\infty\alpha(\omega)\boldsymbol{X}_\omega d\omega \qquad (25.62)$$

where φ, $\boldsymbol{A}$, and $\boldsymbol{X}_\omega$ are all functions of position and time. Nothing is very different: the effect of the material on the field is again mimicked with a reservoir but the amplitudes of the simple harmonic oscillators are now vectors, and the expression for the field now includes the scalar potential. In general a second reservoir should be added to account for losses through the permeability† $\mu(\omega)$, although here we have neglected the magnetic properties of the medium.

---

*Had we dropped this assumption we would have introduced some *spatial dispersion* into the material properties, which is a dependence of the permittivity on wave vector as well as frequency, and amounts to an extra spatial correlation in the field within the medium [1,19].

†See e.g. [5,7].

Applying the Euler–Lagrange equations to (25.60–25.62), we find the equations of motion for the electromagnetic field are given by,

$$\nabla \cdot \left[\varepsilon_0 E + \int_0^\infty \alpha(\omega) X_\omega d\omega \right] = 0 \tag{25.63}$$

and

$$\nabla \times B - \frac{1}{c^2} \frac{\partial E}{\partial t} = \mu_0 \frac{\partial}{\partial t} \int_0^\infty d\omega \, \alpha(\omega) X_\omega. \tag{25.64}$$

The remaining two Maxwell equations listed in (24.1) are identically true when the fields are written in terms of the potentials. The equation for the reservoir is the vector generalisation of the one dimensional case, and again has the solution (25.18–25.19)

$$X_\omega(x, t) = \frac{\alpha(\omega)}{\omega} \int_0^\infty \sin(\omega\tau) E(x, t - \tau) d\tau + C_\omega(x) e^{-i\omega t} + C_\omega^\star(x) e^{i\omega t}. \tag{25.65}$$

Substituting this expression for X_ω into the electromagnetic field Eqs. (25.63–25.64) gives us the behaviour of the electromagnetic field in the medium without reference to the reservoir. These are the macroscopic Maxwell equations,

$$\nabla \cdot D = \rho_f$$
$$\nabla \times H = j_f + \frac{\partial D}{\partial t}. \tag{25.66}$$

The H field is defined as simply proportional to the magnetic field $H = B/\mu_0$, and the *displacement field* D as

$$D(x, t) = \varepsilon_0 \left[E(x, t) + \int_0^\infty d\tau \chi(\tau) E(x, t - \tau) \right]. \tag{25.67}$$

The quantity $\chi(\tau)$ is equal to (25.21) so that the coupling function between the reservoir and the field $\alpha(\omega)$ is given by $([2\omega\varepsilon_0 \text{Im}[\varepsilon(\omega)])/\pi]^{1/2}$ which is the expression we derived in one dimension (25.23). The free charge and current density, ρ_f and j_f in (25.66) — i.e. the current and charge density *not* induced by the field — are equal to

$$\rho_f(x, t) = -\nabla \cdot \int_0^\infty d\omega \, \alpha(\omega) C_\omega(x) e^{-i\omega t} + \text{c.c.}$$
$$j_f(x, t) = -i \int_0^\infty d\omega \, \omega\alpha(\omega) C_\omega(x) e^{-i\omega t} + \text{c.c.} \tag{25.68}$$

and automatically satisfy the continuity equation, $\nabla \cdot j + \partial\rho/\partial t = 0$. The reservoir amplitudes C_ω have the same interpretation as before, which is now explicit in Eq. (25.68): they make up the amplitude of the free electric current density within the medium, which is responsible for the electromagnetic field. For notational brevity we have not indicated any spatial dependence for $\alpha(\omega)$, but all the results given in this section continue to hold when this is a function of position as it is for inhomogeneous media.

To reiterate the point made earlier, remember that we are assuming some degree of dissipation at all frequencies and all points in space, so any field must come from a source. In our case the system is just field plus material, so the source can only be some oscillating current within the material, which is given by j_f. In macro–QED we can thus think of the Casimir effect as being dictated by the interaction of the fluctuating currents within the two semi–infinite plates, across the gap between them. It is worth contrasting this picture with the one that is ordinarily used to understand the Casimir effect — and discussed in Chapter 1 Section 7.1, where the plates simply serve to restrict the allowed modes of the field. Our modification to the traditional understanding is a necessary consequence of properly including dispersion and dissipation.

The derivation of the Hamiltonian from (25.60–25.62) produces the same result as (25.27) and the same canonical momenta as (25.25–25.26), but again with scalar quantities becoming vectors. In three dimensions the quantum mechanical Hamiltonian operator therefore takes the same form as (25.28),

$$\hat{H} = \frac{1}{2} \int d^3x \left\{ \frac{1}{\varepsilon_0} \left(\hat{\Pi}_A + \int_0^\infty d\omega \, \alpha(\omega) \hat{X}_\omega \right)^2 + \frac{1}{\mu_0} \left(\nabla \times \hat{A} \right)^2 \right.$$
$$\left. + \int_0^\infty d\omega \left[\hat{\Pi}_{X_\omega}^2 + \omega^2 \hat{X}_\omega^2 \right] \right\}. \quad (25.69)$$

Gauge condition

Everything is now essentially a matter of listing slightly generalised versions of the operator formulae given in Section 25.1 — except for a slight niggle, which may have already occurred to the reader: *what happened to the scalar potential?* This is not an oddity confined to macro–QED*. The reason for its absence is that the Lagrangian does not contain $\dot{\varphi}$, so the associated canonical momentum is identically zero,

$$\hat{\Pi}_\varphi = 0. \quad (25.70)$$

Equation (25.70) implies that the 'equation of motion' associated with $\hat{\varphi}$, $\nabla \cdot \hat{D} = \hat{\rho}_f$, is not to be understood an equation of motion at all, but must be interpreted as the relationship between $\hat{\varphi}$ and $\hat{A}$, that allows us to eliminate $\hat{\varphi}$ from the Hamiltonian. In order to make sense of this condition on the potentials we fix a gauge, $\nabla \cdot \hat{A} = 0^\dagger$ so that the equation for the divergence of $\hat{D}$ implies,

$$\nabla \hat{\varphi} = \frac{1}{\varepsilon_0} \left(\int_0^\infty \alpha(\omega) \hat{X}_\omega d\omega \right)_L, \quad (25.71)$$

---

*See e.g. [12].

†This is often called the *Coulomb gauge*.

where a subscript 'L' indicates the longitudinal part of the vector*. The expression for the scalar potential given by (25.71) was imposed to obtain (25.69).

Exercise 25.10
Using condition (25.71), derive the Hamiltonian (25.69) from the Lagrangian given by the sum of (25.60–25.62).

Commutation relations

Having fixed a gauge to eliminate the scalar potential from the Hamiltonian, we must also make sure the commutation relations between $\hat{A}$ and $\hat{\Pi}_A$ are consistent with this gauge, i.e. $[\nabla \cdot \hat{A}, \hat{\Pi}_A] = 0$. A consistent set of commutation relations is given by[†]

$$\left[\hat{A}(\boldsymbol{x}, t), \hat{\Pi}_A(\boldsymbol{x}', t)\right] = i\hbar\, \boldsymbol{\delta}_T(\boldsymbol{x} - \boldsymbol{x}'), \tag{25.72}$$

where $\boldsymbol{\delta}_T(\boldsymbol{x} - \boldsymbol{x}')$ is the *transverse delta function* introduced in Chapter 1, Eq. (2.30)[‡]. There is no such complication with the reservoir operators, which are not constrained in this way, and obey the expected generalisation of (25.30),

$$\left[\hat{\boldsymbol{X}}_\omega(\boldsymbol{x}, t), \hat{\Pi}_{\boldsymbol{X}_{\omega'}}(\boldsymbol{x}', t)\right] = i\hbar 1 \delta^{(3)}(\boldsymbol{x} - \boldsymbol{x}')\delta(\omega - \omega'). \tag{25.73}$$

Exercise 25.11
Starting from the following ansatz

$$\left[\hat{A}(\boldsymbol{x}, t), \hat{\Pi}(\boldsymbol{x}', t)\right] = i\hbar \left[1_3 \delta^{(3)}(\boldsymbol{x} - \boldsymbol{x}') - \mathrm{M}(\boldsymbol{x} - \boldsymbol{x}')\right]$$

show that the two constraints $\nabla \cdot \hat{A} = 0$ and $\nabla \cdot \hat{D} = 0$ imply

$$\boldsymbol{k} = \boldsymbol{k} \cdot \tilde{\mathrm{M}}(\boldsymbol{k})$$
$$\boldsymbol{k} = \boldsymbol{k} \cdot \tilde{\mathrm{M}}^T(\boldsymbol{k})$$

where

$$\mathrm{M}(\boldsymbol{x} - \boldsymbol{x}') = \int \frac{d^3 k}{(2\pi)^3} \tilde{\mathrm{M}}(\boldsymbol{k}) e^{i\boldsymbol{k}\cdot(\boldsymbol{x}-\boldsymbol{x}')}.$$

Assuming that both indices of $\tilde{\mathrm{M}}$ are longitudinal, then show that (25.72) must be the correct commutation relation.

---

*The longitudinal part of a vector field $\boldsymbol{V}$ is that part which has divergence, but no curl. In terms of a Fourier expansion of the function this is

$$\boldsymbol{V}_L(\boldsymbol{x}) = \int \frac{d^3 k}{(2\pi)^3} \frac{1}{k^2} \boldsymbol{k}[\boldsymbol{k} \cdot \tilde{\boldsymbol{V}}(\boldsymbol{k})] e^{i\boldsymbol{k}\cdot\boldsymbol{x}} e^{i\boldsymbol{k}\cdot\boldsymbol{x}},$$

where $\tilde{\boldsymbol{V}}$ is the Fourier amplitude of $\boldsymbol{V}$.
[†]See for instance [12, 20, 21].
[‡]The transverse part of a vector field is that part which has curl but zero divergence.

Diagonalising the Hamiltonian

In the quantum mechanical case, the electromagnetic field and reservoir operators still obey the classical equations of motion (see problem 25.13). Their expressions in terms of the creation and annihilation operators for excitations in the reservoir can be found through taking the solutions to the macroscopic Maxwell equations (25.66) in terms of C_ω, and replacing these amplitudes with $\sqrt{\hbar/2\omega}$ times the operator $\hat{C}_\omega$, just as we did in Section 25.1. The operators that result from this process are

$$\hat{E}(x,t) = i\mu_0 \int_0^\infty d\omega\,\omega \int d^3x'\mathrm{G}(x,x',\omega)\cdot\hat{j}_f(x',\omega)e^{-i\omega t} + \text{h.c.}$$

$$\hat{B}(x,t) = \mu_0 \int_0^\infty d\omega \int d^3x'\nabla\times\mathrm{G}(x,x',\omega)\cdot\hat{j}_f(x',\omega)e^{-i\omega t} + \text{h.c.} \qquad (25.74)$$

and

$$\hat{X}_\omega(x,t) = \frac{\alpha(\omega)}{\omega}\int_{-\infty}^t \sin[\omega(t-t')]\hat{E}(x,t')dt' + \sqrt{\frac{\hbar}{2\omega}}\left[\hat{C}_\omega(x)e^{-i\omega t} + \hat{C}_\omega^\dagger(x)e^{i\omega t}\right]. \qquad (25.75)$$

So again the theory works in terms of quanta of current within the medium, from which the field is determined. The electromagnetic Green function $\mathrm{G}(x,x',\omega)$ — defined in Chapters 2 and 3 — now has two vector indices. The operator corresponding to the Fourier amplitude of the free electrical current $\hat{j}_f$ is defined as

$$\hat{j}_f(x,\omega) = -i\sqrt{\frac{\hbar\omega}{2}}\alpha(\omega)\hat{C}_\omega(x). \qquad (25.76)$$

The form of the Hamiltonian in terms of the creation and annihilation operators can again be inferred from the time–dependence of the operators (25.74–25.75) and is the same as (25.45),

$$\hat{H} = \frac{1}{2}\int d^3x \int_0^\infty d\omega\,\hbar\omega\,\hat{C}_\omega(x)\cdot\hat{C}_\omega^\dagger(x) + \text{h.c.} \qquad (25.77)$$

which can be justified in exactly the same way as (25.45), with $\hat{C}_\omega$ and $\hat{C}_\omega^\dagger$ satisfying

$$\left[\hat{C}_\omega(x,t),\hat{C}_{\omega'}^\dagger(x',t)\right] = 1\delta^{(3)}(x-x')\delta(\omega-\omega'). \qquad (25.78)$$

Although this account of the full theory is somewhat cursory, we trust the reader can understand its meaning on the basis of what went before it. The above formulae encompass the theory of light in absorbing media, a theory which we have developed from a Hamiltonian that self–consistently includes the effects of dissipation and dispersion. This theory provides an underlying theoretical framework for the quantum versions of the macroscopic Maxwell equations (24.1), and one that may be extended to unambiguously treat moving objects. The application to moving objects, and the forces between them is the purpose of the second half of this chapter.

Exercise 25.12
Starting from the expression for the electric field operator given by (25.74), show that the correlation of the electric field in the ground state $\langle 0|\hat{\boldsymbol{E}}(\boldsymbol{x},t) \otimes \hat{\boldsymbol{E}}(\boldsymbol{x}',t)|0\rangle$ is the same as predicted by the fluctuation–dissipation theorem given in Chapter 3 Section 21.27,

$$\langle 0|\hat{\boldsymbol{E}}(\boldsymbol{x},t) \otimes \hat{\boldsymbol{E}}(\boldsymbol{x}',t)|0\rangle = \frac{\hbar\mu_0}{\pi} \int_0^\infty d\omega\, \omega^2 \mathrm{Im}\left[\mathrm{G}(\boldsymbol{x},\boldsymbol{x}',\omega)\right].$$

Hint: Use the integral identity for Green functions (25.82).

Exercise 25.13
Starting from the following general expression for a quadratic Hamiltonian,

$$\hat{H} = \sum_{i,j}\left\{\alpha_{ij}\hat{p}_i\hat{p}_j + \beta_{ij}\hat{q}_i\hat{q}_j + \frac{1}{2}\gamma_{ij}\left[\hat{p}_i\hat{q}_j + \hat{q}_j\hat{p}_i\right]\right\} + \sum_i [V_i\hat{q}_i + W_i\hat{p}_i]$$

where α_{ij}, β_{ij} and γ_{ij} are arbitrary symmetric constant matrices, and V_i and W_i are constant vectors, show that the operators obey the classical equations of motion.

Example — a single polariton: Given our rather brief synopsis of macro–QED in three dimensions, an example might be helpful. We could calculate some ground state property of the system, but as established in Section 25.1, the ground state properties of the field come out in agreement with the linear fluctuation–dissipation theorem. However, the theory we have developed is much more general than the fluctuation–dissipation theorem. Unlike the linear fluctuation–dissipation theorem discussed in Chapter 3 Section 19.5, the system does not need to be in a thermal state, and we can apply this theory whenever we have reason to expect the macroscopic Maxwell equations (24.1) to hold.

Consider the simplest possible excitation of the system above the ground state: a single excitation of the medium, with the current aligned along the $\boldsymbol{e}_z$ axis,

$$|\psi\rangle = \int_0^\infty d\omega \int d^3\boldsymbol{x} f(\boldsymbol{x},\omega)\boldsymbol{e}_z \cdot \hat{\boldsymbol{C}}_\omega^\dagger(\boldsymbol{x})|0\rangle, \tag{25.79}$$

where $f(\boldsymbol{x},\omega)$ is a function that is sharply peaked around $\boldsymbol{x}_0$ and ω_0. To ensure that $\langle\psi|\psi\rangle = 1$, the function $f(\boldsymbol{x},\omega)$ must be normalised to one,

$$\int d^3\boldsymbol{x} \int_0^\infty d\omega |f(\boldsymbol{x},\omega)|^2 = 1. \tag{25.80}$$

This is also the lowest level of excitation of the field. We'll call this excitation a *polariton*, which is the name coined by Hopfield [2] for a mixture of electromagnetic and material excitation. The properties of the field in this state can be gleaned from the electric field correlation function which, after a few steps, we find to be

$$\langle\psi|\hat{\boldsymbol{E}}(\boldsymbol{x}_1) \otimes \hat{\boldsymbol{E}}(\boldsymbol{x}_2)|\psi\rangle = \frac{\hbar\mu_0}{\pi}\left[\frac{1}{2}\int_0^\infty d\omega\, \omega^2 \mathrm{Im}[\mathrm{G}(\boldsymbol{x}_1,\boldsymbol{x}_2,\omega)]\right.$$

$$+ \int_0^\infty d\omega \int_0^\infty d\omega' \int d^3\boldsymbol{x}' \int d^3\boldsymbol{x}'' \frac{\omega^2\omega'^2}{c^2}\sqrt{\mathrm{Im}[\varepsilon(\omega)]\mathrm{Im}[\varepsilon(\omega')]}e^{i(\omega'-\omega)t}$$

$$\left. \times f^\star(\boldsymbol{x}',\omega)f^\star(\boldsymbol{x}'',\omega')\mathrm{G}(\boldsymbol{x}_1,\boldsymbol{x}',\omega)\cdot\boldsymbol{e}_z \otimes \boldsymbol{e}_z \cdot \mathrm{G}^\dagger(\boldsymbol{x}_2,\boldsymbol{x}'',\omega')\right] + \text{c.c.} \tag{25.81}$$

where we applied the three dimensional generalisation of result (25.48), which was previously given in Chapter 3 Eq. (21.25),

$$\frac{\omega^2}{c^2}\int d^3\boldsymbol{x}'\mathrm{Im}[\varepsilon(\omega)]\mathrm{G}(\boldsymbol{x}_1,\boldsymbol{x}',\omega)\cdot\mathrm{G}^\dagger(\boldsymbol{x}_2,\boldsymbol{x}',\omega) = \mathrm{Im}[\mathrm{G}(\boldsymbol{x}_1,\boldsymbol{x}_2,\omega)]. \tag{25.82}$$

To proceed we write the expansion coefficient f as a product of Gaussians,

$$f(\boldsymbol{x},\omega) = \left(\frac{1}{\pi(\Delta x)^2}\right)^{3/4} \left(\frac{2}{\pi(\Delta\omega)^2}\right)^{1/4} e^{-\frac{1}{2(\Delta x)^2}(\boldsymbol{x}-\boldsymbol{x}_0)^2} e^{-\frac{1}{2(\Delta\omega)^2}(\omega-\omega_0)^2} \qquad (25.83)$$

and assume that Δx and $\Delta\omega$ are small enough that all the functions in (25.81) are constant over the region where f is significantly different from zero. Carrying out the integrations in (25.81) we obtain the final expression for the correlation function,

$$\langle\psi|\hat{\boldsymbol{E}}(\boldsymbol{x}_1)\otimes\hat{\boldsymbol{E}}(\boldsymbol{x}_2)|\psi\rangle = \frac{\hbar\mu_0}{\pi}\Big[\int_0^\infty d\omega\,\omega^2 \mathrm{Im}[\mathrm{G}(\boldsymbol{x}_1,\boldsymbol{x}_2,\omega)]$$

$$+\mathcal{B}\,\mathrm{Re}\Big[\mathrm{G}(\boldsymbol{x}_1,\boldsymbol{x}_0,\omega_0)\cdot\boldsymbol{e}_z\otimes\boldsymbol{e}_z\cdot\mathrm{G}^\dagger(\boldsymbol{x}_2,\boldsymbol{x}_0,\omega_0)\Big]\Big] \quad (25.84)$$

where $\mathcal{B} = 32\sqrt{2}\pi^2\Delta x^3\,\Delta\omega\,\omega_0^4\,\mathrm{Im}[\varepsilon(\omega_0)]/c^2$. There are several interesting things about (25.84). Firstly, the limit of a point–like excitation $\Delta x \to 0$, or one of infinitesimal bandwidth, $\Delta\omega \to 0$ just gives back the vacuum correlation function. This result has its roots in the normalisation of the state (25.79), and means that the quantum states of light within an absorbing medium must have a finite bandwidth, and originate from a source of non–zero extent. Secondly, notice that the correlation function breaks up into a sum of a vacuum contribution (the three dimensional version of (25.57)) plus an additional term arising from the excitation in the medium. Therefore the correlation of the field is a superposition of the vacuum correlation plus that of the polariton.

While the total intensity of the field diverges in the limit $\boldsymbol{x}_1 \to \boldsymbol{x}_2 = \boldsymbol{x}$, the difference in intensity between the ground and excited states is finite,

$$\langle\psi|\hat{\boldsymbol{E}}(\boldsymbol{x})\otimes\hat{\boldsymbol{E}}(\boldsymbol{x})|\psi\rangle - \langle 0|\hat{\boldsymbol{E}}(\boldsymbol{x})\otimes\hat{\boldsymbol{E}}(\boldsymbol{x})|0\rangle$$

$$= \frac{\hbar\mu_0\mathcal{B}}{\pi}\,\mathrm{Re}\Big[\mathrm{G}(\boldsymbol{x},\boldsymbol{x}_0,\omega_0)\cdot\boldsymbol{e}_z\otimes\boldsymbol{e}_z\cdot\mathrm{G}^\dagger(\boldsymbol{x},\boldsymbol{x}_0,\omega_0)\Big], \quad (25.85)$$

and is exactly what one would obtain for the time average of the classical electric field intensity from a current distributed in space and frequency according to (25.83), with an amplitude proportional to, $(\hbar\,\mathrm{Im}[\varepsilon(\omega)])^{1/2}$. (see problem 25.14)

Exercise 25.14
Consider an electric field due to a current $\boldsymbol{j}(\boldsymbol{x},\omega) = \boldsymbol{e}_z\mathcal{J}(\boldsymbol{x},\omega)$ where $\mathcal{J}(\boldsymbol{x},\omega)$ is sharply peaked around $\boldsymbol{x}_0$ and ω_0. Show that the electric field is of the form

$$\boldsymbol{E}(\boldsymbol{x},t) = A\mathrm{G}(\boldsymbol{x},\boldsymbol{x}_0,\omega_0)\cdot\boldsymbol{e}_z e^{-i\omega_0 t} + \text{c.c.}$$

where A is proportional to the amplitude of the current. From this expression show that the time average of $\boldsymbol{E}\otimes\boldsymbol{E}$ is

$$\langle\boldsymbol{E}(\boldsymbol{x})\otimes\boldsymbol{E}(\boldsymbol{x})\rangle = |A|^2\mathrm{G}(\boldsymbol{x},\boldsymbol{x}_0,\omega_0)\cdot\boldsymbol{e}_z\otimes\boldsymbol{e}_z\cdot\mathrm{G}^\dagger(\boldsymbol{x},\boldsymbol{x}_0,\omega_0) + \text{c.c.}$$

---

§26. VACUUM FORCES BETWEEN MOVING BODIES

The relative motion of macroscopic bodies implies a non–equilibrium situation* and if we were to apply the fluctuation–dissipation theorem, it would in general have

---
*See e.g. [10].

to be with care. In typical calculations of the Casimir force it is imagined that the bodies experiencing the force are held at rest, and we calculate the external force required to maintain this situation. Therefore in such calculations, all the usual equilibrium results apply. However, macro–QED is not restricted to the equilibrium state and can treat quantum forces between objects in relative motion. We finish with a calculation of the quantum force between two bodies in relative motion.

Armed with the theory of the electromagnetic field in realistic media, we now ascertain what effect the quantum field has on the motion of a body. We shall derive a quantum theory of electromagnetic forces through modifying the Lagrangian used in the previous section.

26.1 Moving bodies in 1D macroscopic QED

Returning to electromagnetism in one dimension, imagine that the homogeneous medium that we previously investigated is set into motion with uniform velocity V along the x axis. The *value* of the Lagrangian density will be the same (it is a scalar under Lorentz transformations), but it will *look* different when written in this new reference frame. This modification can be found through rewriting the earlier Lagrangian (25.11–25.13) in a relativistically covariant form*

$$\mathscr{L}_F = \frac{1}{2\mu_0}(\partial_\mu A_z)(\partial^\mu A_z) \tag{26.1}$$

$$\mathscr{L}_I = -V^\mu(\partial_\mu A_z)\int_0^\infty \alpha(\omega)X_\omega d\omega \tag{26.2}$$

$$\mathscr{L}_R = \frac{1}{2}\int_0^\infty \left[(V^\mu\partial_\mu X_\omega)^2 - \omega^2 X_\omega^2\right]d\omega, \tag{26.3}$$

where we have defined $V^\mu = \gamma(c, V)$, $\partial_\mu = (c^{-1}\partial_t, \partial_x)$, $\partial^\mu = (c^{-1}\partial_t, -\partial_x)$, $\gamma = (1 - V^2/c^2)^{-1/2}$, and a repeated Greek index implies summation (the Einstein summation convention). The coupling between the medium and the field takes the rest frame value $\alpha(\omega) = [2\omega\varepsilon_0\text{Im}[\varepsilon'(\omega)]/\pi]^{1/2}$ where $\varepsilon'(\omega)$ is the permittivity measured in the rest frame, and the quantities V^μ and ∂_μ are *four-vectors*, although in this 1D case only the 'x' and 't' components are important. When $V = 0$ then (26.1–26.3) reduce to the Lagrangian density given by (25.11–25.13).

---

*For a recap of relativistic notation see [11].

Exercise 26.1

The Lagrangian is often constructed through taking the difference between the kinetic and potential energy of a system. Consider a particle of mass m at rest. If we associate the energy mc^2 with the particle the action in the rest frame must be $S = -\int mc^2 dt'$ (the minus sign is a matter of convention, and t' is the time in the rest frame). Show that when written in covariant form this is

$$S = -\int m V_\mu dx^\mu = -\int mc^2 \sqrt{1 - V^2/c^2} \, dt$$

which is the relativistic action for a free particle. The equations of motion for the particle can be found through varying this action with respect to V. Now use a similar argument to derive (26.1–26.3) from their rest frame counterparts.

Exercise 26.2

The effect of the motion of the medium appears in (26.1–26.3) as the operator $V^\mu \partial_\mu$. What is the physical meaning of this operator?

The motion of the medium influences the electromagnetic field in two ways: (i) the motion of the reservoir is modified, because absorbed energy is now carried along with the medium rather than remaining at a fixed point; and (ii) the coupling to the reservoir is modified, because a moving dielectric medium polarises in response to both electric and magnetic fields. These modifications are evident in the equations of motion for the field,

$$\left[\frac{\partial^2}{\partial x^2} - \frac{1}{c^2} \frac{\partial^2}{\partial t^2} \right] A_z = -\mu_0 \left(\frac{\partial}{\partial t} + V \frac{\partial}{\partial x} \right) \int_0^\infty \alpha(\omega) X_\omega d\omega \tag{26.4}$$

and the reservoir

$$\left[\left(\frac{\partial}{\partial t} + V \frac{\partial}{\partial x} \right)^2 + \omega^2 \right] X_\omega = -\alpha(\omega) \left(\frac{\partial}{\partial t} + V \frac{\partial}{\partial x} \right) A_z \tag{26.5}$$

where we have assumed the velocity of the medium is slow enough such that $\gamma \sim 1$. The general solution to (26.5) can be written as

$$X_\omega(x, t) = -\alpha(\omega) \int_{-\infty}^\infty dx_0 \int_0^\infty d\tau \mathcal{G}_R(x - x_0, \tau) \left(\frac{\partial}{\partial t} + V \frac{\partial}{\partial x_0} \right) A_z(x_0, t - \tau)$$

$$+ C_\omega(x - Vt) e^{-i\omega t} + C_\omega^*(x - Vt) e^{i\omega t} \tag{26.6}$$

where $\mathcal{G}_R$ is the retarded Green function of Eq. (26.5),

$$\mathcal{G}_R(x - x_0, \tau) = \Theta(\tau) \frac{\sin(\omega \tau)}{\omega} \delta(x - x_0 - V\tau). \tag{26.7}$$

As before, $\mathcal{G}_R$ is the response of X_ω (which mimics the medium) to a sudden force applied at the time $\tau = 0$ at the point $x = x_0$. The argument of the delta function is such that the excitation is carried along with the reservoir, remaining at a point but moving with velocity V.

Exercise 26.3
Show that (26.7) satisfies

$$\left[\left(\frac{\partial}{\partial t} + V\frac{\partial}{\partial x}\right)^2 + \omega^2\right]\mathcal{G}_R(x - x_0, t - t_0) = \delta(t - t_0)\delta(x - x_0).$$

Substituting Eq. (26.6) to (26.4) we obtain the equation for the field which allows us to identify the susceptibility of the moving medium

$$\left[\frac{\partial^2}{\partial x^2} - \frac{1}{c^2}\frac{\partial^2}{\partial t^2}\right]A_z = \frac{1}{c^2}\int_{-\infty}^{\infty}dx_0\int_0^{\infty}d\tau\chi(x - x_0, \tau)\left(\frac{\partial}{\partial t} + V\frac{\partial}{\partial x_0}\right)^2 A_z(x_0, t - \tau),$$

(26.8)

where $C_\omega = 0$ and

$$\chi(x - x_0, \tau) = \frac{2}{\pi}\int_0^{\infty}d\omega\text{Im}[\varepsilon'(\omega)]\sin(\omega\tau)\delta\left(x - x_0 - V\tau\right).$$ (26.9)

Next to the susceptibility in (26.8) we have both spatial and temporal derivatives of the vector potential, which is because a moving material polarises in response to both electric and magnetic fields [1]. From this susceptibility we can identify the effective permittivity as we did in (25.22), which is now a function of both ω and k,

$$\varepsilon(k, \omega) = 1 + \int_{-\infty}^{\infty}dx\int_0^{\infty}d\tau\chi(x - x_0, \tau)e^{i\omega\tau}e^{-ik(x - x_0)}$$

$$= 1 + i\,\text{Im}[\varepsilon'(\omega - Vk)] + \frac{2}{\pi}\text{P}\int_0^{\infty}\frac{\Omega\text{Im}[\varepsilon'(\Omega)]}{\Omega^2 - (\omega - Vk)^2}d\Omega.$$ (26.10)

This is exactly the same as (25.22) but with the frequency shifted from ω to $\omega - Vk$. Our Lagrangian thus describes the physical phenomenon where the dispersion of a moving material is *Doppler shifted* relative to the rest frame (and the theory works for any material). Indeed, we shall find that the same terms in the Lagrangian that are responsible for this Doppler shift in frequency can be used to predict the force on a moving body. The next section will illustrate that in general the Doppler effect is inseparable from the physics of radiation pressure. Note that although we have concentrated our efforts on the case of constant velocity the above approach can be equally well applied to any arbitrary motion of the body, so long as it remains much less than c.

26.2   Computing classical forces

Through making a slight modification to the theory of a uniformly moving medium, we can also describe electromagnetic forces. To do this we simply take the Lagrangian for a uniformly moving homogeneous medium (26.1–26.3), generalise it to the case of an inhomogeneous medium $\alpha(\omega) \rightarrow \alpha(\omega, x - R(t))$ (R is the centre of mass coordinate) and add the centre of mass kinetic energy

$$L = \frac{1}{2}MV^2 + \int dx\,[\mathscr{L}_F + \mathscr{L}_I + \mathscr{L}_R]$$ (26.11)

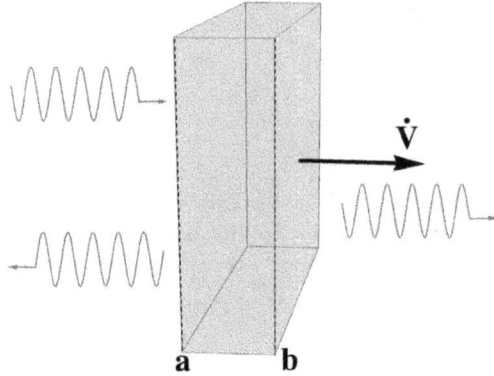

Figure 4.4: Applying the Lagrangian of macro–QED we find that the force on the centre of mass, $M\dot{V}$, is the Lorentz force integrated over the volume of the body. This force can be rewritten in terms of the difference in the electromagnetic stress, σ_{xx} on the two sides, $\sigma_{xx}(b) - \sigma_{xx}(a)$, minus the rate of change of the electromagnetic momentum within the body. In Section 26.3 we show that the operator equivalent of the Lorentz force is the appropriate expression for the quantum mechanical case.

where M is the mass per unit cross–sectional area of the body. Varying the new Lagrangian with respect to R and V then gives us the equation of motion for the position of the body

$$\frac{d}{dt}\left(\frac{\partial L}{\partial V}\right) = \frac{\partial L}{\partial R}$$

which yields an expression that initially looks quite complicated

$$M\dot{V} + \frac{d}{dt}\int dx \int_0^\infty d\omega \left[\frac{\partial X_\omega}{\partial x}\left(\frac{\partial X_\omega}{\partial t} + V\frac{\partial X_\omega}{\partial x}\right) - \frac{\partial A_z}{\partial x}\alpha(\omega, x - R)X_\omega\right]$$

$$= -\int dx \left(\frac{\partial A_z}{\partial t} + V\frac{\partial A_z}{\partial x}\right)\int_0^\infty \frac{\partial \alpha(\omega, x - R)}{\partial R} X_\omega d\omega. \quad (26.12)$$

However, applying the equation of motion for the reservoir (26.5) and enforcing the boundary condition $X_\omega = \Pi_{X_\omega} = 0$ at infinity allows us to simplify this down to

$$M\dot{V} = \int dx \frac{\partial A_z}{\partial x}\int_0^\infty d\omega\, \alpha(\omega, x - R)\left(\frac{\partial X_\omega}{\partial t} + V\frac{\partial X_\omega}{\partial x}\right)$$

$$= -\int dx\, j_z B_y. \quad (26.13)$$

This is simply the total *Lorentz force** on the body with the current density

$$j_z = \int_0^\infty \alpha(\omega, x - R)\left(\frac{\partial X_\omega}{\partial t} + V\frac{\partial X_\omega}{\partial x}\right) \quad (26.14)$$

*In general the Lorentz force density is given by $\boldsymbol{f}_L = \rho\boldsymbol{E} + \boldsymbol{j}\times\boldsymbol{B}$ which reduces to $f_L = -j_z B_y$ in this 1D case.

which is the rest frame current already identified in (25.16), but written in the coordinate system where the medium is in motion (problem 26.5).

Exercise 26.4
Verify that (26.13) follows from (26.12).

Exercise 26.5
Using the fact that $\rho = 0$ in this 1D case, show that the Lorentz transformation of the current (25.16) from the rest frame to one where the material is in motion gives (26.14) when $\gamma \sim 1$. *Hint*: The current transforms between frames as $j_z' = \gamma(j_z - V\rho)$.

Recall that the velocity dependence occurs within the Lagrangian in both the coupling between field and reservoir, and in the dynamics of the reservoir itself. It is the latter of these that is the origin of the modified frequency dependence of the permittivity (26.10). From this we can draw the conclusion that the force on a dielectric body is fundamentally linked to the Doppler effect.

The stress tensor and the Poynting vector

Using the equation for the vector potential (26.4), we can replace the current j_z in (26.13) with $-\mu_0^{-1}[\partial_x^2 - c^{-2}\partial_t^2]A_z$, and the force (26.13) can then be re-written in terms of the fields alone,

$$
\begin{aligned}
M\dot{V} &= \frac{1}{\mu_0} \int dx \left(\frac{1}{c^2} \frac{\partial E_z}{\partial t} - \frac{\partial B_y}{\partial x} \right) B_y \\
&= \int dx \frac{\partial T_{xx}}{\partial x} - \frac{1}{c^2} \frac{\partial}{\partial t} \int S_x dx,
\end{aligned} \tag{26.15}
$$

where the electromagnetic *stress* is identified as

$$
T_{xx} = -\frac{\varepsilon_0}{2} \left[E_z^2 + c^2 B_y^2 \right], \tag{26.16}
$$

which is a single component of the three dimensional *stress tensor** T, and the electric and magnetic fields are defined as in (25.2). Meanwhile the x–component of the *Poynting vector* is

$$
S_x = -\frac{1}{\mu_0} E_z B_y, \tag{26.17}
$$

which represents the electromagnetic power per unit area flowing in the x direction. The two integrands on the right hand side of (26.15) identically cancel in the region of space where there is no medium, as can be verified from the wave equation in the absence of a source. Therefore the integrals may be taken over the body alone, which in this case we assume to extend from $x = a$ to b (as in Fig. 4.4) giving

$$
M\dot{V} = T_{xx}(b) - T_{xx}(a) - \dot{\mathcal{P}}, \tag{26.18}
$$

---

*See, e.g. [11].

where

$$P = \frac{1}{c^2} \int_a^b S_x dx.$$

We can interpret the force given in (26.18) as the difference in radiation pressure on the two sides of the body minus the rate of change of the net electromagnetic momentum P within the body*. After time averaging — which removes P from the equation — this is the result one would obtain from the classical theory of radiation pressure[†], but here we have derived it from an action principle that is set up to self–consistently describe the effects of material dispersion and dissipation. Having constructed the Lagrangian of macro–QED, we got a theory of forces for free. There was no need to postulate a form for the stress tensor; this came automatically from our description of moving media. We can also quantise this theory, thus obtaining a quantum theory of radiation pressure that is not restricted to any particular state of the field or motion of the body, which is a distinct advantage over the Lifshitz theory described in Chapter 3.

Exercise 26.6

Consider a wave incident onto a material occupying the space $0 < x < a$

$$A_z = \frac{E_0}{2i\omega} \begin{cases} \left[e^{i\frac{\omega}{c}x} + re^{-i\frac{\omega}{c}x} \right] e^{-i\omega t} + \text{c.c.} & x < 0 \\ te^{i\frac{\omega}{c}x}e^{-i\omega t} + \text{c.c.} & x > a \end{cases}$$

where r and t are the reflection and transmission coefficients of the interface. Show that for such a field the time average of the stress (26.16) is given by

$$\langle T_{xx} \rangle = -\frac{\varepsilon_0}{2} |E_0|^2 \begin{cases} 1 + |r|^2 & x < 0 \\ |t|^2 & x > a \end{cases}$$

and therefore that the force per unit area imparted by this field is proportional to $1 + |r|^2 - |t|^2$. Can you give an interpretation for this result?

26.3 Quantum theory of radiation pressure

To quantise this theory we take the same approach as in Section 25.1, and first construct the Hamiltonian.

---

*Being concerned with mechanical forces we identify the momentum density in the medium with the Abraham expression $E \times H/c^2$. For further details on the momentum of light in media see [22].

[†]See e.g. [1, 23].

Classical Hamiltonian

The canonical momenta of the field and the reservoir are modified by the motion of the medium

$$\Pi_{A_z} = \frac{\partial \mathscr{L}}{\partial \dot{A}_z} = \varepsilon_0 \frac{\partial A_z}{\partial t} - \int_0^\infty \alpha(\omega, x - R) X_\omega d\omega$$

$$\Pi_{X_\omega} = \frac{\partial \mathscr{L}}{\partial \dot{X}_\omega} = \frac{\partial X_\omega}{\partial t} + V \frac{\partial X_\omega}{\partial x}$$

and the momentum associated with the centre of mass is

$$p = \frac{\partial L}{\partial V} = MV + \mathcal{A}, \tag{26.19}$$

where

$$\mathcal{A} = \int dx \int_0^\infty d\omega \left[\frac{\partial X_\omega}{\partial x} \Pi_{X_\omega} - \frac{\partial A_z}{\partial x} \alpha(\omega, x - R) X_\omega \right]. \tag{26.20}$$

Applying these expressions within the definition of the Hamiltonian, we find

$$H = pV + \int dx \left(\Pi_{A_z} \dot{A}_z + \int_0^\infty d\omega \dot{X}_\omega \Pi_{X_\omega} \right) - L$$

$$= \frac{(p - \mathcal{A})^2}{2M} + \int \mathscr{H}_0 dx, \tag{26.21}$$

where $\mathscr{H}_0$ is given by the earlier expression for a stationary medium (25.27).

The above Hamiltonian (26.21) is of the same *form* as that of a charged particle in an electromagnetic field [11], but it describes the centre of mass motion of a macroscopic body. In this respect the quantity $\mathcal{A}$ is analogous to the vector potential. For a charged particle the vector potential can be thought of as the momentum carried by the charge due to its interaction with the field [24]. Analogously $\mathcal{A}$ is the part of the momentum carried by the centre of mass due to its coupling to both field and material degrees of freedom. In the Hamiltonian formalism the motion of the medium is coupled to the field and the reservoir (the internal degrees of freedom of the material) through the quantity $\mathcal{A}$. As an aside it is worth noting that the term in $\mathcal{A}$ that equals $\partial_x A_z \alpha(\omega) X_\omega$ is the macroscopic version of the Röntgen interaction that occurs between a single moving electric dipole and a magnetic field [25].

Hamiltonian operator

The Hamiltonian operator can be inferred from its classical counterpart (26.21) and takes the form

$$\hat{H} = \frac{(\hat{p} - \hat{A})^2}{2M} + \hat{H}_0, \tag{26.22}$$

where

$$\hat{A} = \int dx \int_0^\infty d\omega \left[\frac{1}{2} \left(\hat{\Pi}_{X_\omega} \frac{\partial \hat{X}_\omega}{\partial x} + \frac{\partial \hat{X}_\omega}{\partial x} \hat{\Pi}_{X_\omega} \right) - \frac{\partial \hat{A}_z}{\partial x} \alpha(\omega, x - \hat{R}) \hat{X}_\omega \right] \tag{26.23}$$

and $\hat{H}_0$ is given by the expression for a stationary body (25.28). The only difference in the form of (26.22–26.23) compared to the classical case is that we have chosen a symmetric ordering of $\hat{X}_\omega$ and $\hat{\Pi}_{X_\omega}$ in $\hat{A}$. For completeness, we note that the commutation relation between the centre of mass $\hat{R}$ and canonical momentum $\hat{p}$ takes the usual value

$$\left[\hat{R}, \hat{p}\right] = i\hbar. \tag{26.24}$$

Operator equations of motion

It is now possible to apply the theory to the problem of quantum electromagnetic forces on objects. The motion of the centre of mass can be determined from the equations of motion for the centre of mass operator, $\hat{R}$, the expectation value of which gives us the average position of the body.

The Hamiltonian gives us both the velocity of the centre of mass

$$\frac{d\hat{R}}{dt} = \frac{i}{\hbar}\left[\hat{H}, \hat{R}\right] = \frac{1}{M}\left(\hat{p} - \hat{A}\right) \tag{26.25}$$

and the acceleration

$$\begin{aligned}\frac{d^2\hat{R}}{dt^2} &= \frac{i}{M\hbar}\left[\hat{H}, \hat{p} - \hat{A}\right] \\ &= -\frac{1}{M}\int dx \int_0^\infty d\omega \frac{\partial \alpha(\omega, x - \hat{R})}{\partial R} \frac{\partial \hat{A}_z}{\partial t}\hat{X}_\omega - \frac{i}{M\hbar}\left[\hat{H}_0, \hat{A}\right]. \end{aligned} \tag{26.26}$$

Evaluating the commutation relations and using the formula $[\hat{A}, \hat{B}\hat{C}] = [\hat{A}, \hat{B}]\hat{C} + \hat{B}[\hat{A}, \hat{C}]$, the acceleration becomes

$$\begin{aligned}\frac{d^2\hat{R}}{dt^2} &= \frac{1}{M}\int dx \int_0^\infty d\omega\, \alpha(\omega, x - \hat{R})\frac{\partial \hat{A}_z}{\partial x}\hat{\Pi}_{X_\omega} \\ &= -\frac{1}{M}\int dx \hat{j}_z \hat{B}_y, \end{aligned} \tag{26.27}$$

which is the operator equivalent of the classical force (26.13), where

$$\hat{j}_z = \int_0^\infty d\omega\, \alpha(\omega, x - \hat{R})\hat{\Pi}_{X_\omega}.$$

This leads us to the conclusion that the force on the centre of mass is determined by the Lorentz force operator. This agrees — for example — with the work of Loudon [26] on quantum mechanical radiation pressure which assumes that the force is given by the expectation value of the Lorentz force operator.

Exercise 26.7
Fill in the steps between (26.26) and (26.27).

As in the classical case, we can write the acceleration of the body entirely in terms of the field. Through applying the equation of motion for the vector potential operator

$$\frac{\partial^2 \hat{A}_z}{\partial x^2} - \frac{1}{c^2}\frac{\partial \hat{A}_z}{\partial t^2} = -\mu_0 \int_0^\infty \alpha(\omega, x - \hat{R})\hat{\Pi}_{X_\omega}$$

we find an expression that is formally identical to the classical result,

$$M\frac{d^2 \hat{R}}{dt^2} = \hat{T}_{xx}(b) - \hat{T}_{xx}(a) - \frac{\partial \hat{\mathcal{P}}}{\partial t} \tag{26.28}$$

where

$$\hat{T}_{xx} = -\frac{\varepsilon_0}{2}\left[\hat{E}_z^2 + c^2 \hat{B}_y^2\right] \tag{26.29}$$

and

$$\hat{\mathcal{P}} = \frac{1}{c^2}\int_a^b \hat{S}_x dx$$

with

$$\hat{S}_x = -\frac{1}{2\mu_0}\left[\hat{E}_z\hat{B}_y + \hat{B}_y\hat{E}_z\right]. \tag{26.30}$$

The force on the centre of mass of a body is thus equal to the expectation value of the operator equivalent of the classical radiation pressure (26.15). This result is true whatever the state of the system. Indeed, one could imagine small objects containing many atoms prepared such that the centre of mass behaves quantum mechanically, and the operator nature of $\hat{R}$ becomes important*.

26.4   The vacuum force

The above theory is now applied to the simplest case of interest: the force on a body initially localised at a point $R = R_0$, with the field and medium in the ground state. The initial wave function of the total system is taken to be of the form,

$$|\psi\rangle = \left(\frac{1}{\pi(\Delta x)^2}\right)^{1/4} e^{-\frac{1}{2(\Delta x)^2}(R-R_0)^2}|0_{R_0}\rangle,$$

where it is assumed that the localisation of the centre of mass Δx is much smaller than any other length scale of interest (i.e. the relevant wavelengths of the field), and $|0_{R_0}\rangle$ is defined as the state where $\hat{C}_\omega(x, R_0)|0_{R_0}\rangle = 0$. The dependence of the creation and annihilation operators on R_0 is necessary because the states of the whole system are different when the body is at different positions. From the

---

*A similar situation has been considered previously, for the case of a perfect mirror interacting with a quantised field [27], with both the position of the mirror and the field imagined to be in a quantum state. The theory developed above is a generalisation of this earlier work to the case of arbitrarily shaped bodies, characterised in terms of a permittivity obeying the Kramers–Kronig relations.

previous section the average value for the force on the body is the expectation value
of (26.28),

$$M\frac{d^2\langle\hat{R}\rangle}{dt^2} = \langle\psi| \left[\hat{T}_{xx}(b) - \hat{T}_{xx}(a) - \frac{\partial\hat{P}}{\partial t}\right] |\psi\rangle. \tag{26.31}$$

Applying our earlier expressions for the field operators (25.37), the expectation value
of the stress and the Poynting vector are found to be

$$\langle\psi|\hat{T}_{xx}|\psi\rangle = -\frac{\varepsilon_0}{2}\lim_{x\to x'}\langle 0_{R_0}| \left[\hat{E}_z(x)\hat{E}_z(x') + c^2\hat{B}_y(x)\hat{B}_y(x')\right]|0_{R_0}\rangle$$

$$= -\frac{\hbar}{2\pi c^2}\lim_{x\to x'}\int_0^\infty d\omega \left\{\omega^2\mathrm{Im}[g(x,x',\omega)] + c^2\frac{\partial^2}{\partial x\partial x'}\mathrm{Im}[g(x,x',\omega)]\right\}$$

$$\tag{26.32}$$

and

$$\langle\psi|S_x|\psi\rangle = 0.$$

To obtain the second line of (26.32) we applied the Green function identity (25.48)
and the same reasoning that led to the ground state correlation function (25.56),
taking the limit $x \to x'$ in the correlation function to obtain the field intensity. The
expectation value for the force is given by the difference in the stress (26.32) on the
two sides of the body, which is the one dimensional version of the *Lifshitz theory*
discussed in Chapter 3 Section 21, restricted to the case of material bodies separated
by vacuum. We have derived this result from a quantum mechanical theory based
on a Hamiltonian derived from a classical action, treating all variables as operators.
In this calculation we have not included the other bodies in the Hamiltonian, but
doing so does not change (26.32). The above result is formally valid for any system
of bodies, and one need only use a Green function that satisfies (25.39) with $\varepsilon(x,\omega)$
defining the configuration of the bodies.

Exercise 26.8
Describe how the above calculation would differ if the centre of mass of the body was prepared in
a quantum mechanical state. Explain why this is usually not important.

Renormalisation

One obvious — and very serious — problem with using the stress (26.32) to calculate the force is that as it stands it is meaningless. As we already discovered
in Section 25.1, these integrals over the Green functions diverge and therefore the
expectation value for the stress (26.32) is infinite. However we know that the force
on the body cannot be infinite so the *difference* between the stress on the two sides
of the body must be a finite number (if it isn't then we are in serious trouble).
To attempt to remove this infinite but spatially uniform contribution to the stress
tensor, we subtract part of the Green function,

$$g(x,x',\omega) \to g(x,x',\omega) - g_0(x-x',\omega) \equiv g^{(S)}(x,x',\omega), \tag{26.33}$$

a process we are referring to as *renormalisation* [28] (recall the convention adopted in the introduction of this book). The function, g_0, is the Green function for a homogeneous medium with the permittivity ε at the point of interest, and $g^{(S)}$ can be thought of as the 'scattered' part of the Green function, as discussed in Chapter 3 Section 21.3. We need to be very careful when we do this, because we are changing the theory by hand — in general *not* a good idea! However, this modification does not change the end result for the force, because in the limit $x \to x'$, g_0 does not depend on x, and therefore (26.33) does not modify the expression for the radiation pressure, $\langle [\hat{\sigma}_{xx}(b) - \hat{\sigma}_{xx}(a)] \rangle$. Although it seems in most cases that (26.33) yields a finite value for the force, at present there is reason to suspect that it can fail in some situations*.

Note that here we have focussed on obtaining a quantum theory of radiation pressure, and (26.33) is merely a trick to extract the finite quantity of interest from the divergent stress. This formal trick is distinct from what is typically referred to as renormalisation in the quantum field theory literature, which is required when the quantity of interest turns out infinite. In that case the divergent contribution has to be absorbed into one or more of the physical constants[†].

26.5   A simple case of quantum friction

When the field and medium are in their ground state then, in addition to the attractive force between two separated objects, there exists a frictional force that serves — for planar media — to bring any relative lateral motion to zero. This phenomenon is known as *quantum friction*[‡], and we shall now consider a simple instance of this effect within our one dimensional theory.

Suppose that the medium is at rest, and the field is initially in the zero particle state $|0\rangle$. Another system is coupled to it, which moves with a fixed velocity $V > 0$, and has an internal degree of freedom ξ which is proportional to its dipole moment (e.g. this could be a moving atom). We imagine that the internal degree of freedom of this object is also in its ground state, and the coupling is switched on at $t = 0$.

To describe this situation we add a new term into the Hamiltonian

$$\hat{H}_I(t) = i\beta\Theta(t)(\hat{a}e^{-i\omega_0 t} - \hat{a}^\dagger e^{i\omega_0 t})\frac{\partial \hat{A}_z(Vt, t)}{\partial t}, \qquad (26.34)$$

where β is proportional to the particle's polarisability, and $\hat{a}$ and $\hat{a}^\dagger$ are the raising and lowering operators associated with the internal degree of freedom ξ. Planck's constant times the transition frequency ω_0 is the energy required to excite the internal degree of freedom. This interaction Hamiltonian (26.34) is written above in the *interaction picture*[§], where the time dependence of all the operators is generated by the Hamiltonian without the interaction term. Adopting a perturbative approach,

---

*See e.g. [29]
[†]See e.g. [12]
[‡]For further details see [30–33].
[§]See e.g. [34].

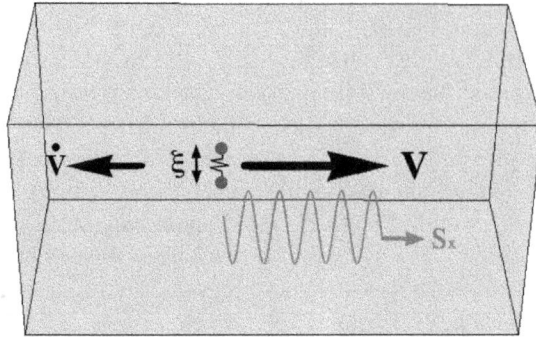

Figure 4.5: A polarisable particle is dragged through a material (e.g. water) at velocity V. Even though both particle and field may initially be in their respective ground states, the motion of the particle leads to an emission of radiation in the same direction as the velocity, and the particle therefore experiences a frictional force $\dot{v}$.

to first order in β the state of the system after the interaction has been switched on is the ground state plus some superposition of excited states,

$$|\psi\rangle = |0\rangle|0\rangle_\xi + \int \frac{dk}{2\pi} \int_0^\infty d\omega c(k,\omega,t)\hat{C}_\omega^\dagger(k)\hat{a}^\dagger|0\rangle|0\rangle_\xi, \qquad (26.35)$$

where $|0\rangle_\xi$ is the ground state of the oscillator, defined as that state which the lowering operator reduces to zero $\hat{a}|0\rangle_\xi = 0$. We have adopted a Fourier transformed representation in which the $\hat{C}_\omega$ operators have been written as a function of k rather than x obeying

$$\left[\hat{C}_\omega(k), \hat{C}_{\omega'}^\dagger(k')\right] = 2\pi\delta(k-k')\delta(\omega-\omega').$$

The vector potential operator (25.37) becomes

$$\hat{A}_z(x,t) = -i\mu_0 \int_0^\infty d\omega \sqrt{\frac{\hbar\omega}{2}}\alpha(\omega) \int \frac{dk}{2\pi}g(k,\omega)\hat{C}_\omega(k)e^{ikx}e^{-i\omega t} + \text{h.c.} \qquad (26.36)$$

where $g(k,\omega)$ is the Fourier transform of (25.38)

$$g(k,\omega) = \frac{1}{k^2 - \frac{\omega^2}{c^2}\varepsilon(\omega)}.$$

In this representation the excitation of the system is encoded in the function $c(k,\omega,t)$ which is given by the matrix element of the interaction Hamiltonian[*],

$$c(k,\omega,t) = -\frac{i}{\hbar}\int_{-\infty}^t \langle 0|_\xi\langle 0|\hat{C}_\omega(k)\hat{a}\hat{H}_I(t')|0\rangle|0\rangle_\xi dt'$$

$$= -i\mu_0\beta\sqrt{\frac{\omega}{2\hbar}}\alpha(\omega)(\omega - Vk)\Theta(t)\frac{e^{i(\omega_0+\omega-Vk)t}-1}{(\omega_0+\omega-Vk)}g^*(k,\omega). \qquad (26.37)$$

---

[*]See e.g. [34].

Already we can see that when $V \neq 0$, the transition amplitude, $c(k, \omega, t)$ is peaked around the point where $\omega_0 + \omega - Vk = 0$. This is the point where a positive frequency, $\omega > 0$, in the rest frame of the medium has been Doppler shifted to a negative value in the rest frame of the particle, $\omega - Vk < 0$. To put it another way, a positive energy excitation of the field and medium, $\hbar\omega$, appears as a negative energy excitation $\hbar(\omega - Vk)$ in the rest frame of the polarisable object, which represents energy available to excite the internal degree of freedom of the particle ξ. After a long time compared to the inverse of the transition frequency, the rate of this transition $\Gamma_{0\to1}(k, \omega)$ is

$$
\begin{aligned}
\Gamma_{0\to1} &= \lim_{t\to\infty} \int_0^\infty d\omega \int_{-\infty}^\infty \frac{dk}{2\pi} \frac{|c(k,\omega,t)|^2}{t} \\
&\to \frac{2\mu_0 \beta^2 \omega_0^2}{\hbar} \int_{\omega_0/V}^\infty \frac{dk}{2\pi} \mathrm{Im}[g(k, Vk - \omega_0)],
\end{aligned} \tag{26.38}
$$

which is zero when $V = 0$. To obtain (26.38) we used the following representation of the delta function

$$
\delta(x) = \frac{1}{\pi} \lim_{\lambda\to\infty} \frac{\sin^2(\lambda x)}{\lambda x^2}.
$$

The result (26.38) has the implication that when a polarisable particle moves though a medium with the field in its ground state, it can become excited. In some sense, what we think of as the vacuum state is not stable to relative motion: both the field and the object can finish in an excited state. The vacuum state of the field in the vicinity of a medium abhors relative motion.

Exercise 26.9
Through examining the frequency of a field ω in two relatively moving frames derive the condition for ω to take different signs in these two frames. Can this occur in free space? Given the discussion given in this section, how might we interpret this sign change?

Emission from the moving particle

During the process of exciting the dipole the field gains momentum, which can be inferred from the Poynting vector (26.31). The non–zero part of the integrated Poynting vector computed from (26.35) is

$$
\begin{aligned}
\int \langle\psi|\hat{S}_x|\psi\rangle dx = &-\frac{1}{2\mu_0} \int \frac{dk'}{2\pi} \int_0^\infty d\omega' \int \frac{dk}{2\pi} \int_0^\infty d\omega c(k,\omega,t) c^*(k',\omega',t) \\
&\times \int \langle 0|\hat{C}_{\omega'}(k') \left[\frac{\partial \hat{A}_z}{\partial x} \frac{\partial \hat{A}_z}{\partial t} + \frac{\partial \hat{A}_z}{\partial t} \frac{\partial \hat{A}_z}{\partial x} \right] \hat{C}_\omega^\dagger(k)|0\rangle dx \quad (26.39)
\end{aligned}
$$

and the matrix element within the integrand is given by

$$\int \langle 0|\hat{C}_{\omega'}(k')\left[\frac{\partial \hat{A}_z}{\partial x}\frac{\partial \hat{A}_z}{\partial t} + \frac{\partial \hat{A}_z}{\partial t}\frac{\partial \hat{A}_z}{\partial x}\right]\hat{C}_\omega^\dagger(k)|0\rangle dx$$

$$= -2\pi\delta(k - k')\mu_0^2\hbar k \mathcal{F}(\omega,\omega')g(k,\omega)g^\star(k,\omega')e^{-i(\omega-\omega')t}, \quad (26.40)$$

where

$$\mathcal{F}(\omega,\omega') = \alpha(\omega)\alpha(\omega')\sqrt{\omega\omega'}(\omega + \omega').$$

Notice that the matrix element (26.40) is an odd function of k, and that the function $c(k,\omega,t)$ is independent of k when $V = 0$. Therefore when $V = 0$ the total Poynting vector is zero, and the coupling of the field to the particle does not result in any net power flow in the field. For times $t \to \infty$, the integrated Poynting vector takes a simple form, which is similar to the transition rate of the particle (26.38):

$$\int \langle \psi|\hat{S}_x|\psi\rangle dx \sim 2\mu_0\beta^2\omega_0^2 \int_{\omega_0/V}^\infty \frac{dk}{2\pi}k(Vk - \omega_0)\text{Im}[g(k, Vk - \omega_0)]^2 \quad (26.41)$$

To obtain this result, the identity $\lim_{\lambda\to\infty}[\exp(i\lambda x) - 1]/x = i\pi\delta(x) - \mathrm{P}(1/x)$ was applied, and the terms involving the principal value were dropped. The quantity (26.41) is positive, meaning that radiation is emitted from the moving particle in the same direction as V. There is thus a force opposing the velocity of the particle, ultimately bringing it to rest. This is the force of quantum friction, and vanishes smoothly as $V \to 0$, when the lower limit of the integration in (26.41) tends to infinity.

26.6　Moving bodies in 3D macroscopic QED

The results in three dimensions are natural generalisations of the one dimensional formulae. In covariant form the earlier Lagrangian density (25.60 — 25.62) is

$$\mathcal{L}_F = -\frac{1}{4\mu_0}F_{\mu\nu}F^{\mu\nu} \tag{26.42}$$

$$\mathcal{L}_I = \frac{c}{2}F_{\mu\nu}P^{\mu\nu} \tag{26.43}$$

$$\mathcal{L}_R = \frac{1}{2}\int_0^\infty d\omega\left[\left(V^\mu\frac{\partial \mathbf{X}_\omega}{\partial x^\mu}\right)^2 - \omega^2\mathbf{X}_\omega^2\right] \tag{26.44}$$

where the *electromagnetic field tensor* is defined as [11]

$$F_{\mu\nu} = \partial_\mu A_\nu - \partial_\nu A_\mu \equiv \begin{pmatrix} 0 & E_x/c & E_y/c & E_z/c \\ -E_x/c & 0 & -B_z & B_y \\ -E_y/c & B_z & 0 & -B_x \\ -E_z/c & -B_y & B_x & 0 \end{pmatrix}$$

and the *polarisation tensor* as [1]

$$
P^{\mu\nu} = \gamma \begin{pmatrix} 0 & P_x/\gamma & P_y & P_z \\ -P_x/\gamma & 0 & VP_y & VP_z \\ -P_y & -VP_y & 0 & 0 \\ -P_z & -VP_z & 0 & 0 \end{pmatrix}.
$$

In this case the velocity of the body is taken along the x–axis and rest frame polarisation is defined as

$$
\boldsymbol{P} = \int_0^\infty d\omega \, \alpha(\omega, \boldsymbol{x} - \boldsymbol{R}) \boldsymbol{X}_\omega.
$$

Note that despite its covariant form, the Lagrangian is a function of the polarisation in the rest frame, and the quantity $\boldsymbol{X}_\omega$ remains a three dimensional vector. To this Lagrangian, $L_0 = \int dx [\mathscr{L}_F + \mathscr{L}_I + \mathscr{L}_R]$, we add the kinetic energy of the centre of mass,

$$
L = \frac{1}{2} M \boldsymbol{V}^2 + L_0, \tag{26.45}
$$

and in the regime $\gamma \sim 1$ this expression forms the basis for the theory of radiation pressure.

Computing the force

From the Lagrangian (26.45) we can immediately calculate the electromagnetic force on a macroscopic body, varying the action with respect to $\boldsymbol{R}$ and $\boldsymbol{V}$

$$
\frac{d}{dt} \left(\frac{\partial L}{\partial \boldsymbol{V}} \right) = \frac{\partial L}{\partial \boldsymbol{R}},
$$

which gives

$$
\frac{d}{dt} \left[M\boldsymbol{V} + \int d^3\boldsymbol{x} \int_0^\infty d\omega \, (\boldsymbol{\nabla} \otimes \boldsymbol{X}_\omega) \cdot \left(\frac{\partial \boldsymbol{X}_\omega}{\partial t} + \boldsymbol{V} \cdot \boldsymbol{\nabla} \boldsymbol{X}_\omega \right) - \int d^3\boldsymbol{x} \boldsymbol{P} \times \boldsymbol{B} \right]
$$
$$
= \frac{\partial}{\partial \boldsymbol{R}} \int d^3\boldsymbol{x} \boldsymbol{P} \cdot (\boldsymbol{E} + \boldsymbol{V} \times \boldsymbol{B}). \tag{26.46}
$$

This can be simplified through applying the equation of motion for the reservoir, $(\partial_t + \boldsymbol{V} \cdot \boldsymbol{\nabla})^2 \boldsymbol{X}_\omega + \omega^2 \boldsymbol{X}_\omega = \alpha(\omega)[\boldsymbol{E} + \boldsymbol{V} \times \boldsymbol{B}]$ (c.f. (26.5)), and after the application of a few vector identities we find the expression for the force (26.46) becomes

$$
M\dot{\boldsymbol{V}} = \int d^3\boldsymbol{x} \, [\rho \boldsymbol{E} + \boldsymbol{j} \times \boldsymbol{B}], \tag{26.47}
$$

which is the total Lorentz force on the body, with charge density

$$
\rho = -\boldsymbol{\nabla} \cdot \boldsymbol{P}
$$

and current density

$$
\boldsymbol{j} = -\boldsymbol{V}(\boldsymbol{\nabla} \cdot \boldsymbol{P}) + \int_0^\infty \alpha(\omega, \boldsymbol{x} - \boldsymbol{R}) \left(\frac{\partial}{\partial t} + \boldsymbol{V} \cdot \boldsymbol{\nabla} \right) \boldsymbol{X}_\omega.
$$

To obtain this expression the reader might find the following formula useful:

$$(\boldsymbol{\nabla} \otimes \boldsymbol{P}) \cdot \boldsymbol{V} \times \boldsymbol{B} = [(\boldsymbol{V} \times \boldsymbol{B}) \times \boldsymbol{\nabla}] \times \boldsymbol{P} + (\boldsymbol{\nabla} \cdot \boldsymbol{P})\boldsymbol{V} \times \boldsymbol{B}.$$

Applying the Maxwell equations $\boldsymbol{\nabla} \cdot \boldsymbol{E} = \rho/\varepsilon_0$ and $\boldsymbol{\nabla} \times \boldsymbol{B} = \mu_0 \boldsymbol{j} + c^{-2}\dot{\boldsymbol{E}}$ the force (26.47) can be re–written as

$$M\dot{\boldsymbol{V}} = \int d^3x \left[\boldsymbol{\nabla} \cdot \mathrm{T} - \frac{1}{c^2}\frac{\partial \boldsymbol{S}}{\partial t} \right] \tag{26.48}$$

where we have recovered the *stress tensor*,

$$\mathrm{T} = \varepsilon_0 \left[\boldsymbol{E} \otimes \boldsymbol{E} + c^2 \boldsymbol{B} \otimes \boldsymbol{B} - \frac{1}{2}\mathbb{1}\left(\boldsymbol{E}^2 + c^2 \boldsymbol{B}^2 \right) \right],$$

and the *Poynting vector*,

$$\boldsymbol{S} = \frac{1}{\mu_0}\boldsymbol{E} \times \boldsymbol{B}.$$

As we have already established in the one–dimensional case, the time average of the equation of motion (26.48) reproduces the known result from the classical theory of radiation pressure: the average force is given by the integral of the stress tensor over the surface of the body. This comes as a natural consequence extending the Lagrangian of macroscopic QED to moving media.

Quantum theory

From (26.42–26.44) the canonical momenta of this system are

$$\boldsymbol{\Pi}_{\boldsymbol{A}} = \varepsilon_0 \left(\dot{\boldsymbol{A}} + \boldsymbol{\nabla}\varphi \right) - \int_0^\infty d\omega\, \alpha(\omega, \boldsymbol{x} - \boldsymbol{R})\boldsymbol{X}_\omega$$

$$\boldsymbol{\Pi}_{\boldsymbol{X}_\omega} = \frac{\partial \boldsymbol{X}_\omega}{\partial t} + (\boldsymbol{V} \cdot \boldsymbol{\nabla})\boldsymbol{X}_\omega \tag{26.49}$$

and

$$\boldsymbol{p} = \frac{\partial L}{\partial \boldsymbol{V}} = M\boldsymbol{V} + \boldsymbol{\mathcal{A}},$$

where

$$\boldsymbol{\mathcal{A}} = \int d^3x \int_0^\infty d\omega\, [(\boldsymbol{\nabla} \otimes \boldsymbol{X}_\omega) \cdot \boldsymbol{\Pi}_{\boldsymbol{X}_\omega} - \alpha(\omega, \boldsymbol{x} - \boldsymbol{R})\boldsymbol{X}_\omega \times \boldsymbol{B}], \tag{26.50}$$

a quantity which again plays the role of an effective vector potential for the motion of the centre of mass. In terms of these canonical variables the Hamiltonian is

$$H = \boldsymbol{p} \cdot \boldsymbol{V} + \int d^3x \left[\dot{\boldsymbol{A}} \cdot \boldsymbol{\Pi}_{\boldsymbol{A}} + \int_0^\infty d\omega \dot{\boldsymbol{X}}_\omega \cdot \boldsymbol{\Pi}_{\boldsymbol{X}_\omega} \right] - L$$

$$= \frac{(\boldsymbol{p} - \boldsymbol{\mathcal{A}})^2}{2M} + H_0, \tag{26.51}$$

where the Lagrangian is given by (26.45), and H_0 equals the stationary result (25.69). In quantum mechanics the Hamiltonian takes the same form, which — as expected — is very similar to the one dimensional result (26.22),

$$\hat{H} = \frac{\left(\hat{\boldsymbol{p}} - \hat{\boldsymbol{A}}\right)^2}{2M} + \hat{H}_0, \qquad (26.52)$$

where the operator $\hat{\boldsymbol{A}}$ only formally differs from (26.50) because we must symmetrise the ordering of the operators

$$\hat{\boldsymbol{A}} = \int d^3x \int_0^\infty d\omega \left\{ \frac{1}{2} \left[(\boldsymbol{\nabla} \otimes \hat{\boldsymbol{X}}_\omega) \cdot \hat{\boldsymbol{\Pi}}_{X_\omega} + \hat{\boldsymbol{\Pi}}_{X_\omega} \cdot (\hat{\boldsymbol{X}}_\omega \otimes \overleftarrow{\boldsymbol{\nabla}}) \right] \right.$$
$$\left. - \alpha(\omega, \boldsymbol{x} - \hat{\boldsymbol{R}})\hat{\boldsymbol{X}}_\omega \times \hat{\boldsymbol{B}} \right\}. \quad (26.53)$$

The Hamiltonian (26.52) describes the quantum mechanical motion of a polarisable body in the electromagnetic field. The quantity $\hat{H}_0$ is equal to the Hamiltonian for the field and medium for a body at rest (25.69), at a position determined by $\hat{\boldsymbol{R}}$, and the results for a stationary body can be reclaimed if we take the limit $M \to \infty$.

Quantum forces

Our main concern is vacuum forces, so the first thing is to determine the average acceleration of the centre of mass in response to a quantum state of the electromagnetic field. Both the time derivative of the centre of mass operator, $\hat{\boldsymbol{R}}$, and its acceleration are given by expressions that are formally identical to the classical ones, with the velocity given by

$$\frac{d\hat{\boldsymbol{R}}}{dt} = \frac{i}{\hbar} \left[\hat{H}, \hat{\boldsymbol{R}} \right] = \frac{(\hat{\boldsymbol{p}} - \hat{\boldsymbol{A}})}{M} \equiv \hat{\boldsymbol{V}}, \qquad (26.54)$$

where we have applied $[\hat{\boldsymbol{R}}, \hat{\boldsymbol{p}}] = i\hbar \mathbb{1}$. The acceleration is then equal to

$$M\frac{d^2\hat{\boldsymbol{R}}}{dt^2} = \frac{i}{\hbar} \left[\hat{H}, \hat{\boldsymbol{p}} - \hat{\boldsymbol{A}} \right]$$
$$= \int d^3x \left[\hat{\rho}\hat{\boldsymbol{E}} + \hat{\boldsymbol{j}} \times \hat{\boldsymbol{B}} \right], \qquad (26.55)$$

where the charge and current density operators are

$$\hat{\rho} = -\boldsymbol{\nabla} \cdot \hat{\boldsymbol{P}}$$
$$\hat{\boldsymbol{j}} = \int_0^\infty \alpha(\omega, \boldsymbol{x} - \boldsymbol{R})\hat{\boldsymbol{\Pi}}_{X_\omega} + \frac{1}{2}\left(\hat{\boldsymbol{V}}\hat{\rho} + \hat{\rho}\hat{\boldsymbol{V}} \right).$$

Exercise 26.10
Fill in the steps between the two lines of (26.55).
Hint: Don't forget that in this vector case $\hat{\boldsymbol{p}} - \hat{\boldsymbol{A}}$ does not commute with $(\hat{\boldsymbol{p}} - \hat{\boldsymbol{A}})^2$.

The right hand side of (26.55) is formally the same as (26.47), and as the field operators obey the classical Maxwell equations, we can also re–express this force in terms of the fields alone, finding the operator analogue of (26.48)

$$M\frac{d^2\hat{\boldsymbol{R}}}{dt^2} = \int d^3x \left[\boldsymbol{\nabla} \cdot \hat{\mathrm{T}} - \frac{1}{c^2}\frac{\partial \hat{\boldsymbol{S}}}{\partial t} \right],$$

where we have defined the *stress tensor* operator

$$\hat{\mathrm{T}} = \varepsilon_0 \left[\hat{\boldsymbol{E}} \otimes \hat{\boldsymbol{E}} + c^2 \hat{\boldsymbol{B}} \otimes \hat{\boldsymbol{B}} - \frac{1}{2}\mathbb{1}\left(\hat{\boldsymbol{E}}^2 + c^2\hat{\boldsymbol{B}}^2 \right) \right] \qquad (26.56)$$

and the *Poynting vector* operator

$$\hat{\boldsymbol{S}} = \frac{1}{2\mu_0} \left[\hat{\boldsymbol{E}} \times \hat{\boldsymbol{B}} - \hat{\boldsymbol{B}} \times \hat{\boldsymbol{E}} \right].$$

We have thus found that the centre of mass of a body obeys the operator equivalent of the classical theory of radiation pressure. For a stationary localised object at position $\boldsymbol{R}_0$ with the field and medium in the ground state — as we considered in Section 26.4 — the average value of the force on a body is given by

$$M\frac{d^2\langle\hat{\boldsymbol{R}}\rangle}{dt^2} = \int_{\partial V} d\boldsymbol{s} \cdot \langle 0_{\boldsymbol{R}_0}|\hat{\mathrm{T}}|0_{\boldsymbol{R}_0}\rangle, \qquad (26.57)$$

where ∂V stands for the surface of the body, and

$$\langle 0_{\boldsymbol{R}_0}|\hat{\mathrm{T}}|0_{\boldsymbol{R}_0}\rangle = \frac{\hbar}{\pi} \lim_{\boldsymbol{x}_1 \to \boldsymbol{x}_2} \int_0^\infty d\omega \left\{ \boldsymbol{\theta}(\boldsymbol{x}_1, \boldsymbol{x}_2, \omega) - \frac{1}{2}\mathbb{1}\mathrm{Tr}\left[\boldsymbol{\theta}(\boldsymbol{x}_1, \boldsymbol{x}_2, \omega)\right] \right\} \qquad (26.58)$$

with

$$\boldsymbol{\theta}(\boldsymbol{x}_1, \boldsymbol{x}_2, \omega) = \frac{\omega^2}{c^2}\mathrm{Im}\left[\mathrm{G}^{(S)}(\boldsymbol{x}_1, \boldsymbol{x}_2, \omega)\right] + \boldsymbol{\nabla}_1 \times \mathrm{Im}\left[\mathrm{G}^{(S)}(\boldsymbol{x}_1, \boldsymbol{x}_2, \omega)\right] \times \overleftarrow{\boldsymbol{\nabla}}_2. \qquad (26.59)$$

The superscript '(S)' implies the scattered part of the Green function, as discussed in Section 26.4 and Chapter 3. To obtain this expression for the force we used the field operators given by (25.74), as well as the integral identity for Green functions given by (25.82). Equation (26.57) establishes that the ground state electromagnetic field imparts an acceleration to a body that, on average, is equal to the expression used in Lifshitz theory (see Chapter 3), in the case of a body surrounded by vacuum. We have recovered Lifshitz theory as a natural consequence of self–consistently applying macro–QED to moving media. But this only turns out to be a special case, and there is no restriction on what the state of the system is, besides the assumption that macroscopic electromagnetism is valid. Although this formalism reproduces the fluctuation–dissipation theorem, it does not assume it.

26.7 Quantum friction between sliding plates

To conclude the chapter, we'll apply the above theory to the phenomenon of quantum friction, which was already partially discussed in Section 26.5. In our 1D treatment the friction phenomenon was inferred from the behaviour of a polarisable particle moving at a constant velocity through a medium. We now aim to show directly that when two separated bodies slide past one another, there is a force that serves to bring them to relative rest even for perfectly smooth bodies at zero temperature. As we shall see, this is a situation when standard Lifshitz theory (and its derivatives) cannot be applied. Indeed, there has been some controversy over the existence of this force that came from comparing results derived within the formalism of Lifshitz theory to results derived within a different framework [32, 37]. Using macro–QED we shall show that the usual fluctuation–dissipation theorem that underlies Lifshitz theory is not applicable to bodies in relative motion, and we find an extra term in the correlation function which turns out to be the source of the controversy. Using macro–QED we shall show that it is then possible to reclaim the expression for the frictional force given by J. B. Pendry [32].

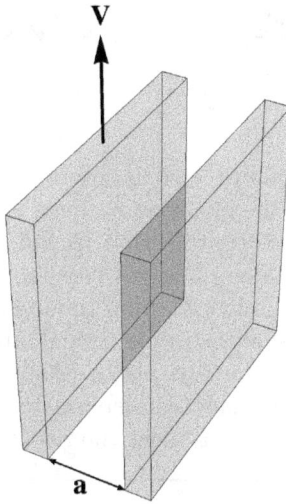

Figure 4.6: Two planar materials are separated by a distance a and slide relative to one another with velocity $\boldsymbol{V}$. The ground state field in the gap between the two bodies can be thought of as being generated by currents within each of them. The interaction between these currents serves to bring the relative motion to zero, a phenomenon known as quantum friction.

Consider two semi–infinite planar media separated by a distance a, with one of the bodies (say the one on the left) moving at a velocity $\boldsymbol{V}$ lying in the y–z plane (see Fig. 4.6). We assume that the bodies are massive enough that the velocity operator can be replaced with the vector $\boldsymbol{V}$. We have already established in Eq. (25.76) that the current operator associated with excitations of electrical current in the stationary body is given by

$$\hat{\boldsymbol{j}}_R(\boldsymbol{x},t) = \frac{\partial \hat{\boldsymbol{P}}_R(\boldsymbol{x},t)}{\partial t}, \tag{26.60}$$

where

$$\hat{\boldsymbol{P}}_{L,R}(\boldsymbol{x},t) = \int_0^\infty d\omega \int \frac{d^2\boldsymbol{k}_\parallel}{(2\pi)^2} \sqrt{\frac{\hbar}{2\omega}} \alpha_{L,R}(\omega,x) \hat{\boldsymbol{C}}_\omega(x,\boldsymbol{k}_\parallel) e^{i(\boldsymbol{k}_\parallel \cdot \boldsymbol{x} - \omega t)} + \text{h.c.} \tag{26.61}$$

and

$$\alpha_{L,R}(\omega,x) = \alpha(\omega) \begin{cases} \Theta(-x) & L \\ \Theta(x-a) & R. \end{cases}$$

Meanwhile the operator for current in the moving plate takes the form of a Lorentz transformation of the rest frame value

$$\begin{aligned} \hat{\boldsymbol{j}}_L(\boldsymbol{x},t) &= \frac{\partial \hat{\boldsymbol{P}}_L(\boldsymbol{x}-\boldsymbol{V}t,t)}{\partial t} - \nabla \times \left(\boldsymbol{V} \times \hat{\boldsymbol{P}}_L(\boldsymbol{x}-\boldsymbol{V}t,t)\right) \\ &\sim \frac{\partial \hat{\boldsymbol{P}}_L(\boldsymbol{x}-\boldsymbol{V}t,t)}{\partial t}, \end{aligned} \tag{26.62}$$

an expression which is valid only to leading order and neglects all relativistic effects, barring the Doppler shift. In the final line we have neglected the curl of a quantity, which is an entirely transverse contribution to the current density. This is valid because in the end we shall take a low velocity limit where the plates are closely spaced. In such a limit only the longitudinal (nonretarded) components of the field are relevant. A more careful analysis would have distinguished between the creation and annihilation operators in the two plates (26.60) and (26.62), since from the perspective of the Hamiltonian they correspond to different reservoirs. However, making such a distinction would not affect the result of the present calculation.

Exercise 26.11
Show that — for low velocities — the current transforms as (26.62).

The total current in the system is given by the sum of these two contributions, $\hat{\boldsymbol{j}} = \hat{\boldsymbol{j}}_L + \hat{\boldsymbol{j}}_R$, and this is the source of the electric field

$$\hat{\boldsymbol{E}}(\boldsymbol{x},t) = -\mu_0 \int d^3\boldsymbol{x}' \int_{-\infty}^t dt' \mathrm{G}(\boldsymbol{x},\boldsymbol{x}',t-t') \cdot \frac{\partial \hat{\boldsymbol{j}}(\boldsymbol{x}',t')}{\partial t'}, \tag{26.63}$$

where the Green function in the above formula is for the whole system and is written in the time rather than frequency domain. From this expression for the electric field operator (26.63), we can calculate the stress tensor* (26.56). The expression for the force (26.57) shows that only the off diagonal components of the stress tensor are relevant for the lateral (frictional) force, which is of interest here. To determine these off diagonal components we evaluate the electric field correlation function,

$$\langle 0|\hat{\boldsymbol{E}}(\boldsymbol{x},t) \otimes \hat{\boldsymbol{E}}(\boldsymbol{x}',t)|0\rangle = \mu_0^2 \int d^3\boldsymbol{x}_1 \int_{-\infty}^{t} dt_1 \int d^3\boldsymbol{x}_2 \int_{-\infty}^{t} dt_2$$

$$\times \mathrm{G}(\boldsymbol{x},\boldsymbol{x}_1,t-t_1) \cdot \langle 0|\frac{\partial \hat{\boldsymbol{j}}(\boldsymbol{x}_1,t_1)}{\partial t_1} \otimes \frac{\partial \hat{\boldsymbol{j}}(\boldsymbol{x}_2,t_2)}{\partial t_2}|0\rangle \cdot \mathrm{G}^T(\boldsymbol{x}',\boldsymbol{x}_2,t-t_2). \quad (26.64)$$

The correlation in the electric field is thus determined by the ground state correlation in the electrical current within the two plates, which is modified by the relative motion. Using their representation in terms of the creation and annihilation operators (26.60–26.62) we find the vacuum correlation between the electrical currents is given by

$$\langle 0|\frac{\partial \hat{\boldsymbol{j}}(\boldsymbol{x}_1,t_1)}{\partial t_1} \otimes \frac{\partial \hat{\boldsymbol{j}}(\boldsymbol{x}_2,t_2)}{\partial t_2}|0\rangle = 1\delta(\boldsymbol{x}_1-\boldsymbol{x}_2) \int_0^{\infty} d\omega \int \frac{d^2\boldsymbol{k}_{\|}}{(2\pi)^2} \frac{\hbar\varepsilon_0}{\pi} \mathrm{Im}[\varepsilon(\omega)] e^{i\boldsymbol{k}_{\|}\cdot(\boldsymbol{x}_1-\boldsymbol{x}_2)}$$

$$\times \left\{ \Theta(x_1-a)\omega^4 e^{-i\omega(t_1-t_2)} + \Theta(-x_1)\omega_+^4 e^{-i\omega_+(t_1-t_2)} \right\} \quad (26.65)$$

where $\omega_{\pm} = \omega \pm \boldsymbol{V} \cdot \boldsymbol{k}_{\|}$, and $\boldsymbol{k}_{\|}$ is a wave vector lying in the y–z plane. Inserting the electric current correlation function (26.65) into (26.64) gives us the electric field correlation function written without reference to the operators

$$\langle 0|\hat{\boldsymbol{E}}(\boldsymbol{x},t) \otimes \hat{\boldsymbol{E}}(\boldsymbol{x}',t)|0\rangle = \frac{\hbar\mu_0}{\pi c^2} \int_0^{\infty} d\omega \int \frac{d^2\boldsymbol{k}_{\|}}{(2\pi)^2} \mathrm{Im}[\varepsilon(\omega)] e^{i\boldsymbol{k}_{\|}\cdot(\boldsymbol{x}-\boldsymbol{x}')}$$

$$\times \left\{ \omega^4 \int_a^{\infty} dx_1 \mathrm{G}(x,x_1,\boldsymbol{k}_{\|},\omega) \cdot \mathrm{G}^{\dagger}(x',x_1,\boldsymbol{k}_{\|},\omega) \right.$$

$$\left. + \omega_+^4 \int_{-\infty}^{0} dx_1 \mathrm{G}(x,x_1,\boldsymbol{k}_{\|},\omega_+) \cdot \mathrm{G}^{\dagger}(x',x_1,\boldsymbol{k}_{\|},\omega_+) \right\} \quad (26.66)$$

where $\mathrm{G}^{\dagger} = (\mathrm{G}^T)^{\star}$.

Integral identities for the Green function: To this order we are neglecting all effects of the motion except the Doppler shift within the dispersion of the medium. Therefore the differential equation satisfied by the Green function is

$$\boldsymbol{\nabla} \times \boldsymbol{\nabla} \times \mathrm{G}(x,x',\boldsymbol{k}_{\|},\omega) - \frac{\omega^2}{c^2}[\varepsilon(\omega_-)\Theta(-x) + \varepsilon(\omega)\Theta(x-a)]\mathrm{G}(x,x',\boldsymbol{k}_{\|},\omega) = 1\delta(x-x'), \quad (26.67)$$

where $\boldsymbol{\nabla} = \boldsymbol{e}_x\partial_x + i\boldsymbol{k}_{\|}$, and $\omega_{\pm} = \omega \pm \boldsymbol{V} \cdot \boldsymbol{k}_{\|}$. Meanwhile the Hermitian conjugate of the Green function obeys

$$\boldsymbol{\nabla} \times \boldsymbol{\nabla} \times \mathrm{G}^{\dagger}(x',x,\boldsymbol{k}_{\|},\omega) - \frac{\omega^2}{c^2}[\varepsilon^{\star}(\omega_-)\Theta(-x) + \varepsilon^{\star}(\omega)\Theta(x-a)]\mathrm{G}^{\dagger}(x',x,\boldsymbol{k}_{\|},\omega) = 1\delta(x-x') \quad (26.68)$$

---

*The magnetic field operator can be inferred from the Maxwell equation $\boldsymbol{\nabla} \times \hat{\boldsymbol{E}} = -\frac{\partial \hat{\boldsymbol{B}}}{\partial t}$.

Multiplying (26.68) on the left by the Green function, $G(x, x'', \boldsymbol{k}_{\|}, \omega)$, integrating over x, and then subtracting the Hermitian conjugate of the resulting expression with x' and x'' reversed, we obtain a generalisation of (25.82),

$$\frac{G(x', x'', \boldsymbol{k}_{\|}, \omega) - G^{\dagger}(x'', x', \boldsymbol{k}_{\|}, \omega)}{2i} = \frac{\omega^2}{c^2}\left[\text{Im}[\varepsilon(\omega_-)] \int_{-\infty}^{0} dx G(x', x, \boldsymbol{k}_{\|}, \omega) \cdot G^{\dagger}(x'', x, \boldsymbol{k}_{\|}, \omega)\right.$$
$$\left. + \text{Im}[\varepsilon(\omega)] \int_{a}^{\infty} dx G(x', x, \boldsymbol{k}_{\|}, \omega) \cdot G^{\dagger}(x'', x, \boldsymbol{k}_{\|}, \omega)\right]. \quad (26.69)$$

For integration over only the region $x < 0$ we have instead

$$\frac{\omega^2}{c^2}\text{Im}[\varepsilon(\omega_-)] \int_{-\infty}^{0} dx G(x', x, \boldsymbol{k}_{\|}, \omega) \cdot G^{\dagger}(x'', x, \boldsymbol{k}_{\|}, \omega) = \frac{G(x', x'', \boldsymbol{k}_{\|}, \omega) - G^{\dagger}(x'', x', \boldsymbol{k}_{\|}, \omega)}{2i}$$
$$- \frac{1}{2i}\left[G(x', 0, \boldsymbol{k}_{\|}, \omega) \cdot \boldsymbol{e}_x \times \boldsymbol{\nabla} \times G^{\dagger}(x'', 0, \boldsymbol{k}_{\|}, \omega) - \text{h.c.}\right]. \quad (26.70)$$

Applying (26.69) to (26.66), gives us the electric field correlation function, and after taking the limit, $\boldsymbol{x} \to \boldsymbol{x}'$, we obtain the electric contribution to the stress tensor

$$\langle 0|\hat{\boldsymbol{E}}(\boldsymbol{x}, t) \otimes \hat{\boldsymbol{E}}(\boldsymbol{x}, t)|0\rangle = \frac{\hbar\mu_0}{\pi} \int \frac{d^2\boldsymbol{k}_{\|}}{(2\pi)^2} \int_{0}^{\infty} d\omega\, \omega^2 \text{Im}[\bar{G}^{(S)}(x, x, \boldsymbol{k}_{\|}, \omega)]$$
$$- \int_{-\infty}^{0} dx_1 \int \frac{d^2\boldsymbol{k}_{\|}}{(2\pi)^2} \int_{0}^{\boldsymbol{V}\cdot\boldsymbol{k}_{\|}} d\omega \bar{Q}(x, x_1, \boldsymbol{k}_{\|}, \omega), \quad (26.71)$$

where we have introduced symmetrised quantities, e.g. $\bar{G} = \frac{1}{2}[G + G^T]$, which makes explicit the independence of the right hand side from the order of the electric field operators. The quantity Q within the correlation function (26.71) is given by

$$Q(x, x_1, \boldsymbol{k}_{\|}, \omega) = \frac{\hbar\mu_0}{\pi} \frac{\omega^4}{c^2} \text{Im}[\varepsilon(\omega_-)] G(x, x_1, \boldsymbol{k}_{\|}, \omega) \cdot G^{\dagger}(x, x_1, \boldsymbol{k}_{\|}, \omega). \quad (26.72)$$

The result (26.71) is that which enters Lifshitz theory (26.59), plus an additional velocity dependent term equal to an integral over low frequency excitations within the moving plate. The new term Q comes from an imbalance between the frequency spectrum of the electrical current in the stationary plate and the one in motion, which in turn is due to the Doppler effect. This is the origin of the friction between the plates. The imbalance of frequencies is similar to the effect discussed in Section 26.5, where the Doppler effect causes a sign change of some of the frequencies between reference frames, which can slow a moving particle.

Integrating the term on the second line of (26.71) over the moving plate and applying the result (26.70) results in two terms,

$$\int_{-\infty}^{0} dx_1 Q(x, x_1, \boldsymbol{k}_{\|}, \omega) = \frac{\hbar\mu_0\omega^2}{\pi} \frac{G(x, x, \boldsymbol{k}_{\|}, \omega) - G^{\dagger}(x, x, \boldsymbol{k}_{\|}, \omega)}{2i}$$
$$- \frac{\hbar\mu_0\omega^2}{2\pi i}\left\{G(x, 0, \boldsymbol{k}_{\|}, \omega) \cdot \boldsymbol{e}_x \times \left(\boldsymbol{\nabla} \times G^{\dagger}(x, 0, \boldsymbol{k}_{\|}, \omega)\right) - \text{h.c.}\right\} \quad (26.73)$$

where the curl in the second line is taken with respect to the second spatial coordinate. The term on the second line came from an integration by parts, and can be thought of as a flux of momentum leaving the moving plate (it takes the form of an electric field crossed with a magnetic field, like the Poynting vector). This is the term responsible for the frictional force between the plates, which we now examine in detail, using the explicit form for the Green function.

The Green function between relatively moving plates: In empty space the Green function satisfies the equation given in Chapter 3, Section 21.3 with $\varepsilon = 1$. When written in terms of x, x', $\boldsymbol{k}_\|$ and ω this equals

$$G_0(x, x', \boldsymbol{k}_\|, \omega) = \left[1 - \frac{\boldsymbol{k}_\pm \otimes \boldsymbol{k}_\pm}{k_0^2}\right] \frac{i e^{i\sqrt{k_0^2 - k_\|^2}|x - x'|}}{2\sqrt{k_0^2 - k_\|^2}}.$$

where $k_0 = \omega/c$ and

$$\boldsymbol{k}_\pm = \text{sign}(x - x')\boldsymbol{e}_x \sqrt{k_0^2 - k_\|^2} + \boldsymbol{k}_\|,$$

which is simply the 1D Green function (25.38) with a bit of dressing to take account of polarisation and angle of incidence. For realistic sliding velocities (on the order of metres per second), Q is evaluated in a regime where $k_\| \gg k_0$ and the field exponentially dies away from the source*. In this regime only the first few reflections (we consider two) of the field from the two plates contribute significantly to the Green function. Moreover when the permeability $\mu = 1$ the Fresnel reflection coefficient for s–polarised radiation is zero in this limit[†]. Therefore the Green function can be approximated by

$$G(x, x', \boldsymbol{k}_\|, \omega) \sim G_0(x, x', \boldsymbol{k}_\|, \omega) + \frac{1}{2\kappa}\left[\boldsymbol{u} \otimes \boldsymbol{u}^* r_p(\omega_-)e^{-\kappa(x+x')} + \boldsymbol{u}^* \otimes \boldsymbol{u}\, r_p(\omega)e^{-\kappa(2a-x-x')}\right.$$

$$\left. + \boldsymbol{u}^* \otimes \boldsymbol{u}^* r_p(\omega)r_p(\omega_-)e^{-\kappa(2a+x'-x)} + \boldsymbol{u} \otimes \boldsymbol{u}\, r_p(\omega)r_p(\omega_-)e^{-\kappa(2a+x-x')}\right],$$

$$(26.74)$$

where $\kappa = \sqrt{k_\|^2 - k_0^2}$, $r_p(\omega)$ is the p–polarised reflection coefficient for a body at rest, and the p–polarised and s–polarised unit vectors are respectively given by

$$\boldsymbol{u} = \frac{1}{k_0 k_\|}[k_\|^2 \boldsymbol{e}_x - i\kappa \boldsymbol{k}_\|]$$

$$\boldsymbol{v} = \frac{1}{k_\|}\boldsymbol{e}_x \times \boldsymbol{k}_\|.$$
$$(26.75)$$

The curl of the Hermitian conjugate of the Green function (26.74) with respect to the second index is

$$\boldsymbol{\nabla}' \times G^\dagger(x, x', \boldsymbol{k}_\|, \omega) \sim \boldsymbol{\nabla}' \times G_0(x, x', \boldsymbol{k}_\|, \omega) - \frac{ik_0}{2\kappa}\boldsymbol{v} \otimes \left[\boldsymbol{u}^* r_p^\star(\omega_-)e^{-\kappa(x+x')} + \boldsymbol{u} r_p^\star(\omega)e^{-\kappa(2a-x-x')}\right.$$

$$\left. + \boldsymbol{u} r_p^\star(\omega)r_p^\star(\omega_-)e^{-\kappa(2a+x'-x)} + \boldsymbol{u}^* r_p^\star(\omega)r_p^\star(\omega_-)e^{-\kappa(2a+x-x')}\right]. \quad (26.76)$$

---

*This is known as the *electrostatic*, or *nonretarded* limit by those working on the physics of the electromagnetic field close to surfaces. See, e.g. [35].
    [†]See e.g. [1]

Inserting expressions (26.74) and (26.76) for the Green function into the surface term in (26.73) we find

$$\frac{1}{2i}\boldsymbol{e}_x \cdot \left[G(x,0,\boldsymbol{k}_{\|},\omega) \cdot \boldsymbol{e}_x \times (\boldsymbol{\nabla} \times G^{\dagger}(x,0,\boldsymbol{k}_{\|},\omega)) - \text{h.c.}\right] \cdot \boldsymbol{e}_y$$

$$= \frac{k_y e^{-2\kappa a}}{k_0^2}\text{Im}[r_p(\omega)]\text{Im}[r_p(\omega_-)] \quad (26.77)$$

where terms of order $\exp(-4\kappa a)$ and smaller have been neglected and only the symmetric part of the tensor has been retained. The x–y component of the electric field correlation function (26.71) is

$$\langle 0|\hat{E}_x(\boldsymbol{x},t)\hat{E}_y(\boldsymbol{x},t)|0\rangle = \frac{\hbar\mu_0}{\pi}\int\frac{d^2\boldsymbol{k}_{\|}}{(2\pi)^2}\int_{Vk_y}^{\infty}d\omega\omega^2\text{Im}[\bar{G}_{xy}^{(S)}(x,x,\boldsymbol{k}_{\|},\omega)]+$$

$$\frac{\hbar\mu_0}{2\pi i}\int\frac{d^2\boldsymbol{k}_{\|}}{(2\pi)^2}\int_0^{Vk_y}d\omega\omega^2\boldsymbol{e}_x\cdot\left[G(x,0,\boldsymbol{k}_{\|},\omega)\cdot\boldsymbol{e}_x\times(\boldsymbol{\nabla}\times G^{\dagger}(x,0,\boldsymbol{k}_{\|},\omega)) - \text{h.c.}\right]\cdot\boldsymbol{e}_y,$$

$$(26.78)$$

where only the symmetric part of the tensor on the second line is included. In this 'electrostatic' regime the magnetic correlation function does not contribute to the stress*. Furthermore it was shown in [37] that the first term on the right of (26.78) is zero, although we shall not prove this here. Therefore the frictional force is ε_0 times the second term in (26.78) which after we apply (26.77) is

$$\langle 0|\hat{T}_{xy}|0\rangle = \frac{\hbar}{\pi}\int\frac{d^2\boldsymbol{k}_{\|}}{(2\pi)^2}\int_0^{Vk_y}d\omega k_y\text{Im}[r_p(\omega)]\text{Im}[r_p(\omega_-)]e^{-2k_{\|}a}, \quad (26.79)$$

where we have taken the limit $k_{\|} \gg k_0$. The above off diagonal element of the stress tensor is the limiting expression found by Volokitin and Persson [33] and Pendry [36]. As we already mentioned, this limit is appropriate for low sliding velocities (relative to the speed of light), and we can see that the frictional force is determined by the imaginary parts of the reflection coefficients at low frequencies and large wave vectors. The integral over ω is such that $\omega_- \leq 0$ and therefore $\text{Im}[r_p(\omega_-)] \leq 0$ and the stress is negative. This means that the force on the left hand body acts against its motion, and the force on the right hand body acts in the opposite direction; i.e. the relative motion of the bodies is being reduced. Roughly speaking, for materials with a large degree of dissipation this frictional force is large. The vacuum field close to a dissipative body can thus be thought of as being like a viscous fluid that inhibits motion parallel to the surface.

---

*See [32].

Exercise 26.12

Show that the reflection coefficient for p–polarised radiation

$$r_p = \frac{\varepsilon(\omega)\sqrt{k_0^2 - k_\parallel^2} - \sqrt{\varepsilon(\omega)k_0^2 - k_\parallel^2}}{\varepsilon(\omega)\sqrt{k_0^2 + k_\parallel^2} + \sqrt{\varepsilon(\omega)k_0^2 - k_\parallel^2}}$$

approaches $(\varepsilon(\omega) - 1)/(\varepsilon(\omega) + 1)$ as $k_\parallel/k_0 \to \infty$. Using this result show that in the same limit

$$\mathrm{Im}[r_p(\omega)] \to \frac{2\mathrm{Im}[\varepsilon(\omega)]}{|\varepsilon(\omega) + 1|^2}.$$

For the simple case of a constant conductivity σ the permittivity is $\varepsilon(\omega) = \varepsilon_b + i\sigma/\omega\varepsilon_0$ (ε_b assumed constant). For this case show that the stress tensor (26.79) is given by

$$\langle 0|\hat{T}_{xy}|0\rangle = \frac{4\sigma^2\hbar}{\pi\varepsilon_0^2} \int \frac{d^2\mathbf{k}_\parallel}{(2\pi)^2} \int_0^{Vk_y} d\omega k_y \frac{\omega\omega_-}{|i\sigma/\varepsilon_0 + (1 + \varepsilon_b)\omega|^2 |i\sigma/\varepsilon_0 + (1 + \varepsilon_b)\omega_-|^2} e^{-2k_\parallel a}.$$

Using your favourite software or programming language numerically evaluate this integral as a function of V and a and plot the results.

We complete this chapter having developed the theory of quantum friction alongside ordinary Lifshitz theory, and all within the formalism of macroscopic QED. This formalism is a very general way of treating the quantum mechanics of macroscopic bodies and the electromagnetic field and may be applied to any problem in Casimir physics. Not only does this justify the formulae of Lifshitz theory on the basis of a complete quantum mechanical theory, but it opens up the possibility of exploring new effects that might arise, for example, when the centre of mass of a macroscopic body is prepared in a quantum mechanical state.

§27. PROBLEMS

Problem 4.1
Taking the Hamiltonian (26.52) show that in the limit of a very massive body in uniform motion it becomes

$$H = -\mathbf{V} \cdot \int d^3x \int_0^\infty d\omega \left[(\boldsymbol{\nabla} \otimes \mathbf{X}_\omega) \cdot \mathbf{\Pi}_{\mathbf{X}_\omega} - \alpha(\omega, \mathbf{x} - \mathbf{R})\mathbf{X}_\omega \times \mathbf{B} \right] + H_0 + \text{const.} \qquad (27.1)$$

where $\mathbf{V} = \mathbf{p}/M$. Show that if the constant is neglected then this Hamiltonian can have an arbitrarily low energy (which could become negative), while the original Hamiltonian (26.52) is always positive. What is the resolution of this apparent contradiction?

Problem 4.2
Find the equations of motion for the electromagnetic field and the reservoir from the Hamiltonian (27.1), and find expressions for the electromagnetic field in terms of the $\mathbf{C}_\omega(k_x, y, z)$ and $\mathbf{C}_\omega^\star(k_x, y, z)$ of the reservoir. Assume that the medium is homogeneous and moves along the x–axis.

Problem 4.3
Extend the solutions of the above problem to the quantum case and show that a suitable Hamiltonian to describe the evolution of the operators is

$$\hat{H} = \int \frac{dk_x}{2\pi} \int_0^\infty d\omega \hbar(\omega + Vk_x)\hat{\mathbf{C}}_\omega^\dagger(k_x, y, z) \cdot \hat{\mathbf{C}}_\omega(k_x, y, z).$$

What are the eigenvalues of this Hamiltonian and how do they differ from the $V = 0$ case?

Problem 4.4
Consider a detector coupled to the electromagnetic field with a Hamiltonian

$$\hat{H}_D = \hbar\omega_0 \left[\hat{a}^\dagger \hat{a} + \frac{1}{2}\right]$$

and an interaction

$$\hat{H}_I = i\beta \left[\hat{a} - \hat{a}^\dagger\right] e \cdot \hat{E}(x_0, t),$$

where β is a constant determining the strength of the interaction, and e is a unit vector determining its orientation. Suppose that the detector is initially in the first excited state and find an expression (valid to first order in perturbation theory) for the transition rate into the ground state in terms of the electric field operator. Using the expression for the electric field operator given in the text evaluate this in terms of the Green function.

Problem 4.5
Find an expression for the electromagnetic Green function in a region of space outside of a dielectric half space ($x > 0$) characterised by reflection coefficients $r_\lambda(\omega, k)$ for the two polarisations $\lambda = 1, 2$. Use the result of problem (4.4) to find the transition rate of the detector as a function of the distance from the surface.

Hint: Start from the free space Green function, decompose it into a sum of freely propagating waves, and add in the reflected waves necessary to fulfil the boundary conditions.

§28. Bibliography

[1] L. D. Landau, E. M. Lifshitz, and L. P. Pitaevskii, *Electrodynamics of Continuous Media*, Vol. VIII (Butterworth Heinemann, Oxford, 2004).

[2] J. J. Hopfield, "Theory of the contribution of Excitons to the Complex Dielectric Constant of Crystals" *Phys. Rev.*, **112**, 1555 (1958).

[3] B. Huttner and S. M. Barnett, "Quantization of the electromagnetic field in dielectrics" *Phys. Rev. A*, **46**, 4306 (1992).

[4] L. G. Suttorp and A. J. van Wonderen, "Fano diagonalization of a polariton model for an inhomogeneous absorptive dielectric" *Euro. Phys. Lett.*, **67**, 766 (2004).

[5] F. Kheirandish and M. Soltani, "Extension of the Huttner–Barnett model to a magnetodielectric medium" *Phys. Rev. A*, **78**, 012102 (2008).

[6] S. Scheel and S. Y. Buhmann, "Macroscopic quantum electrodynamics - Concepts and applications" *Act. Phys. Solv.*, **58**, 675 (2008).

[7] T. G. Philbin, "Canonical quantization of macroscopic electromagnetism" *New J. Phys.*, **12**, 123008 (2010).

[8] L. D. Landau and E.M. Lifshitz, *Mechanics*, Vol. I (Butterworth Heinemann, Oxford, 2004).

[9] C. Lanczos, *The Variational Principles of Mechanics* (Dover Publications, New York, 2013).

[10] L. D. Landau and E. M. Lifshitz, *Statistical Physics (Part 1)*, Vol. V (Butterworth Heinemann, Oxford, 2005).

[11] L. D. Landau and E.M. Lifshitz, *The Classical Theory of Fields*, Vol. II (Butterworth Heinemann, Oxford, 2003).

[12] L. H. Ryder, *Quantum Field Theory* (Cambridge University Press, Cambridge, 2003).

[13] H. Kleinert, *Path Integrals in Quantum Mechanics, Statistics, Polymer Physics and Financial Markets* (World Scientific, Singapore, 2006).

[14] P. A. M. Dirac, *Quantum Mechanics* (Oxford University Press, London, 1958).

[15] W. Eckhardt, "First and second fluctuation–dissipation–theorem in electromagnetic fluctuation theory" *Opt. Comm.*, **41**, 305 (1982).

[16] U. Fano, "Effects of Configuration Interaction on Intensities and Phase Shifts" *Physical Review*, **124**, 1866 (1961).

[17] S. A. R. Horsley, "Canonical quantization of the electromagnetic field interacting with a moving dielectric medium" *Phys. Rev. A*, **86**, 023830 (2012).

[18] J. W. Dettman, *Applied Complex Variables* (Dover Publications, New York, 1984).

[19] S. A. R. Horsley and T. G. Philbin, "Canonical quantization of electromagnetism in spatially dispersive media" *New J. Phys.*, **16**, 013030 (2014).

[20] S. Weinberg, *The Quantum Theory of Fields (Vol. 1)* (Cambridge University Press, Cambridge, 1995).

[21] U. Leonhardt, *Essential Quantum Optics*, (Cambridge University Press, New York, 2010).

[22] S. M. Barnett, "Resolution of the Abraham–Minkowski Dilemma" *Phys. Rev. Lett.*, **104**, 070401 (2010).

[23] L. Novotny and B. Hecht, *Principles of Nano Optics* (Cambridge University Press, Cambridge, 2006).

[24] M. D. Semon and J. R. Taylor, "Thoughts on the magnetic vector potential" *Am. J. Phys.*, **64**, 1361 (1996).

[25] M. Wilkens, "Significance of Röntgen current in quantum optics: Spontaneous emission of moving atoms" *Phys. Rev. A*, **49**, 570 (1994).

[26] R. Loudon, "Theory of the radiation pressure on dielectric surfaces" *J. Mod. Opt.*, **49**, 821(2002).

[27] C. K. Law, "Interaction between a moving mirror and radiation pressure: A Hamiltonian formulation" *Phys. Rev. A*, **51**, 2537 (1995).

[28] E. M. Lifshitz and L. P. Pitaevskii, *Statistical Physics (Part 2)*, Vol. IX (Butterworth Heinemann, Oxford, 2003).

[29] W. M. R. Simpson, S. A. R. Horsley, and U. Leonhardt, "Divergence of Casimir stress in inhomogeneous media" *Phys. Rev. A*, **87**, 043806 (2013).

[30] L. S. Levitov, "Van Der Walls' Friction" *Europhys. Lett.*, **8**, 499 (1989).

[31] V. B. Braginsky and F. Y. Khalili, "Friction and fluctuations produced by the quantum ground state" *Phys. Lett. A*, **161**, 197 (1991).

[32] J. B. Pendry, "Shearing the vacuum - quantum friction" *J. Phys. Cond. Mat.* **9**, 10301 (1997).

[33] A. I. Volokitin and B. N. J. Persson, "Theory of the interaction forces and the radiative heat transfer between moving bodies" *Phys. Rev. B*, **78**, 155437 (2008).

[34] V. B. Berestetskii, L. P. Pitaevskii, and E. M. Lifshitz, *Quantum Electrodynamics*, Vol. IV (Butterworth Heinemann, Oxford, 2004).

[35] I. D. Mayergoyz, *Plasmon Resonances in Nanoparticles*, (World Scientific, Singapore, 2013).

[36] J. B. Pendry, "Quantum friction - fact or fiction?" *New J. Phys.* **12**, 033028 (2010).

[37] T. G. Philbin and U. Leonhardt, "No quantum friction between uniformly moving plates" *New J. Phys.* **11**, 033035 (2009).

Chapter 5

Measuring Casimir phenomena

RICARDO S. DECCA

> I have no way of knowing whether the events that I am about to
> narrate are effects or causes.
>
> > – Jorge Luis Borges, *Collected Fictions*

The fact that two macroscopic bodies in vacuum interact with each other due to the modification of the vacuum is, perhaps, one of the most intriguing consequences of quantum physics. From its origins in 1948 [1], the field now known as *Casimir physics* has shown remarkable progress. As demonstrated in Section 7.1, the force per unit area for two *perfect mirrors* separated by a distance z is $P_C = -\pi^2 \hbar c / 240 z^4$. This interaction becomes dominant at small separations, the pressure between two such plates separated by $z = 10$ nm being $\sim 10^5$ Pa. The strong separation dependence makes the interaction rather small at larger z. Two 1 cm^2 plates separated by 1 μm experience a force $\sim 10^{-7}$ N.

Despite the strength of the interaction at small separations, early attempts to observe the phenomenon, made relatively soon after its prediction [2], proved difficult: separating two macroscopic bodies by submicron distances is a gargantuan task. Consequently, these early experiments, while inconclusive, left the door open for further development. Things were pretty quiet for some time until 1997 [3] when the *modern era* for the Casimir effect got started. With his measurements Lamoreaux showed that not only were Casimir's predictions qualitatively correct, it was possible to extract *quantitative* information out of the experiments. Following these first modern experiments, a significant number of groups contributed to a rapid increase in the quality and quantity of experiments, developing new approaches to measure the interaction. In this regard, many aspects of the interaction have since then been tested, such as the effects of the physical properties of the materials, finite temperature, and geometry.

In this chapter, after a general description of equipment needed to do the experiments (Section 29), we will summarise relevant *experimental* results obtained in the area of Casimir physics. This Casimir physics subdivision is understood as the interaction of macroscopic bodies separated by distances sufficiently large that the effect of the finite speed of light must be taken into account. These results are contained in Section 30. Other equally important areas, like the Casimir-Polder or the van der Waals regime, are described in Chapters 1 and 2. In order to better understand the experiments done to date and to help prepare the reader to understand the minutiae of the measurements of the Casimir interaction, further details and considerations on the measurement of the Casimir interaction, equipment calibrations and problems and shortcomings of the technique are described in subsequent sections in this chapter, and in Section 38.

§29. GENERAL EXPERIMENTAL CONSIDERATIONS

All experimental determinations of the Casimir interaction require some *common attributes*:

1. A device able to measure the *interaction* between the bodies. The device that has been used is always a mechanical transducer, where the interaction produces a change in its physical characteristics, being either the position of the transducer or its resonant frequency.

2. The device used needs to be *calibrated* against a known force. Most calibrations are done in the geometry of the experiment using the electrostatic interaction between the bodies.

3. In ensuring the quality of the measurement, care needs to be taken to reduce the *noise* in the device. Consequently most experimentalists take great care to reduce vibrations from the environment (by mounting the experiment on vibration isolation systems and working in places where vibrations are naturally small).

4. *Residual forces* between the bodies need to be minimised and properly accounted for. This means that these forces (for example, the electrostatic and gravitational attraction, and magnetic forces) need to be understood and, if possible, made smaller than the experimental error on the measurement of the Casimir interaction. The magnitude of the attractive *electrostatic force* F_e between two metallic coated surfaces, assumed to be equipotentials, is given by

$$F_e = \frac{1}{2}\frac{\partial C(z)}{\partial z}\Delta V^2. \tag{29.1}$$

In this expression $C(z)$ is the *capacitance* between the bodies when separated by a distance z, $\Delta V = V_1 - V_2 + V_o$, where V_i the external electric potential applied to body i and V_o a *residual potential difference* between the bodies which appears even when $V_1 = V_2$. When the bodies are covered with different metals V_o can be identified with the difference in the work function between them. It is worth mentioning here, however, that even when the exposed surfaces are covered with one and the same metal, details of the fabrication, geometry, and the materials used to close the circuit between the two surfaces make $V_o \neq 0$. Details on how F_e is minimised and how it is used for calibration purposes are presented in Section 34. For now, it suffices to say that if $C(z)$ is known, then a calibration can be obtained.

If the surfaces cannot be considered equipotentials then there are remnant effects of the electrostatic interaction. These effects are presented in Section 38.2. Although it does not receive the same degree of attention in the Casimir community, the electrostatic interaction with the surroundings is equally important. It is critical that all surrounding components, either mobile or stationary, are made out of good conductors which are properly grounded. The presence of, for example, aluminium with a layer of aluminium oxide at distances on the order of cm from the interaction detecting mechanism is sufficient to induce enough systematics to confound the experimental interpretation. Similarly, samples need to be kept clean and unaltered during the experimental measurements. Partially to fulfil this requirement most experiments are carried out in vacuum.

While the gravitational force between the plates cannot be nullified, it is small (usually within the experimental error), and a judicious selection of the geometry makes it almost constant with separation within the experimental z-range.

Magnetic materials are usually avoided. When they are used, care has to be exercised to ensure they do not have a permanent magnetisation, since the magnetic interaction would swamp the Casimir force.

5. The *separation* between the bodies needs to be determined. While the separation between the interacting bodies can be controlled by screws, piezoelectric actuators, or electrostatic actuation, it has to be measured at every position where the experiment is performed.

§30. Representative experimental results

In this section a few experimental results are highlighted. A brief description of the approach used is provided, but without an in-depth discussion of the specifics of the experiments and techniques used. A more complete description of some of the approaches used and their limitations are given in Section 32.

Since the determination of the Casimir interaction depends on the modes supported by the cavity, defined by the two surfaces, any factor that changes these modes will modify the interaction. Hence, effects of the materials forming the cavity (both the mirrors and the medium between them), the geometry of the cavity, and the flatness of the mirrors need to be taken into account. Furthermore, the interaction between the surfaces is given by zero-point fluctuations but also the thermal population of the different modes. While most (but not all) measurements are done at small separations, when the separation becomes larger than the thermal wavelength ($\lambda_T \sim 10~\mu$m) thermal effects also become important. At such large separations, however, the Casimir force becomes very small and such experiments require large areas of interaction between the bodies. On the other hand, even for small separations, the interpretation of precise enough experiments requires the effects of finite temperature to be taken into account.

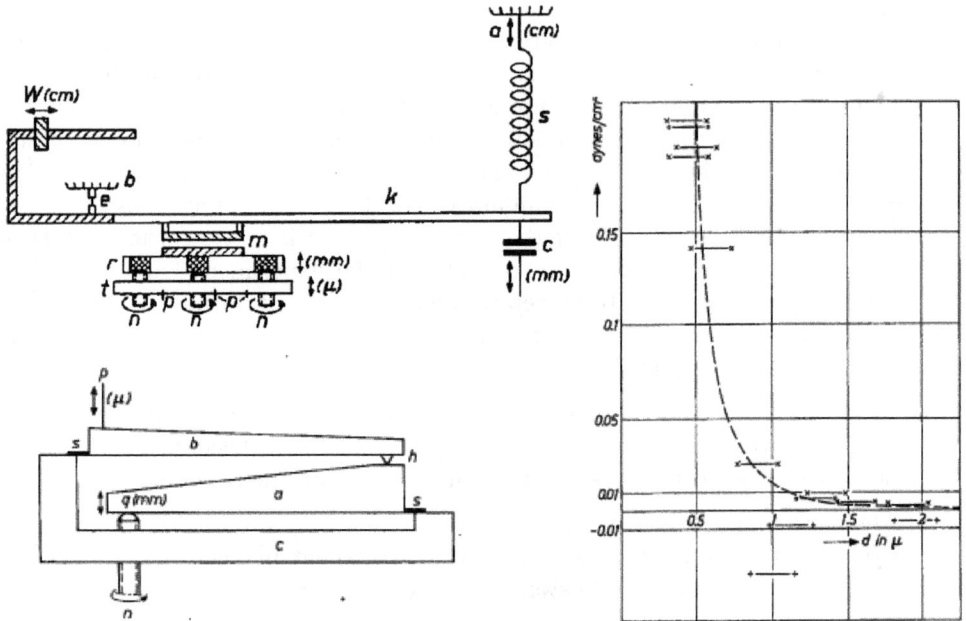

Figure 5.1: *Left panel, top*: Schematic of one of the apparatus used by Sparnaay. The top plate was balanced by adjusting the position a of the vertical spring S and the position and values of the weights W, which controlled the pivot of the lever arm around the point e. *Left panel, bottom*: The bottom plate's r vertical and angular position were adjusted by turning the three double lever system screws shown in the insert. *Right panel*: Data obtained.

30.1   Early results

Although the predicted Casimir force per unit area between two parallel plates is large for very small separations, an experimental determination of its strength was a very difficult task in the 1950s due to technological limitations from the difficulty of placing two parallel plates at short distances. Whilst techniques for positioning bodies with precision have improved enormously since then, it still remains a complex task to try to align two flat surfaces. Nevertheless, one of the first experiments was carried out in 1958 by Marcus Sparnaay at Philips in Eindhoven [2] (see

Fig. 5.1). He used a spring balance to measure the force between parallel plates made from different materials (aluminium, chromium and steel), determining the extension of the spring by measuring the capacitance between the plates. Since a relatively small residual charge in each plate would overcome the Casimir interaction, Sparnaay made the plates touch before each determination. He was also able to align the mirrors to a large degree at separations greater than 1 μm. In his own words Marcus Sparnaay concluded that his results "did not contradict Casimir's theoretical prediction". Even less conclusive were the experiments performed by Derjaguin and Abrikosova [4, 5] performed on fused silica, where systematic effects from charging obscured the interpretation of the data.

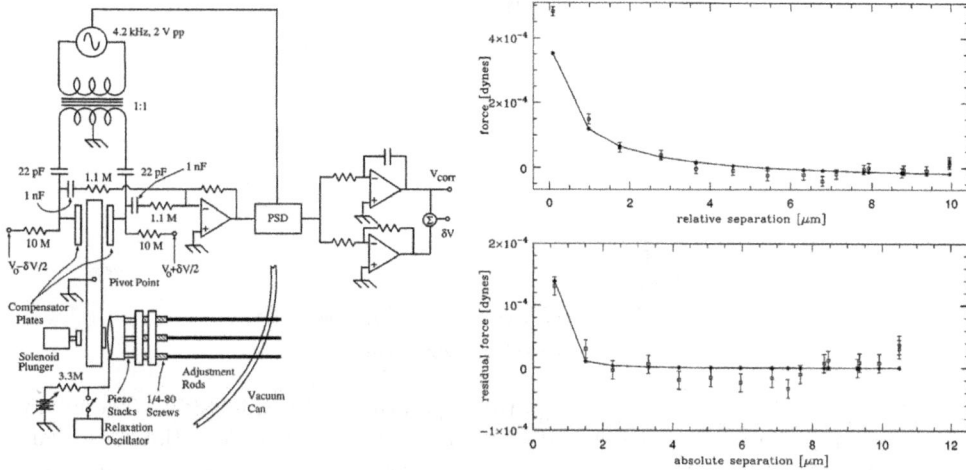

Figure 5.2: *Left panel*: Schematic of the apparatus used by Lamoreaux. The bar can oscillate around the tungsten wire defining the pivot point. The Casimir interaction occurs at the bottom, and the position of the bar is adjusted by the electrostatic force provided by the two compensating plates, together with the feedback control loop shown. *Right panel*: the top shows the data obtained (points) and the calculated Casimir force (line), once the electrostatic contribution has been subtracted. The bottom shows the data when the calculated Casimir force (without thermal corrections) is subtracted.

30.2 Beginning of the "modern Casimir era"

Torsional pendulum measurements

Measurements made by Lamoreaux, then at the University of Washington, Seattle, revolutionised the experimental landscape. Lamoreaux [3] used a *torsional pendulum* in a vacuum chamber. The pendulum was made of a tungsten wire with a perpendicular bar attached to it (see Fig. 5.2). Glued to the bar was a $R = 2$ cm

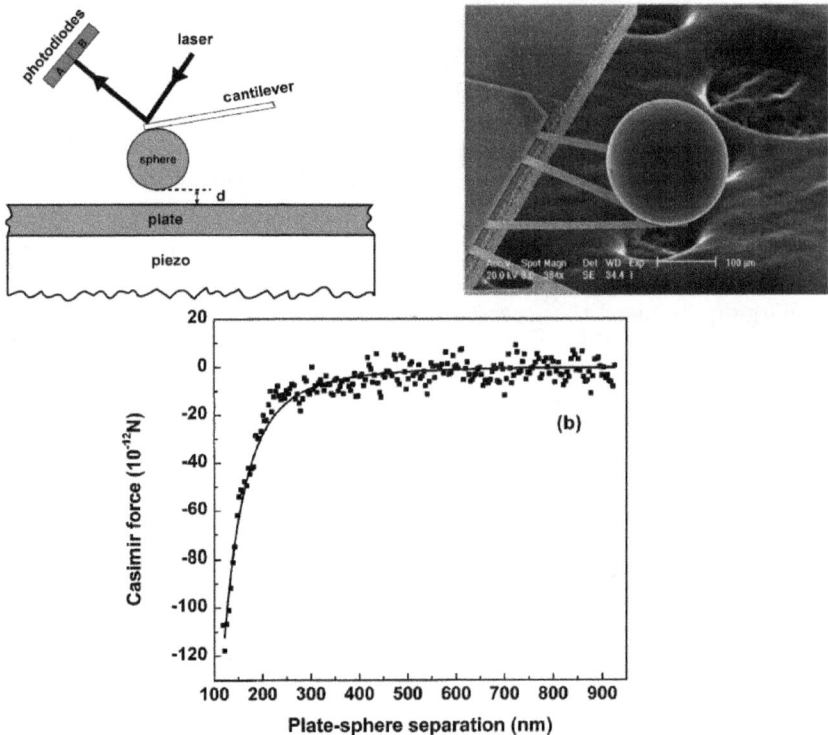

Figure 5.3: *Top left panel*: Schematic of the apparatus used by Mohideen. The interaction between the sphere and the plate deflects the cantilever. This deflection is measured by the reflection of the laser beam at the back of the cantilever. *Top right panel*: scanning electron micrography of the polysterene sphere glued to the tip-less AFM cantilever. *Bottom panel*: The points represent Casimir interaction (obtained after minimising the electrostatic interaction) between the sphere and the plate. The solid line is the calculation of the force with finite conductivity, finite temperature and roughness taken into account.

spherical silica lens covered with gold. The distance between the gold covered lens and a gold covered quartz plate was changed by means of piezoelectric actuators. The angular deviation induced in the torsional pendulum due to the Casimir torque was used as an error signal in a feedback control loop. The control loop applied a counteracting known force to the lens to balance the effect of the Casimir torque. Using a sphere in the presence of a plate allowed Lamoreaux to solve the alignment problems (there is always the closest point between them), but introduced some extra complications. The experimental complication is that, loosely speaking, the strength of the force is reduced due to the reduced *area of interaction* between the bodies, given by $A_{\text{eff}} \sim \pi R z$ instead of the physical area. On the theoretical part, using a sphere requires a modification of the expression for the interaction. In the limit when $R >> z$ the use of the *proximity force approximation* is extensive [6].

Succinctly, in the case of the interaction between a perfectly conducting sphere and a perfectly conducting plate, $F_C = -\pi^3 R\hbar c / 360 z^3$.

Exercise 30.1
Determine the number of degrees of freedom (both translational and rotational) to bring into alignment a pairwise combinations of spheres, cylinders and planes (i.e. a sphere with a sphere a sphere, a sphere with a cylinder, ...). Rank in increasing order the effective area of interaction for each configuration.

Modified atomic force microscope measurements

Soon after Lamoreaux's experiments Mohideen and co-workers at the University of California at Riverside [7] used a modified *atomic force microscope* (AFM) to measure the Casimir force (see Fig. 5.3). Mohideen attached a polystyrene sphere 200 μm in diameter to a tip-less AFM cantilever. The sphere tip was then sputter-coated with a metal. In a series of experiments performed in vacuum, they used piezoelectric actuators to bring the sphere to within 100 nm of a flat disk, which was also coated with the same metal. As in AFM microscopy, the resulting attraction between the sphere and the disk was monitored by the deviation of a laser beam on the back of the cantilever. In order to obtain a calibration for the cantilever's spring constant (29.1) was used. The point of contact was defined by touching the sphere against the disk.

Also an AFM was used by Ederth at the Royal Institute of Technology in Stockholm, Sweden, [8] (see Fig. 5.4). He measured the force between two $R \sim 10$mm fuse silica gold-coated cylinders that were arranged at $90°$ to each other and that were as little as 20 nm apart. The gold-coated cylinders were also coated by a 2.1 nm hydro-carbon layer to prevent degradation of the exposed surface, since the experiments were performed under normal atmospheric conditions. The two surfaces were connected by a thin, loose gold wire ensuring an all–gold circuit, and no net mechanical force between the cylinders. Separation was controlled by piezoelectric actuators, with its motion measured by a LVDT device. The attractive force was measured by the deformation induced on a bimorph cantilever to which one of the cylinders was glued. Deformations of the surfaces had to be taken into account, and the lack of a direct measurement of the separation between them hindered the interpretation of the results.

Parallel plate configuration

The only recent experiment to replicate Casimir's original set-up of two plane parallel mirrors was carried out by Carugno, Onofrio and co-workers at the University of Padova in Italy [9]. They measured the change in the *resonance frequency* of a cantilever as it is approached by another flat surface. The origin of this shift is described in detail in Section 32.2 below. It is based in the fact that, for small oscillations, the mechanical response of the oscillator is modified due to the presence of the interaction with the other surface, which, for the attractive Casimir interaction, softens the spring constant of the cantilever. The separations were maintained between 0.5 and 3 μm. The surfaces were made of chromium covered silicon. The experiments were made in vacuum, and an electrostatic calibration was used to characterise and calibrate the system. Once again, significant difficulties arising mainly from the problems in obtaining sufficient parallelism between the plates yielded discrepancies when comparing the results with theoretical calculations.

Microelectromechanical torsional oscillator measurements

Also in 2001 Chan, working in Capasso's group at Bell Labs [10], used a silicon *microelectromechanical torsional oscillator* to determine the Casimir interaction between a $R = 100$ μm gold-coated polystyrene sphere and a gold-coated plate, which is part of the micromechanical torsional oscillator. The experiment was done at low pressure, $P < 10^{-3}$ torr, to improve the response of the oscillator. The separation between the interacting surfaces was provided by piezoelectric actuators.

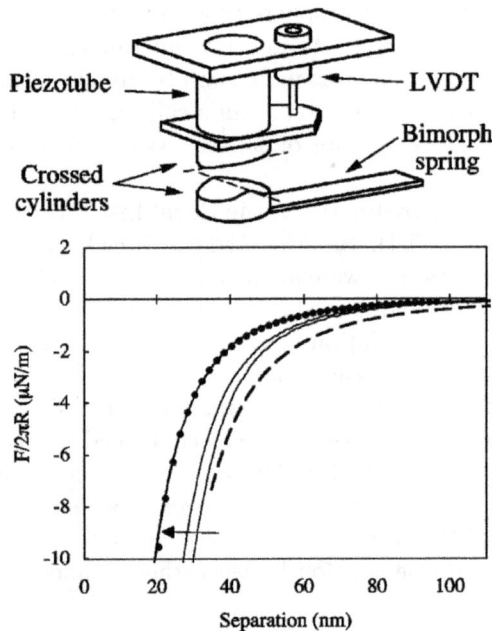

Figure 5.4: *Top panel*: Schematic of the apparatus used by Ederth. The interaction between the cylinders deflect the bimorph spring. The linear voltage to distance transducer (LVDT) measures the extension of the piezoelectric tube. *Bottom panel*: Measurement of the interaction. The two solid center lines are two sets of 5 measurements each for different samples. When taking into account deformations of the samples (as a fitting parameter), these two curves collapse into the one to the left, which agrees well with the Casimir interaction when finite conductivity corrections are taken into account. The dashed line to the right is the Casimir interaction without finite conductivity corrections.

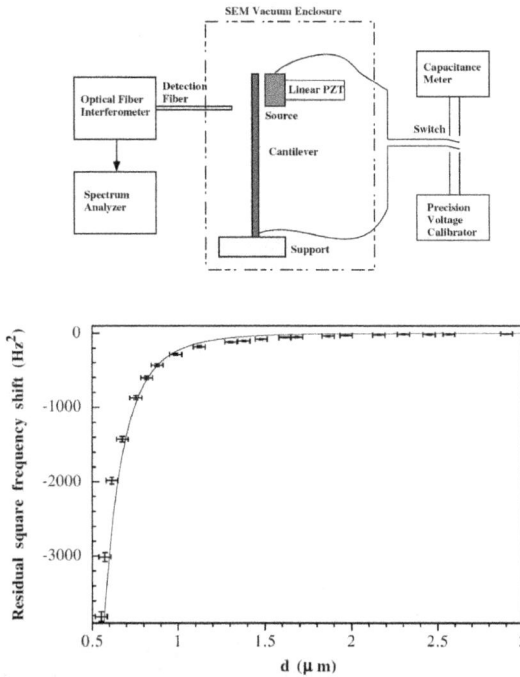

Figure 5.5: *Top panel*: Schematic of the apparatus used by Bressi and co-workers. One of the plates, denoted as the source, is moved by means of the piezoelectric transducer (PZT). The Casimir interaction modifies the resonant frequency of the cantilever. The spectral response of the motion of the cantilever at room temperature is obtained by looking into the interferogram obtained by the reflections of the light at the end of the optical fibre (reference mirror) and the cantilever, which is analysed by the spectrum analyser. The calibration is made using the electrostatic interaction between the source and the cantilever. The experiment is made within a scanning electron microscope (SEM). *Bottom panel*: Square of the resonance frequency change (which is related to the Casimir interaction) as a function of separation. The solid curve is the expected result using the Casimir interaction between perfect conductors.

The system was calibrated using the electrostatic force between the sphere and the plate. This calibration was done to obtain the torsional constant of the oscillator and the distance between the plate and the sphere. The deflection of the plate was determined by measuring the capacitance between the plate and two underlying electrodes. The same group also measured the non-linear effects of the Casimir interaction on the microelectromechanical torsional oscillator [11].

30.3   Effect of material properties

In most cases the experimental configuration preferred has involved the interaction between two gold-coated surfaces (gold is an inert metal, easy to deposit and durable) with either air or vacuum between the surfaces. As was described in Section 6.1, the modes in the cavity (and hence its zero-point energy) are modified

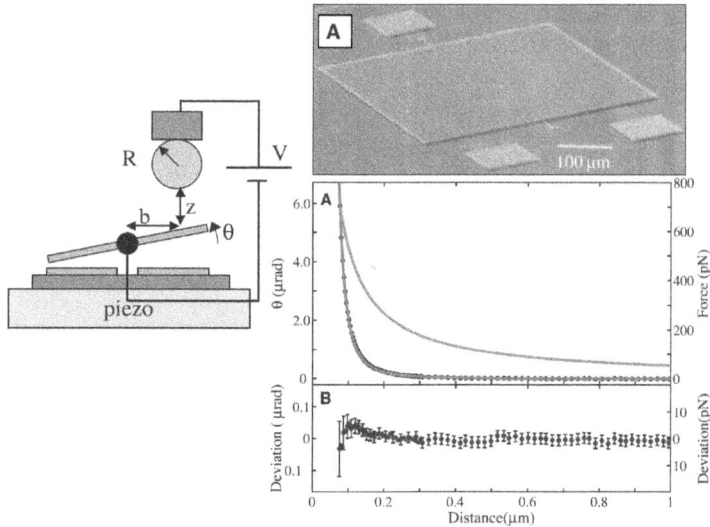

Figure 5.6: *Left panel*: Schematic of the apparatus used by Chan and co-workers. The interaction between the sphere and the plate deflects the torsional oscillator by θ due to the torque caused by the force at a lever arm b. This deflection is measured by the difference in capacitance between the plate and the left and right electrodes. *Top right panel*: scanning electron micrography of the torsional oscillator. The square pads are the places where the contacts are established for the oscillator (top) and the electrodes (two at the bottom). A potential V can be applied between the plate and the sphere. This is used to calibrate the system and then to minimise the electrostatic force to obtain the Casimir interaction. *Bottom right panel*: Measurement of the Casimir force (A) with the fit of the Casimir force between a sphere and a plate when finite conductivity and roughness corrections are taken into account. (B) shows the residual differences between the measured and calculated Casimir interactions. The top curve in (A) shows the residual electrostatic force (normalised) when a residual potential between the sphere and the plate is present. The difference in the dependence of the electrostatic and Casimir interactions is clear.

by the properties of the materials forming the cavity. Consequently, it is natural to experimentally investigate the effects arising from the use of different materials. There have been many experiments where the materials forming the surfaces are not gold, and also where the properties of materials have been changed *in situ*. In this section we provide a brief description of a few such experiments. Some other experiments where magnetic materials have been used are discussed in Section 38.1.

Measurements using different metals

Decca and co-workers [12] used a gold-coated sapphire sphere and a copper-coated plate of micromachined torsional oscillators. These oscillators were similar to the ones used by Chan. The schematic of the apparatus and the results obtained are

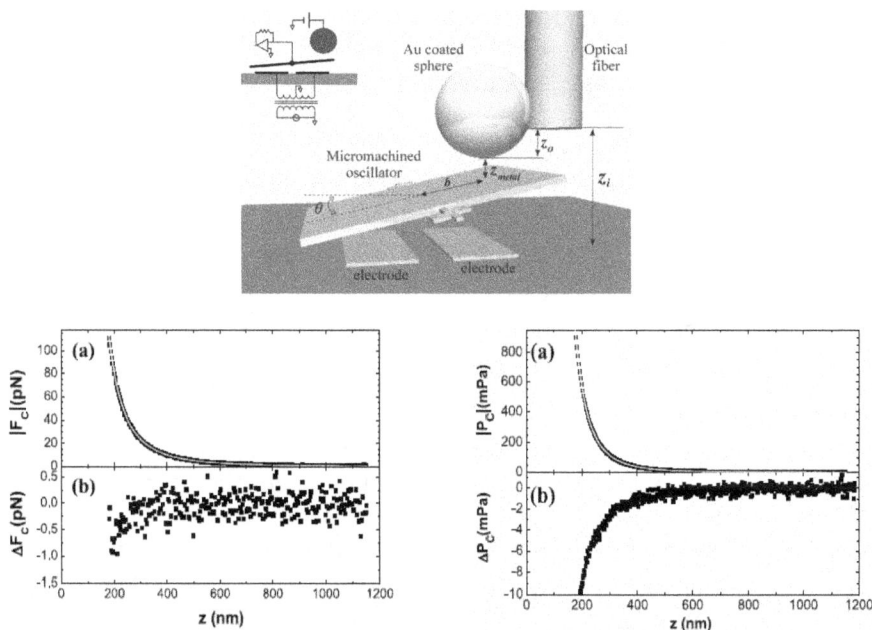

Figure 5.7: *Top panel*: Schematic of the apparatus used by Decca and co-workers. The optical fiber is part of an optical interferometer used to keep the distance z_i constant. θ is measured during the experiment by measuring the difference in capacitance between the oscillator and the two underlying electrodes by means of the circuit shown in the inset. z_o is obtained from the electrostatic calibration. All other parameters are measured independently. *Bottom left panel*: (a) $F_C(z)$ obtained by measuring $\theta(z)$, together with a theoretical prediction; (b) Difference between the data and the theory. *Bottom right panel*: Same as bottom left, but for $P_C(z)$.

shown in Fig. 5.7. The measurements shown are the force between the interacting surfaces and the pressure between them. This requires a little explanation, which is offered in Section 32. Basically two things can be measured on the oscillators, which are the deviation produced by the force and the change in the resonant frequency produced by the extra interaction as the surfaces become closer together. This change in the resonance frequency can be associated with the gradient of the interaction

$$
\omega_r(z) = \omega_o \left[1 - \frac{b^2}{2I\omega_o^2} \frac{\partial F_C(z)}{\partial z} \right],
\tag{30.1}
$$

where $\omega_r(z)$ is the resonant angular frequency (at separation z), b is the lever arm (distance between the point of maximum force and the axis of rotation of the plate), I is the moment of inertia of the plate, and ω_o is the resonant angular frequency of the oscillator when no interaction is present. This expression will be derived in

Section 32.2. Using the proximity force approximation [6], which states that for a sphere in front of a plate $F_C(z) = 2\pi R E_C(z)$, where $E_C(z)$ the energy per unit area for the plate-plate configuration, the pressure $P_C(z) = \partial E_C(z)/\partial z$ is

$$P_C(z) = \frac{1}{2\pi R}\frac{\partial F_C(z)}{\partial z}, \tag{30.2}$$

as shown in Section 32.2.

Measurements using semiconductors

Lamoreaux and collaborators [13] used his torsional pendulum to measure the interaction between crystalline germanium. The experiments were performed using a Ge sphere ($R = (15.10 \pm 0:05)$ cm) and a plate. Interestingly, they found that they had to apply a position-dependent potential difference between the two surfaces to minimise the force between them. Even at the minimum, they concluded that the measured force was a contribution of a residual electrostatic force and the Casimir interaction. Once the residual force was subtracted, the remnant Casimir interaction was observed to agree with the Lifshitz prediction (21.6) when the properties of germanium are used. A more detailed description of the effects of a non-homogeneous potential distribution in the sample is further described in Section 38.2.

Mohideen and co-workers [14] used the modified AFM system to measure the Casimir interaction between the Au-coated sphere and a silicon plate. When the Si surface is properly passivated to minimise its charging, they found excellent agreement between the data and the results predicted by the theory.

Measurements where the material's properties are changed *in situ*

Ianuzzi and co-workers [15] were the first to use an approach where the dielectric properties of the material are changed *in-situ* by changing the external conditions. They used an MTO (similar to the one used by Chan [10, 11]) where the plate was gold-coated. The $R = 100$ μm polystyrene sphere was coated with an engineered sample consisting of a seven repetitions of layers of Mg (10 nm thick) and Ni (2 nm), which were followed by an evaporation of a thin film of Pd (5 nm). This engineered material is metallic as deposited, but becomes semi-transparent when exposed to a hydrogen rich atmosphere, as shown in Fig. 5.8. Although the experiment was not successful in showing a change in the Casimir interaction between the two states of the hydrogen switchable mirror, it is instructive since it experimentally points out that changes in the reflection coefficients have to be made over a large frequency range to be detectable by a measurement of the Casimir interaction, as shown in (21.6).

Following the same line of thought, in which the optical properties of the sample are changed by adjusting an external parameter, Mohideen and his co-workers and Palasantzas and his collaborators have also made significant contributions. They both used a modified AFM setup as schematically shown in Fig. 5.3. Mohideen's group has done experiments changing the properties of indium-tin oxide (ITO) and

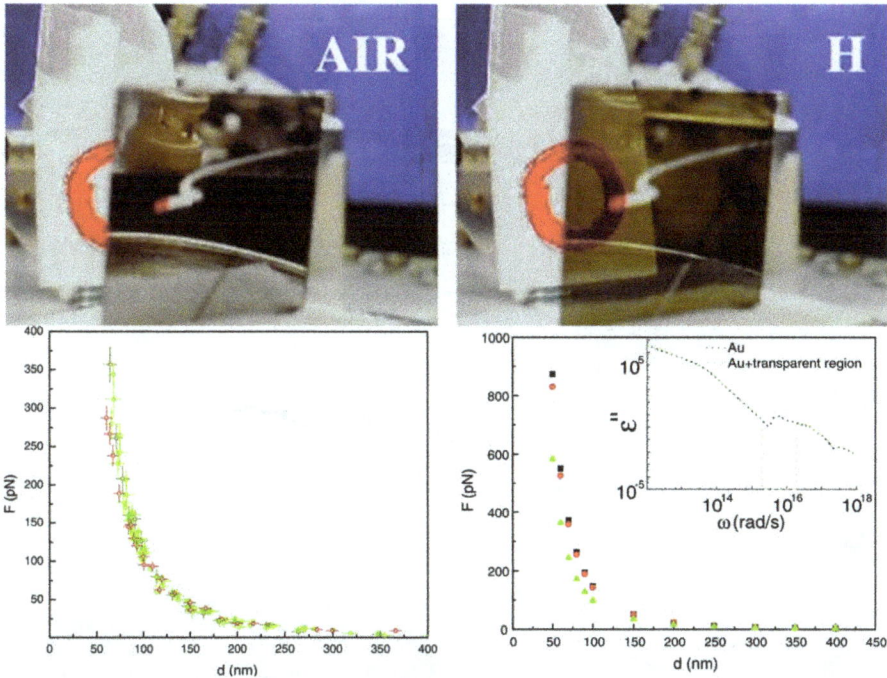

Figure 5.8: *Top panel*: Optical properties of the hydrogen switchable mirror used on the sphere. The material is reflective in air and semi-transparent in a hydrogen-rich atmosphere. *Left bottom panel*: Casimir interaction as a function of separation. There is no experimental difference on the measured Casimir interaction between Au and the two different states of the switchable mirror (metallic in green, semi-transparent in red). *Right bottom panel*: Results of the calculation of the Casimir force by using the Lifshitz formula for a material with dielectric function equal to that of gold with the exception of a frequency range where the material is supposed to be transparent. Black squares, pure gold; red circles, transparent window from 0.2 to 2.5 μm; green triangles, transparent window from 1 to 200 μm. The inset shows the region in energy where the material is made transparent, i.e. Im $\epsilon = 0$.

silicon. Palansantzas' group worked on phase change materials (PCM) similar to those used in CDs and bluerays.

On the experiments done in ITO, Mohideen [3] used his setup with an Au-coated sphere and an ITO plate. It was found that ultraviolet illumination of the sample led to a separation-dependent decrease of the interaction of up to 35 % in contrast to the untreated sample. Naively, one might expect these changes to come from observable variations in Re ϵ and Im ϵ. However, ellipsometry measurements yielded no significant difference of the Im ϵ for the treated and untreated samples. Consequently, using the Kramers–Kronig relations (19.14), one would expect that this lack of variation in ϵ would translate into the measured force. The authors found that the experimental data agrees well with Lifshitz theory in the case of the untreated sample. For the UV-treated sample, in order to obtain a good agreement

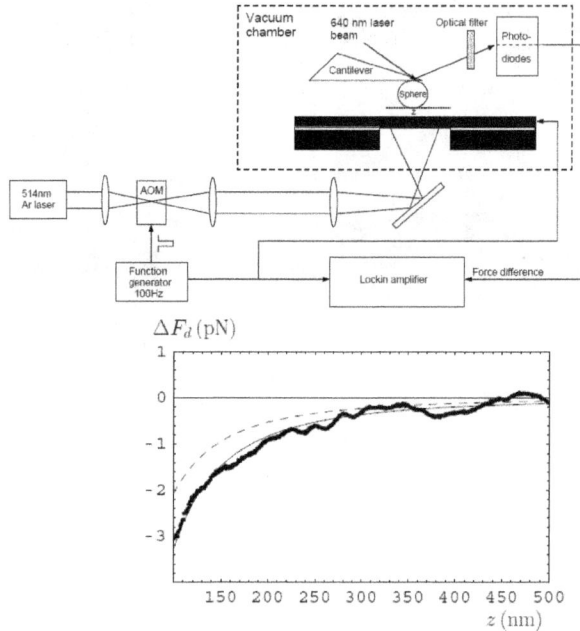

Figure 5.9: *Top panel*: Schematic of the experimental setup used to measure the Casimir interaction between a gold-coated sphere and a Si plate. The dotted line represents the vacuum chamber where the modified AFM apparatus interacts with a back-thinned Si membrane. The Si sample is illuminated at a rate of 100 Hz by a 514 nm laser. The excitation of carriers from the valence to the conduction band provides a variation of the dielectric permittivity which translates into a change in the measured force. *Bottom panel*: difference between the measured forces. The illuminated state is more metallic (hence a more attractive force). The dots are the experimental data. The solid curve is the prediction by Lifshitz theory when the DC conductivity of Si is not taken into account. The dashed curve takes into account the DC conductivity of Si.

between theory and experiment, the authors have to disregard the contribution of free carriers to the electric permittivity. They hypothesise this is due to a phase transition in the ITO sample where the effect of UV treatment induces disorder in the sample which translates into a Mott–Anderson localisation, with little or no effect in the dielectric permittivity.

The same group performed experiments using a back-thinned Si membrane [17]. The Si crystalline wafer was thinned to $\sim 4~\mu m$, thick enough to ensure that no light from the Ar laser will hit the sphere, yet thin enough to have a $10^{19}~cm^{-3}$ density of carriers optically excited from the valence to the conduction band. As shown in Fig. 5.9 there is good agreement between theory and experiment only if the DC conductivity of the non-illuminated sample is disregarded. This issue pertains to an open question in the community that is discussed later (see Section 38.1).

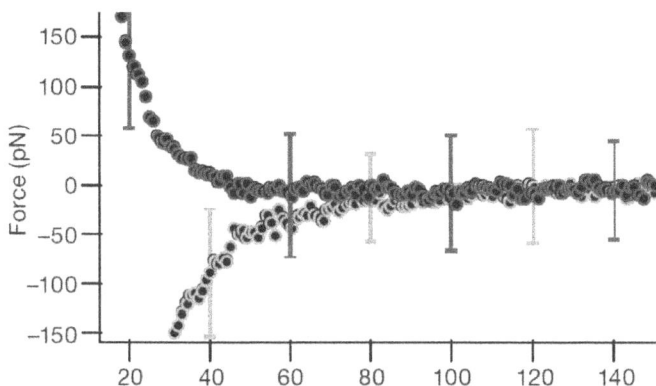

Figure 5.10: Experimental measurement of the Casimir interaction in a bromobenzene filled cell, as a function of separation in nm. The lighter points and error bars show the interaction between a Au-coated polystyrene sphere and a gold coated plate. Dark points and error bars correspond to the use of a silica plate. The vertical bars are the standard deviation of the measurements.

of the dielectric properties of a sample. The group deposited amorphous Ag-In-Sb-Te (AIST) thin films onto standard Al-coated Si wafers. Of the prepared AIST samples half were thermally annealed to the crystalline state. The dielectric permittivity was determined between 0.04 and 8.9 eV. The amorphous film is transparent in the infrared region, while the crystalline material is metallic. The group found good qualitative agreement between theory and experiment. Quantitative comparison was hindered by a small but measurable variation of the residual electrostatic force as a function of separation (see Section 38.2).

Measurements with repulsive Casimir interaction: gold–bromobenzene–silica

As a final example, we show an experiment where a *repulsive interaction* is observed within a cavity. For materials with purely electric response (i.e. magnetic permeability of unity), this is only possible (with reference to Fig. 3.3) if the vacuum is replaced by a medium with dielectric permittivity $\epsilon_2(\omega)$ such that $\epsilon_1(\omega) > \epsilon_2(\omega) > \epsilon_3(\omega)$ for all ω. Munday and collaborators [19] used a modified AFM apparatus suited to work in a fluid cell. They did the measurement in a bromobenzene(Bz)-filled cell, where an Au-coated sphere was glued to the AFM cantilever. Two different experiments were done, as shown in Fig. 5.10: for one case, the plate completing the cavity was silica (SiO); for the other case, it was gold. In the former the condition $\epsilon_{Au}(\omega) > \epsilon_{Bz}(\omega) > \epsilon_{SiO}(\omega)$ is satisfied. However, when the silica is replaced with gold it clearly cannot be satisfied. The group found that good agreement is found with Eq. (21.6) when the proper modifications for using three media are introduced.

30.4   Effects of temperature

The great majority of Casmir force experiments have been done at room temperature. As described in Section 21.4, the effect of temperature on the pressure between
two parallel plates is accounted for by performing a summation over Matsubara frequencies, $\zeta_l = l4\pi^2 k_B T/h$, where l is the order of the Matsubara frequency, and at
$T = 300$ K, $\zeta_1 \sim 3 \times 10^{14}$ rad/s. In order to modify the range of frequencies relevant
for the experiment, the temperature needs to be changed by large amounts.

Experiments in which the temperature has been reduced have been performed
by Decca and co-workers [20] and by Mohideen's group [4, 5]. Recently Chan and
co-workers [23] have been working on a system where the experiments, performed
completely on microelectromechanical systems, are done at 4 K (these experiments,
which pioneered a new direction for the Casimir interaction detection, are described
in Section 30.5).

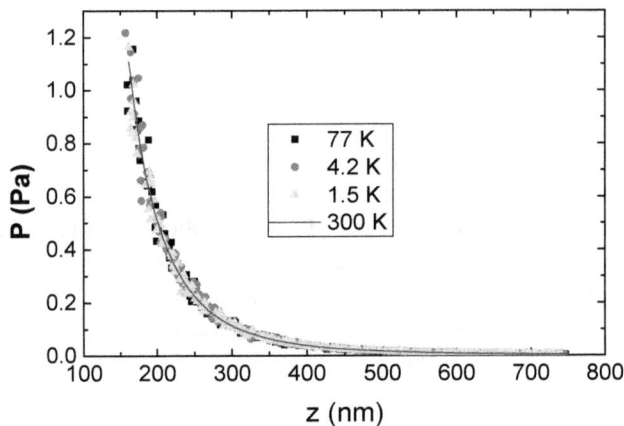

Figure 5.11: Casimir interaction for different temperatures. All the measurements are
consistent with the 300 K measurements, but the noise in the measurements increases at
lower temperatures.

Decca used a torsional oscillator to measure the Casimir interaction between a
gold-coated $R \sim 150$ μm sphere and the gold-coated plate of the oscillator. The
main difference with the other experiments performed by his group is that it was

performed in a vessel surrounded by the refrigerating liquid (either liquid helium or liquid nitrogen), which in turn was immersed in a vessel with liquid nitrogen. The experiment was performed below the superfluid transition for ^{4}He, and then at the fixed points of boiling helium and boiling nitrogen, i.e. 4.2 and 77 K respectively. The results were less precise than the results obtained at room temperature. It was hypothesised that the introduction of noise due to the pumping of the liquid helium and the continuous bubbling-induced noise from the liquid nitrogen vessel were detrimental for the experiment. Furthermore, the optical interferometer used to control the separation suffered from drifts associated with the temperature gradient due to the change in the index of refraction of the optical fiber, as well as for the stress induced changes in the polarisation of light in the fiber due to the same temperature variation. These problems precluded obtaining a firm conclusion on the effect of dissipation on the Casimir interaction. Results from this experiment are shown in Fig. 5.11.

The experiments done by Mohideen's group were better designed and implemented. In these experiments the group used its standard approach based on a gold-coated sphere glued to a AFM cantilever. The results obtained are in Fig. 5.12. These experiments shed very useful light on the most pressing controversy existing in the area of precise measurements of the Casimir force, i.e. what is the relevant model for the description of the optical properties of the Casimir effect? Unlike the previous experiment, these measurements were precise enough to show that decreasing the temperature, which causes a reduction in the electron-phonon scattering, did not affect the Casimir interaction. The full extent of the controversy is presented in Section 38.1.

While the described experiments provided some insight into the temperature-dependence of the Casimir interaction, it would seem that the most obvious manifestation of a temperature dependent effect would be more easily observed by increasing the temperature. This is, however, quite dificult to achieve. When increasing the temperature, the differences between thermal expansion coefficients between the different materials, and the fact that the temperature is varying as a function of time, severely conspire against a reliable measurement of the temperature-dependence of the interaction. A system properly aligned at room temperature will suffer severe displacements when warmed up. A mechanical arm of 10 cm in a system made of two different materials with a difference in their coefficients of thermal expansion as low as 1×10^{-6} 1/K will shift by ~ 15 μm when its temperature is increased to 400 K. Since the actuators will have to be kept colder, there would be gradients of temperatures which would produce severe drifts in the system. Furthermore, for the case of thin layers of the force sensitive transducer (for example the gold-covered 3.5 μm thick silicon oscillators shown in Fig. 5.6) the difference in the expansion coefficients between the materials used in the structure would produce enough stress to warp it. Consequently, while increasing the temperature does not imply violating any fundamental constraint, the complexity of its implementation has kept its realisation at bay.

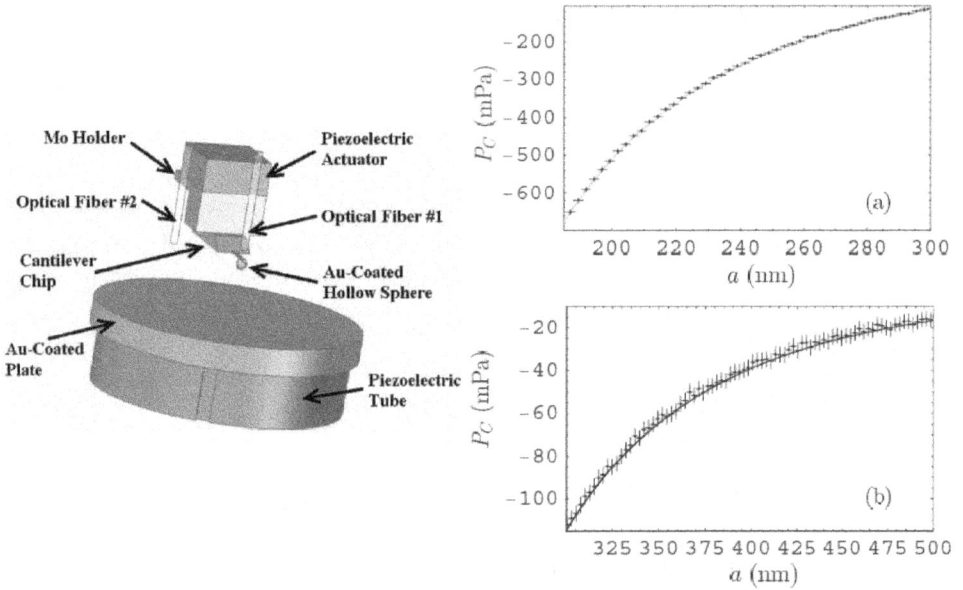

Figure 5.12: *Left panel*: Schematic of the experimental setup. The materials were selected for their low thermal expansion coefficient. One optical fibre is used to measure the vibrations of the cantilever, the other one is part of an optical interferometer to measure the distance to a reference plate. *Right panel*: Experimentally determined Casimir interaction at 77 K (crosses) compared with the theoretical prediction. (a) and (b) represent the interaction at different separations.

30.5   Recent directions

One innovative approach was recently developed by Chan to measure the Casimir interaction [23]. In this method, the vast repository of knowledge concerning the processing of silicon sequestered in the semiconducting industry has been successfully translated into the development of a fully *integrated microelectromechanical system* for the measurement of the Casimir interaction, as shown in Fig. 5.13. In this approach the system is made out of silicon, with the measurement of the interaction between parallel plates defined by the movable electrode and the beam in Fig. 5.13. These plates have typical dimensions of 100 μm long by ~ 1.42 μm wide. The Casimir interaction is sensed by the change in the resonance frequency of oscillation of the beam. To measure this frequency of oscillation a small alternating current is circulated through the beam at its resonant frequency $\omega_r > 7 \times 10^6$ rad/s. In the presence of a 5 T magnetic field perpendicular to the substrate, the beam is subjected to a periodic Lorentz force. Vibrations of the beam in the magnetic field generate an induced electromotive force that is detected with a current amplifier. Like other systems using micromechanical components, the experiments are done at low pressure $(P < 10^{-5}$ Torr). Since the magnet used is a superconducting magnet, Chan elected to perform the experiments at 4.2 K. The motion of the movable electrode is obtained by applying a voltage V_{comb} to the comb actuator. The presence of V_{comb} causes the movable electrode (blue in the figure) to be pulled in towards the stationary part of the comb (grey in the figure). An independent potential V_e can be applied between the movable electrode and the beam to do the required calibration of the system and to minimise electrostatic contributions when performing the Casimir interaction measurements.

This approach has important advantages: by construction the interacting surfaces can be made parallel and the shapes of the surfaces can be selected from the lithography process. The miniaturisation of the system implies that the mechanical arm (typically tens of cm in other systems) is only a few hundred μm, and the high resonant frequency provides an almost complete isolation from environmental vibrations. Furthermore, the distance actuation is built into the system and the whole setup is made out of a single material, silicon. These features provide much better mechanical and thermal stability.

§31. Techniques for measurement

Now that a significant number of experiments have been presented, it is time to start analysing how the measurements are done. All experiments described in Section 30 do not measure a force directly but rely on the use of a force *transducer*. This transducer is, in all cases, a mechanical device. As in the simple spring-mass system, the presence of the interaction produces a *deflection* on the system, which is ultimately what is observed. For the small forces measured in the experiments this description using a *driven damped simple harmonic oscillator* is sufficient to describe the observed results. Under these conditions we can then obtain what is the smallest interaction that can be detected with the system. This is known as the *minimum detectable interaction*, and it is perhaps the most critical parameter in understanding the potential of the system. In order to understand the minimum detectable interaction, we need to start with the mechanical oscillator equation of motion (which we describe here as a torsional oscillator, but can easily be translated to the case of a cantilever), which is represented by

$$I\frac{d^2\theta}{dt^2} + b\frac{d\theta}{dt} + \kappa\theta = \tau_{ext}(t) + \tau_{th}(t). \tag{31.1}$$

Here θ is the angular deviation of the oscillator, I is its moment of inertia, $-b\frac{d\theta}{dt}$ is the damping torque (assumed to oppose the angular velocity), κ is the elastic restitution constant of the torsional oscillator, τ_{ext} is the external torque applied to the oscillator, and τ_{th}, which is the thermal noise, represents the effect the environment at a temperature T has in the oscillator. In the absence of τ_{ext}, $\theta(t)$ is given by the response of the oscillator to the incoherent driving torque representing the oscillator

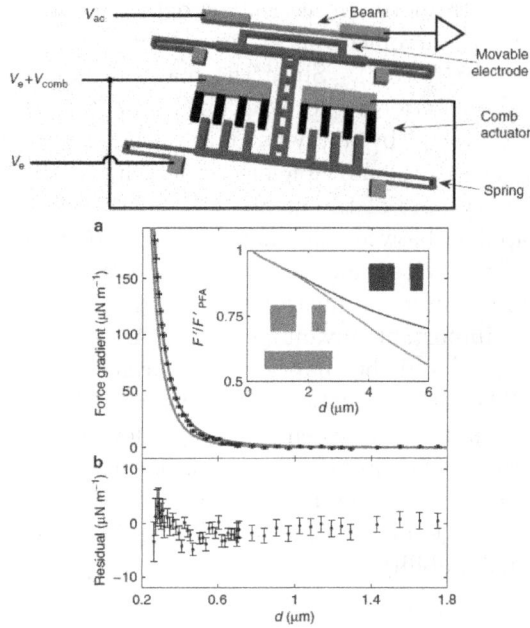

Figure 5.13: *Top panel*: Schematic of the experimental setup. *Bottom panel*: Measurements of the gradient of the Casimir force between the beam and the movable electrode as a function of separation d after compensating for the residual voltage. (a) The lower line represents the calculated Casimir force gradients between an electrode and a beam made of silicon using a boundary-element method [16]. The top line includes possible contributions from patch potentials. *Inset*: the ratios of the calculated Casimir force between the beam and the electrode to the forces given by the PFA are plotted as the bottom (with substrate) and top lines (without substrate). (b) Deviations of the measured force gradient from the top line in a. Error bars represent ±1 standard deviation.

in thermal contact with a bath of modes at a definite temperature. Under these conditions, the form of the torque can be derived using the equipartition theorem of statistical mechanics.

A harmonic torsional oscillator in equilibrium with a bath of temperature T has an energy expectation value for each mode equal to $k_B T/2$, where $k_B = 1.3810^{23}$ J/ K is Boltzmann's constant. Thus

$$\frac{1}{2}\kappa\langle\theta^2\rangle = \frac{1}{2}k_B T, \tag{31.2}$$

where $\langle\theta^2\rangle$ is the mean-square angular displacement θ^2_{rms}. Hence if θ^2_{rms} can be measured the elastic constant of the oscillator can be inferred from (31.2). Usually time-domain observations of θ are not accurate with the required sensitivity and time resolution. Hence, the determinations are done in the frequency domain. Recalling that

$$\langle \theta(t)^2 \rangle = \int_0^\infty S_\theta^2(f) df, \tag{31.3}$$

where $S_\theta^2(f)$ is the one-sided deflection *power spectral density*, κ can be determined. Since the thermal noise is white, and in the limit considered here where damping is small compared with the elastic force (i.e. the mechanical $Q = f_o/\Delta f >> 1$, where f_o is the resonant frequency of the oscillator and Δf is the width of the resonance), it follows that

$$S_\theta(f) = \frac{S_\tau}{\kappa} \frac{f_o^2}{\sqrt{(f_o^2 - f^2)^2 + f^2 f_o^2/Q^2}}. \tag{31.4}$$

In (31.4) S_τ^2 is the constant (since thermal noise is white) one-sided torque power spectral density.

31.1   Minimum detectable interaction

Combining Eqs. (31.4), (31.3) and (31.2), we obtain

$$S_\theta^2(0) = \frac{2k_B T}{\pi \kappa Q f_o}. \tag{31.5}$$

The power spectral density at all frequencies for a harmonic oscillator at thermal equilibrium is obtained by substituting $S_\theta^2(0)$ into (31.4)

$$S_\theta(f) = \sqrt{\frac{2k_B T}{\pi \kappa Q f_o}} \frac{f_o^2}{\sqrt{(f_o^2 - f^2)^2 + f^2 f_o^2/Q^2}}, \tag{31.6}$$

with $S_\tau^2 = 2\kappa k_B T/\pi Q f_o$. Thermal oscillator deflection fluctuations can be treated as if due to a force fluctuation of this spectral density. The deflection thermal-noise power in a narrow bandwidth δf centered at f_*, such as would be measured with a lock-in amplifier, is

$$\theta_{min}^2(f_*) = \int_{f_* - \delta f/2}^{f_* + \delta f/2} S_\theta^2(f) df, \tag{31.7}$$

and $\theta_{min}(f_o) = \sqrt{\frac{2Q\delta f k_B T}{\pi \kappa f_o}}$ while $\theta_{min}(0) = \sqrt{\frac{2\delta f k_B T}{\pi \kappa Q f_o}}$. On the other hand, and by doing a similar analysis, the minimum detectable torque is found to be

$$\tau_{min} = \sqrt{\frac{2\kappa \delta f k_B T}{\pi Q f_o}}, \tag{31.8}$$

which is independent of frequency, provided the only source of noise is thermal.

Exercise 31.1
Derive the expressions for $\theta_{min}(f_o)$, $\theta_{min}(0)$, and τ_{min}

Exercise 31.2
Rewrite expressions 31.1–31.8 for the case of a linear oscillator.

Equation (31.8) indicates that the best sensitivity occurs with

 i. lowest possible spring constant κ,

 ii. lowest possible temperature T,

 iii. highest possible quality factor Q,

 iv. highest possible resonance frequency f_o.

Unfortunately not all these quantities can be addressed independently, and a detailed analysis of the apparatus is needed to achieve optimal results. More importantly, if additional *non-thermal* noise is present, its effect has to be added into (31.1), and τ_{min} will no longer be frequency-independent. Even in the case that non-thermal noise sources have been taken into account, electronic limits in the determination of θ impose more severe conditions away from resonance (where the amplitude of the deflection is small) than at resonance. Consequently, the value obtained in (31.8) is more closely achieved when the system is near resonance.

An experimental situation where the effects of a residual force is observed is shown in Fig. 5.14. In the experiment [1], performed at the resonant frequency, the measured signal was obtained in a Casimir-like configuration where the driving force was expected to be zero. It is clear that there is a separation-dependent residual force which prevents us from observing the behaviour described by (31.8), except when the separation between the interacting surfaces is large (which was verified as corresponding to the thermodynamic limit using (31.7) and (31.8)). As a general rule, it is important to understand that either the experiment is thermal noise limited, following the behaviour described in (31.7) and (31.8), or that the sources of noise are properly characterised.

§32. MEASUREMENT OF THE INTERACTION

For the force transducers in Section 30 two major distinctions can be made. In the one case, the change in the equilibrium position of the transducer, induced by the interaction, is detected. We call this a static method and it is described in Section 32.1. The other case uses the fact that a small extra attractive interaction effectively softens the spring constant of the transducer, causing a reduction in its resonant frequency. The change of the resonant frequency is then measured. We call this a dynamic method, and it is described in Section 32.2.

32.1 Static method

Different categories of experiments can also be distinguished within this class. One of them corresponds to AFM based experiments. Another consists of experiments where a torsional balance or an oscillator are used as transducers. In both these cases (which are the most popular approaches), by the end of the detection process the deflection has been transformed into an electric signal, which is directly measured.

In AFM based experiments, the motion of the cantilever is detected by reflecting a laser beam off the cantilever and using a position-sensitive detector to measure the motion of the beam (see Fig. 5.3) [26]. This deflection is measured by a position-sensitive detector consisting of two photodiodes. The normalised difference in the photocurrent I_p generated between the photodiodes 1 and 2, $(I_{p1} - I_{p2})/(I_{p1} + I_{p2})$, is proportional to the deflection of the beam. In the torsional balance or oscillator systems, the motion is generally detected by measuring the capacitance difference between the transducer and two stationary plates (see Figs. 5.2 and 5.6). This capacitance difference, for the small amplitudes of motion involved, is proportional to the motion of the transducer, and hence the interaction. Alternatively an optical interferometer could be used to detect the motion of the transducer. The main conclusion is that, independently of the approach used (whether the photocurrent, in AFM and optical interferometry approaches, or the capacitance) the minimum signal that can be detected by the electronic components ultimately limits the sensitivity of the apparatus.

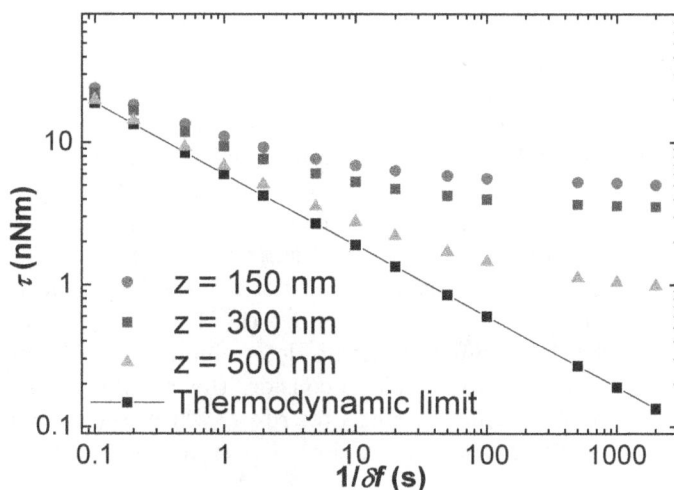

Figure 5.14: Effect of integration time in the minimum detectable torque. The thermodynamic limit is obtained when the two surfaces are far away. As the separation between the samples is decreased a finite residual interaction is observed.

As an example, let's consider how the *electronic noise* enters into the determination of the minimum detectable signal. To stay within the notation developed in Section 31, we will assume an angle is measured (i.e. a torsional balance is used). We will assume that the conversion factor between the measured capacitance and the angular displacement is not a function of frequency, i.e. the amplifier has a frequency-independent response*. In this case we find that (31.6) needs to be modified by adding the response of the electronic system. Since the sources of noise are independent, the minimum detectable deviation becomes

$$\theta_{min}^* = \sqrt{\frac{2k_B T \delta f}{\pi \kappa Q f_o} + \theta_{el}^2},$$

(32.1)

where θ_{el} is the minimum detectable angular deviation due to the electronic noise and the measurement is done at a fixed frequency with a bandwidth δf. Consequently, for sufficiently long integration times, the minimum detectable deviation (and the minimum detectable interaction) is limited by the electronic noise.

As an introduction to the dynamic methods of measurement, it is important to note that while θ_{el} is in general approximately constant as a function of frequency, θ_{min} from (31.7) is not. If an interaction modulated at f_o is present, then the amplitude of motion of the oscillator is very large. In this case, at resonance, the minimum detectable torque is still very close to τ_{min} from (31.8). On the other hand, far from resonance, $\tau_{min} \sim \kappa \theta_{el}$.

32.2 Dynamic method

It is advisable to work at resonance to improve the sensitivity of the Casimir interaction detection. Unfortunately, it is difficult to modulate the Casimir interaction at the resonance frequency. One possibility is to modulate the separation between the bodies. However, this is not feasible for high-Q mechanical oscillators, since if there are parts moving at resonance the oscillator will start vibrating at that frequency (even for large separations where the interaction is not measurable). Another possibility is to modulate the effective cavity (i.e. change the dielectric permittivity) at resonance. In the context of Casimir interaction determination, this approach has not been implemented. Another approach is to consider the effect of the interaction itself on the properties of the oscillator. Let's rewrite (31.1) in the presence of an interaction – for example the Casimir interaction. For simplicity we will omit the thermal bath

$$I \frac{d^2\theta}{dt^2} + b \frac{d\theta}{dt} + \kappa \theta = \tau_C(z),$$

(32.2)

and $\tau_C(z)$ is the torque due to the Casimir interaction at a separation z. For the sphere in front of a plane this is simply the product of the Casimir force F_C by

---

*To clarify things, the deflection of the transducer is static. The method to measure the deflection is, in general, done at a fixed frequency far from resonance using standard lock-in amplification techniques. While the capacitance is fixed, it is measured using an AC signal.

the lever arm b (see Fig. 5.6). Since the Casimir interaction is small, (32.2) can be Taylor expanded around the equilibrium position. For an attractive Casimir interaction this yields a change in the resonant frequency as noted in (30.1): the gradient of the interaction decreases the resonant frequency of the oscillator.

Exercise 32.1
Show that (30.1) is obtained when the Taylor's expansion of $\tau_C(z)$ if truncated at first order.

In this approach, the gradient of the interaction is then directly determined if the resonant frequency is measured. To measure the resonant frequency two different equivalent approaches can be used. In one, the system is permitted to vibrate under the influence of thermal fluctuations and the resonant frequency is measured by looking into the power spectrum of the motion of the oscillator. The other approach, which is more common and yielded the most precise measurements of the Casimir interaction to date, is to use a phase-locked loop [27] (PLL) circuit to maintain the system at resonance. In this case, the errors in the determination of the resonant frequency mainly arise from the noise in the phase induced by the PLL scheme [28].

With the gradient of the interaction determined, one can use the proximity force approximation to obtain the pressure between two parallel plates at the same distance z. The proximity force approximation states that the total energy $E_{tot}(z)$ can be obtained as a summation (integral) over all the individual infinitesimal areas, treating these infinitesimal areas as planes with an energy per unit area $E_C(z)$. For the case of the sphere-plane configuration

$$E_{tot} = 2\pi R \int_0^{\pi/2} E_C(z+R(1-\cos\theta))Rd(\cos\theta) = -2\pi R \int_0^R E_C(z+R-y)dy, \quad (32.3)$$

where $y = R\cos\theta$ is the horizontal distance along the sphere, and the direction of y is taken to be perpendicular to the axis of rotation of the torsional oscillator. Hence each infinitesimal area is at a distance $b_y = b + y$ and the torque is

$$\tau_C(z) = 2\pi R \left[\int_{-R}^0 y \frac{\partial E_C(z+R+y)}{\partial y}dy + \int_0^R y\frac{\partial E_C(z+R-y)}{\partial y}dy \right.$$
$$\left. + b\int_0^R \frac{\partial E_C(z+R-y)}{\partial y}dy \right]. \quad (32.4)$$

The first two integrals cancel each other out, and the last integral is proportional to $F_C(z)$. Hence, $F_C(z) = 2\pi RE_C(z)$, and as mentioned earlier, (30.2), $P_C(z) = \frac{1}{2\pi R}\frac{\partial F_C(z)}{\partial z}$ follows.

§33. MEASUREMENT OF THE SEPARATION

The measurement of the separation between the interacting bodies is as important as the measurement of the interaction itself if quantitative information is going to be extracted. The relative motion between the bodies is accomplished by means of piezo-electric materials. In general, the measurement of the separation requires the knowledge of one specific separation point (obtained, for example, when the two bodies are in contact, or during the calibration of the apparatus – see Section 34) and the size of the displacement of the piezo elements under a given applied voltage, which require a calibration of the piezo element. Most groups either calibrate the piezo-elements using an optical interferometer or, alternatively, continuously measure the separation between the interacting bodies during the Casimir interaction experiment. Using these approaches the separation between the bodies can routinely be found to better than 1 nm.

Exercise 33.1

Assuming an error in the position of 1 nm, estimate the error propagated into the pressure if two plates made from ideal metals are placed in front of each other.

---

§34. Calibration of the apparatus

The most common method for *calibrating* the apparatus is to use a known interaction existing between the bodies under consideration. In general, this interaction is the electrostatic force between the bodies. If the capacitance between the bodies is known either analytically or numerically, then the electrostatic interaction can be obtained. In what follows we will concentrate on the most common geometry (i.e. sphere-plane geometry) but extensions to other configurations are easily achieved. The electrostatic force between a sphere and a plate separated by a distance z is given by

$$F_e(z, V) = -2\pi\epsilon_o(V - V_o)^2 \sum_{n=0}^{\infty} \frac{\coth(u) - n\coth(nu)}{\sinh(nu)}$$

$$\simeq \Xi(z)(V - V_o)^2, \tag{34.1}$$

$$\Xi(z) = -2\pi\epsilon_o \sum_{m=0}^{7} A_m t^{m-1}, \tag{34.2}$$

where

$$\sum_{n=0}^{\infty} \frac{\coth(u) - n\coth(nu)}{\sinh(nu)} = \frac{\partial C}{\partial z} \tag{34.3}$$

is the derivative of the capacitance between a sphere and an infinite plane*. In (34.1) and (34.2), ϵ_o is the permittivity of free space, V is an applied potential difference and V_o is a residual potential difference between the plate and the sphere, $u = 1 + t$, A_m are fitting coefficients, and $t = z/R$. While the full expression is exact, the series is slowly convergent, and it is easier to use the approximation

---

*The derivation of the capacitance between a sphere and a plane is a very instructive exercise. See Ref. [29] for an approach to obtain it using the method of images.

shown (developed in Refs. [30]). The values of the different parameters are A_0=0.5, A_1=-1.18260, A_2=22.2375, A_3=-571.366, A_4=9592.45, A_5=-90200.5, A_6=383084, and A_7=-300357. Using this expression errors smaller than 1 part in 10^5 are made. In (34.1) it has been assumed that the contact potential V_o is independent of separation. If this is not the case a more involved analysis, where the $V_o(z)$ dependence is taken into account, would be needed [31]. The origin of V_o resides in the use of many different metals along the circuit that connects the plate and the sphere, the use of polycrystalline surfaces, and the fact that the surfaces are contaminated (most measurements are performed at a pressures where water vapor, nitrogen and carbon compounds remain on the metallic surfaces).

The expression shown in (34.1) shows that the electrostatic force can be nullified by a judicious choice of the potential difference between the sphere and the plate. While customarily F_e is measured at constant z to find the position of the minimum of the parabolic dependence, it is found that by taking the derivative $\frac{\partial F_e}{\partial V}$ the potential V_o is more accurately determined [32]. In order to find $\frac{\partial F_e}{\partial V}$, a harmonic potential $V(t) = V_{ac}\cos(\omega t)$ with small amplitude $V_{ac} \leq 0.1$ mV is added to a DC electrostatic potential, V_{DC}, and a standard lock-in detection technique is used to measure the derivative of F_e at V_{DC}. The value of ω is selected to coincide with the resonant frequency of the system. It is worth mentioning that while the values obtained for V_o agree with those obtained from finding the minimum in the magnitude of F_e, ΔV_o is a factor of 2/3 using the method described here when compared to the minimum in the parabolic dependence of F_e on V. In general V_o needs to be determined from the minimum separation where the Casimir interaction is to be measured up to separations of several microns, where the Casimir interaction presents a negligible contribution.

Once it has been determined that there is no systematic dependence of V_o on position, separation, or time, (34.1) is used to provide a calibration of the system. The calibration of the system can be done both by measuring F_e or the change in the resonant frequency of the oscillator due to the electrostatic interaction. These determinations are done at z large enough such that the Casimir interaction does not have a measurable contribution. In both cases the actual separation z is known to within a constant Δz (i.e. the differences in z are well determined interferometrically, but a fixed separation is not known). For the case of $F_e(z)$, using

$$\theta = \frac{b}{\kappa} \Xi(z + \Delta z)(V - V_o)^2, \tag{34.4}$$

and using Δz as a fitting parameter, the best fits for κ/b and D_1 are obtained. In reality, since θ enters in z through the $b\theta$ term (see Fig. 5.6), this process has to be repeated until no further changes in the values of Δz are observed. A similar procedure is used when instead of using the force the change on the resonant frequency is employed.

§35. Comparison with theory

Once the separation dependence of the interaction has been obtained, it is desirable to establish quantitative comparisons between the data and the theoretical predictions. This can be done, for example, by using (21.6). There are, however, many factors that need to be taken into account:

i. The proximity force approximation is used, since data is not obtained for a plane-plane configuration where (21.6) is valid. We saw that in general this is not a problem.

ii. The reflection coefficients for both polarisations of light need to be known at all frequencies. This implies that the dielectric permittivity and the magnetic permeability need to be known at all frequencies, or at the very least for frequencies sufficiently large such that they are both seen to approach unity. As will be discussed in Section 38, this represents a problem, particularly for the value to be used at zero frequency.

iii. Samples are not infinitely smooth. Whilst care is taken to ensure that the materials are as smooth as possible, in reality they present some surface roughness. In typical samples the *roughness* is a just a few nm. When the measurements are done at separations of 100 nm and above, the contribution of the corrections due to surface roughness are a fraction of a percent. This, however, is not the case when measurements are performed at separations on the order of 10 nm. The corrections employed range from the use of a perturbative approach to the proximity force approximation. In this approach the topography of the sample is measured by AFM and the information is used to determine the fraction of the sample that is at a given separation.

iv. Precision measurements executed at room temperature need to include finite temperature corrections, which is achieved by performing a summation over Matsubara frequencies, as shown in Section 21.4.

As was mentioned in Section 30.3, the comparison between theory and experiment is not always successful. Section 38 describes in more detail where those problems arise from.

§36. Problems

Hint: Both problems require the use of numerical integrations.

Problem 5.1
Consider a mechanical transducer well described by a driven, damped, simple harmonic oscillator as represented by (31.1), where $\tau_{ext}(t)$ is harmonic at angular frequency ω. Consider the electronic circuit to provide a noise, which translated into angular displacement has a frequency independent power spectrum density of 10^{-9} rad^2/Hz. (i) Obtain the relevant parameters for κ, b, etc from one of the cited papers using torsional oscillators. (ii) Calculate the sensitivity of the apparatus as a function of ω for different δf of 10 Hz, 1 Hz, 0.1 Hz, 0.01 Hz

Problem 5.2

The measured interaction data as a function of separation $P_C(z)$ obtained in [2] are available online http://physics.iupui.edu/people/ricardo-decca/. Use (21.6) at room temperature with the dielectric function given by a Drude model with a plasma frequency $\omega_P = 8.9$ eV and a dissipation $\gamma = 0.035$ eV. Plot the difference between the two as a function of separation. How does this difference change as the metal is made a "better metal" (higher ω_p and lower γ)? When it is a worse metal?

§37. BIBLIOGRAPHY

[1] H.B.G. Casimir, "On the attraction between two perfectly conducting plates", *Proc. Kongl. Ned. Akad. v. Wetensch.* **B51**, 793 (1948).

[2] M.J. Sparnaay, "Measurements of attractive forces between flat plates", *Physica* **24**, 751 (1958).

[3] S.K. Lamoreaux, "Demonstration of the Casimir force in the 0.6 to 6 μm range", *Phys. Rev. Lett.* **78**, 5 (1997).

[4] B.V. Deriagin and I.I. Abrikosova, "Direct measurement of the molecular attraction of solid bodies. 1. Statement of the problem and method of measuring forces by using negative feedback", *Sov.Phys.JETP* **3**, 819 (1957) (*Zh.Eksp.Teor.Fiz.* **30**, 993 (1956)).

[5] B.V. Deriagin and I.I. Abrikosova, "Direct measurement of the molecular attraction of solid bodies. 2. Method for measuring the gap. Results of experiments", *Sov.Phys.JETP* **4**, 2 (1957) (*Zh.Eksp.Teor.Fiz.* **31**, 3 (1956)).

[6] J. Blocki, J. Randrup, W. Świątecki, and C. Tsang, "Proximity forces", *Ann. Phys.* **105**, 427 (1977).

[7] U. Mohideen and A. Roy, "Precision measurement of the Casimir force from 0.1 to 0.9 μm", *Phys. Rev. Lett.* **81**, 4549 (1998).

[8] T. Ederth, "Template-stripped gold surfaces with 0.4-nm rms roughness suitable for force measurements: Application to the Casimir force in the 20–100-nm range", *Phys. Rev. A* **62**, 062104 (2000).

[9] G. Bressi, G. Carugno, R. Onofrio, and G. Ruoso, "Measurement of the Casimir force between parallel metallic surfaces", *Phys. Rev. Lett.* **88**, 041804 (2002).

[10] H.B. Chan, V.A. Aksyuk, R.N. Kleiman, D.J. Bishop, and F. Capasso, "Quantum mechanical actuation of microelectromechanical systems by the Casimir force", *Science* **291**, 1941 (2001).

[11] H.B. Chan, V.A. Aksyuk, R.N. Kleiman, D.J. Bishop, and F. Capasso, "Nonlinear micromechanical Casimir oscillator", *Phys. Rev. Lett.* **87**, 211801 (2001).

[12] R.S. Decca, D. López, E. Fischbach and D.E. Krause, "Measurement of the Casimir force between dissimilar metals", *Phys. Rev. Lett.* **91**, 050402 (2003).

[13] W.J. Kim, A.O. Sushkov, D.A.R. Dalvit, and S.K. Lamoreaux, "Measurement of the short-range attractive force between Ge plates using a torsion balance", *Phys. Rev. Lett.* **103**, 060401 (2009).

[14] F. Chen, U. Mohideen, G.L. Klimchitskaya, and V.M. Mostepanenko, "Investigation of the Casimir force between metal and semiconductor test bodies", *Phys. Rev. A* **72**, 020101 (2005).

[15] D. Iannuzzi, M. Lisanti, and F. Capasso, "Effect of hydrogen-switchable mirrors on the Casimir force", *Proc. Nat. Acad. Sci.* **101**, 4019 (2004).

[16] A.A. Banishev, C.-C. Chang, R. Castillo-Garza, G.L. Klimchitskaya, V.M. Mostepanenko, and U. Mohideen, "Modifying the Casimir force between indium tin oxide film and Au sphere", *Phys. Rev. B* **85**, 045436 (2012).

[17] F. Chen, G.L. Klimchitskaya, V.M. Mostepanenko, and U. Mohideen, "Demonstration of optically modulated dispersion forces", *Opt. Lett.* **15**, 4823 (2007).

[18] G. Torricelli, P.J. van Zwol, O. Shpak, C. Binns, G. Palasantzas, B.J. Kooi, V.B. Svetovoy, and M. Wuttig, "Switching Casimir forces with phase-change materials", *Phys. Rev. A* **82**, 010101 (2010).

[19] J.N. Munday, F. Capasso, and V.A. Parsegian, "Measured long-range repulsive CasimirLifshitz forces", *Nature* **457**, 170 (2009).

[20] R.S. Decca, D. López, and E. Osquiguil, "New results for the Casimir Interaction: Sample characterization and low temperature measurements", *Int. J. Mod. Phys. A* **25**, 2223 (2010).

[21] R. Castillo-Garza and U. Mohideen, " Variable-temperature device for precision Casimir-force-gradient measurement", *Rev. Sci. Instrum.* **84**, 025110 (2013).

[22] R. Castillo-Garza, J. Xu, G.L. Klimchitskaya, V.M. Mostepanenko, and U. Mohideen, "Casimir interaction at liquid nitrogen temperature: Comparison between experiment and theory", *Phys. Rev. B* **88**, 075402 (2013).

[23] J. Zou, Z. Marcet, A.W. Rodriguez, M.T.H. Reid, A.P. McCauley, I.I. Kravchenko, T. Lu, Y. Bao, S.G. Johnson, and H.B. Chan, "Casimir forces on a silicon micromechanical chip", *Nat. Commun.* **4**, 1845 (2013).

[24] M.T.H. Reid, A.W. Rodriguez, J. White, and S.G. Johnson, "Efficient computation of Casimir interactions between arbitrary 3D objects", *Phys. Rev. Lett.* **103**, 040401 (2009).

[25] R.S. Decca, D. López, H.B. Chan, E. Fischbach, D.E. Krause, and J.R. Jamell, "Constraining new forces in the Casimir regime using the isoelectronic technique", *Phys. Rev. Lett.* **94**, 240401 (2005).

[26] G. Haugstaad, *Atomic Force Microscopy: Understanding Basic Modes and Advanced Applications* (Wiley, Hoboken, NJ, 2012).

[27] R.C. Richardson and E. N. Smith, Eds., *Experimental Techniques in Condensed Matter Physics at Low Temperatures* (Addison-Wesley, Reading, MA, 1998).

[28] E. Rubiola, *Phase Noise and Frequency Stability in Oscillators*, (Cambridge University Press, Cambridge, UK, 2009).

[29] A. López Dávalos and D. Zanette, *Fundamentals of Electromagnetism: Vacuum Electrodynamics, Media, and Relativity*, (Springer-Verlag, Berlin, Germany, 1999).

[30] F. Chen, U. Mohideen, G.L. Klimchitskaya, and V.M. Mostepanenko, "Experimental test for the conductivity properties from the Casimir force between metal and semiconductor", *Phys. Rev. A* **74**, 022103 (2006).

[31] W.J. Kim, M. Brown-Hayes, D.A R. Dalvit, J.H. Brownell, and R. Onofrio, "Anomalies in electrostatic calibrations for the measurement of the Casimir force in a sphere-plane geometry", *Phys. Rev. A* **78**, 020101 (2008).

[32] R.S. Decca and D. López, "Measurement of the Casimir force using a micromechanical torsional oscillator: Electrostatic calibration", *Int. J. Mod. Phys. A* **24**, 1748 (2009).

[33] R.S. Decca, D. López, E. Fischbach, G.L. Klimchitskaya, D.E. Krause, and V.M. Mostepanenko"Tests of new physics from precise measurements of the Casimir pressure between two gold-coated plates", *Phys. Rev. D* **75**, 077101 (2007).

Chapter 6

Casimir forces at the cutting edge

WILLIAM SIMPSON, ULF LEONHARDT, RICARDO DECCA, STEFAN Y. BUHMANN

> The scientist is not a person who gives the right answers, he's one
> who asks the right questions.
>
> – Claude Lévi-Strauss, *Le Cru et le cuit* (1964)

It would be a pity if you read this far, decided that it all made perfect sense, and put the book down, supposing that scientists have got the Casimir effect 'pretty much figured out'. They haven't, and physics only thrives when people ask questions. Here are just a few examples of some fairly fundamental problems that have been exercising physicists in the Casimir community. We begin by reviewing two unresolved issues in experimental Casimir physics, following the previous chapter, and conclude by raising a number of questions in the theory of the Casimir effect.

§38. Experimental issues in Casimir physics

One of the most contentious issues remaining in the Casimir community concerns the precise form of the frequency dependent dielectric permittivity $\varepsilon(\omega)$ that should be used for modelling the interacting bodies. As discussed in Section 30, most disagreements between theory and experiments can be "solved" (i.e. good agreement between the model and data can be obtained) by selecting a particular response of the medium to electromagnetic radiation, especially in regions where it cannot be measured. In general, it is experimentally observed that, when using small spheres ($R \leq 150\mu$m), the measurements are limited to separations smaller than 1μm, the effects of residual electrostatic interactions are negligible, and a plasma model for the electric permittivity needs to be used to obtain good agreement between experiment and theory. On the other hand, when using larger spheres, a residual electrostatic force is present, measurements can be carried out to larger separations, and a Drude model needs to be used to minimise the differences between experimental data and theoretical calculations.

The second issue concerns the fact that, in reality, the interacting bodies do not possess equipotential surfaces. It can be shown that there will always be a residual electrostatic force as a result. The controversy concerns the strength of this interaction: if it is smaller than the minimum detectable interaction, then it does not contribute to the signal; if this is not the case, then it must be accounted for.

38.1   Drude or plasma?

The title of this section follows an animated controversy that started in 2005 and has not yet been resolved. In 2005 Decca and co-workers [1] used their sphere-plane configuration and showed that there was a better agreement between the experimental data and the theoretical computations obtained using (21.6) when the low frequency permittivity of Au was modeled using a plasma response

$$\varepsilon(\omega) = 1 - \frac{\omega_P^2}{\omega^2}, \tag{38.1}$$

instead of the accepted Drude response

$$\varepsilon(\omega) = 1 - \frac{\omega_P^2}{\omega(\omega + i\gamma)}. \tag{38.2}$$

In these equations, presented in Section 1.2, ω_P is the plasma frequency of the material and γ is the relaxation of free electrons in the medium.

Before presenting a list of experiments supporting one or the other approach, let's briefly describe the problem. On the one hand, we all know that gold has a finite resistivity at low frequencies, consequently a description using (38.1), where the relaxation parameter is zero, does not seem appropriate. On the other hand, the experimental data in many experiments seem to favour the plasma model. The way in which the model enters the calculation was described in Section 21.2 in Eq. (21.6). In this approach the electromagnetic response of the material must be known for all frequencies. However, the response cannot be measured at all frequencies, for rather obvious reasons. This presents us with a problem, particularly at low frequencies, where the electric permittivity needs to be extrapolated from the value measured at the lowest possible frequency to *zero* frequency. After this extrapolation is achieved, the Kramers–Kronig relations must be applied to obtain the full response at all frequencies, followed by the usual *Wick rotation* $\xi = i\omega$, before finally performing the summation over Matsubara frequencies.

After checking the contributions to the Casimir interaction arising from different Matsubara frequencies, it has been observed that, at $T = 300$ K, the discrepancies originate in the zeroth-order Matsubara term at $\zeta_0 = 0$ [1] . For other ζ_l there is minimal difference on $\varepsilon(i\zeta_l)$ and, consequently, the value of the interaction.

Exercise 38.1

Calculate $\varepsilon(i\zeta_l)$ for a plasma model and a Drude model for gold at 300 K. Use $\omega_P = 8.9$ eV and a dissipation $\gamma = 0.035$ eV

Plasma model support

Results supporting the plasma model were more firmly established when better measurements were obtained by the same group [2] and later corroborated by Mohideen's group [3]. These groups used a dynamic method with $R \sim 100\mu$m spheres in their respective micromechanical torsional oscillator and AFM setups. It is quite remarkable that these two independent experiments yield data which, within their experimental errors, completely agree in all the measured separation range. Furthermore, the temperature dependent experiments performed in Mohideen's group [4,5] also favour the plasma model. The data is seen to agree with the plasma model at all temperatures, and only agrees with the Drude model at low temperatures, where the effects of $\gamma(T)$ are negligible.

It is worth mentioning that all these results yield an interaction that is larger than that which is expected from a Drude response, since the behaviour of the plasma model is closer to the response of an ideal reflector. Hence, making the metal forming the cavity a "worse metal" (i.e. decreasing ω_P or increasing γ) only exacerbates the observed differences.

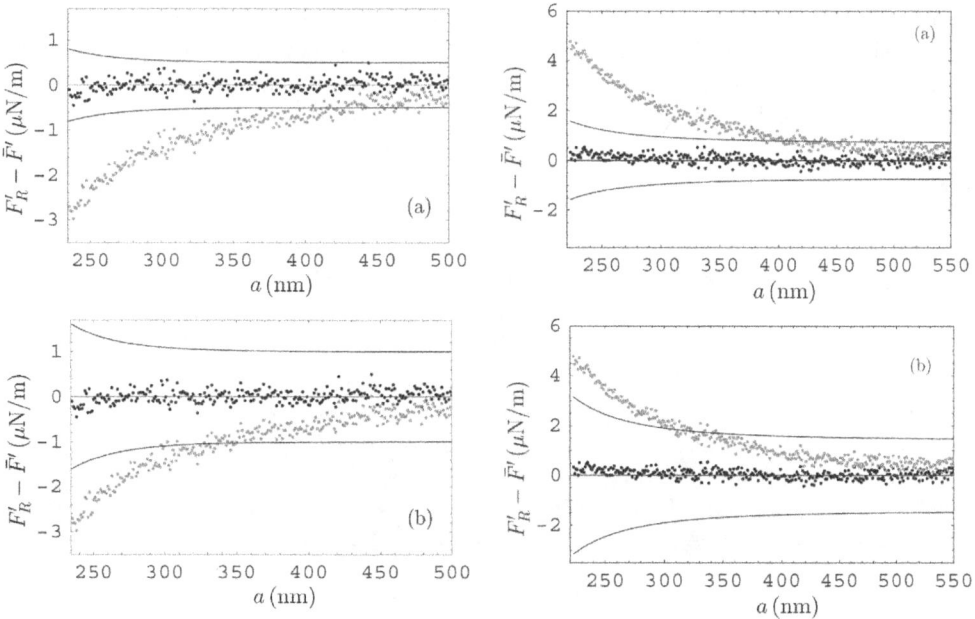

Figure 6.1: *Left panel:* Differences between the theoretical and mean experimental gradients of the Casimir force found at the experimental separations between a gold-coated plate and a gold-coated sphere. Plasma model results are shown in black. Drude model in grey. The solid lines indicate the borders of the (a) 67% and (b) 95% confidence intervals. *Right panel:* Same as the left panel but for both bodies being nickel-coated.

The last piece of evidence to support the plasma model also arises from experiments done by Mohideen's group [6]. In their series of experiments they measured the interaction between (1) two gold-coated bodies, (2) a nickel-coated body and a gold coated one, and (3) two nickel-coated bodies. All these experiments were done using the dynamic method with $R \sim 100\mu m$ spheres with an AFM approach. Since nickel is magnetic, its frequency dependent magnetic permeability $\mu(\omega)$ needs to be taken into account. As observed in the experiments, the plasma model predicts a stronger attraction than the Drude model in experiment (1), the same for experiment (2) and a weaker attraction for experiment (3). The results of experiment (1) and (3) are shown in Fig. 6.1.

Drude model support

Contrary to the previous experiments, first Lamoreaux and his coworkers [7] and then Tang's group [8] found experimental evidence in support of the Drude model. Lamoreaux used his torsional pendulum to measure the interaction between a $R = 15.6\mu m$ sphere and a flat plate, both covered with gold. Since the radius of the curvature is large, it allowed for the experiments to be performed at large separations, up to $z = 7\mu m$. The group found that, after subtracting a residual electrostatic force, the data was explained by using the Drude and not the plasma model. Furthermore, since the separations involved are comparable with thermal lengths at $T = 300$ K, the thermal effects associated with the Casimir effect can be observed. The results show that they need to be included into the theoretical description to explain the results.

Tang and co-workers reported measurements on a Casimir force sensor that bridges the measurement at microscale and macroscale by utilising a nanomembrane of millimeter lateral dimensions as a sensitive force transducer. A separate millimeter-sized gold coated sphere is used in a sphere-plate configuration to approach the nanomembrane. Since both surfaces have relatively large areas, a sizeable Casimir force can be measured even at larger separation distances. Like other oscillator based systems, the change in the resonance frequency of the oscillator is determined and from here the gradient of the Casimir force between the sphere and the plate is extracted. In the experiment it was found that, after subtracting a residual electrostatic contribution, the Drude model provides a better fit than the plasma model for separations between 1 and 2 μm.

38.2   Patch potentials

Let us turn now to our second controversial issue. As described in Section 38.1, experiments between large spheres and plates require the subtraction of a large residual electrostatic contribution, even after the electrostatic interaction has been minimised. The existence of this residual electrostatic interaction has also been proposed to explain the measurement of a more attractive interaction than the one predicted by the Drude model. While this is an interesting argument for the gold-gold cavity, it is clearly not sufficient to explain the results of the nickel-nickel cavity experiments. Nevertheless, it is important to understand the origin of the force and how it is possible for it to be present in some experiments whilst not being observable in others.

Non-equipotential metallic structures

The metallic structures used to form the Casimir cavity are not equipotentials, as described in Section 34. The reasons reside in using polycrystalline surfaces with different crystalline orientations (and work functions) exposed, and performing experiments under conditions where the surfaces cannot be made free of contaminants. Dalvit and co-workers [9] have provided a theoretical framework to calculate the electrostatic interaction between two surfaces with an arbitrary electrostatic potential distribution, if the distribution is known. While the details of the paper are beyond the scope of this book, a number of points can be stated clearly:

i. If the electric potential in the surface is not constant, there is a residual interaction.

ii. This residual interaction can be *minimised* by applying a potential difference V_o between the surfaces, but it cannot be made zero.

iii. The electric potential V_o is a function of separation.

These conclusions have to be qualified under two opposite conditions. Let's assume that the characteristic size of the potential distribution on the sample is l. In the case in which the effective area of interaction $A_{eff} \simeq \pi R \times z$ is either much larger or much smaller than l^2, it turns out that V_o is effectively constant as a function of separation (from an experimental point of view) and the residual interaction is too small to be measured:

1. if $A_{eff} << l^2$, then the surfaces are effectively equipotentials.

2. if $A_{eff} >> l^2$, averaging greatly reduces the strength of the residual force to the point where it becomes undetectable.

A recently published article [10] shows that the effect of patches does not explain the observed differences between the data and the Drude model prediction for experiments performed by Decca's and Mohideen's groups on gold-coated surfaces. However, there is still work to be done to completely understand the role of contamination, pressure, the nature of the contaminants, the size of the patches, among other factors, if we are to completely understand the significance of patch potentials in measurements of Casimir phenomena.

§39. THEORETICAL QUESTIONS IN CASIMIR PHYSICS

In addition to the goal of empirical adequacy, theorists are concerned with questions of self-consistency: Casimir physics is not free from conundrums, and we conclude our tour by raising a number of questions in the theory of the Casimir effect that turn out to be closely related.

39.1   Repulsive shells and Casimir's electron

In 1953 Casimir suggested an intriguing model that could potentially explain the stability of charged particles and the value of the fine-structure constant [11]. The argument goes as follows:

Imagine a particle, like the electron, as an electrically charged hollow sphere. Two forces are acting upon it: the Coulomb force (producing electrostatic repulsion) and the force of the quantum vacuum (presumed to be attractive). The energy of the quantum vacuum on a spherical shell of radius r must be given by a dimensionless constant α times $\hbar c/r$ on purely dimensional grounds:

$$E_{Casimir} = -\alpha \frac{\hbar c}{2r}. \tag{39.1}$$

Now, the electrostatic energy of a uniformly charged sphere is proportional to the square e^2 of its charge and is also inversely proportional to its size r [12]:

$$E_{Coulomb} = \frac{e^2}{8\pi\varepsilon_0 r}. \tag{39.2}$$

It follows that the Casimir force should balance the electrostatic repulsion, regardless of how small the distance parameter r may be, as long as α assumes a certain value given by the strength of the Casimir force. This strength depends on the internal structure of the particle – the fact that it is a spherical shell – but not on its size, which could be imperceptibly small. Thus Casimir's model, albeit crude, suggests a seductively simple explanation for both the fine-structure constant and the stability of charged elementary particles:

$$E_{Coulomb} = -E_{Casimir} \implies \alpha = \frac{e^2}{4\pi\varepsilon_0 \hbar c}. \tag{39.3}$$

On this analysis, the stabilising constant α turns out to be none other than the fine-structure constant.

Boyer's repulsive Casimir energy

Unfortunately Casimir's hopes for connecting the quantum vacuum to the stabilising Poincare stresses required in the classical electron model proved to be short-lived. In a seminal paper by Boyer in 1968 it was argued (in a QED calculation) that the sign of the zero-point energy is, in fact, the opposite to that proposed in Casimir's model:

$$E = \frac{0.09\hbar c}{2r}, \tag{39.4}$$

where r is the radius of the sphere. It follows that the Casimir force on a spherical shell is repulsive, not attractive; the force of the vacuum works with the Coulomb force to *expand* the sphere, rather than stabilise it [13]. Moreover, the multiplicative constant in (39.4) is not remotely close to the value of the fine-structure constant $\alpha \approx 1/137$. In Boyer's own words, this seemed 'a most melancholy' conclusion, subsequently corroborated by more sophisticated attempts at the same calculation [14]. Indeed, the spherical shell has become an archetype for a shape that causes Casimir repulsion.

The Kenneth-Klich Contradiction

However, repulsive Casimir forces remain the disputed exception (dispersion energies associated with polarisable media are typically attractive) and Boyer's conclusion is not beyond question. In 2006, two theorists produced a general theorem claiming that the Casimir force between two dielectric objects, related by reflection, is necessarily *attractive* [15]. This powerful result affords a 'no-go theorem' for repulsive Casimir forces in symmetric configurations. It is applicable, for instance, to the case of two slightly separated hemispheres – that is, to a spherical shell sliced with a subtle knife. It proves that the pieces will attract, not repel.

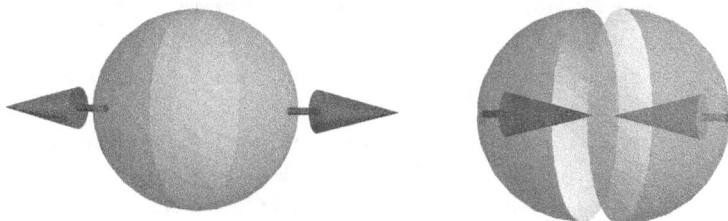

Figure 6.2: According to Boyer, the outward pressure on a conducting shell is repulsive (*left*). But according to Kenneth and Klich's theorem, two slightly separated hemispheres must attract (*right*). How is this possible?

The validity of the theorem is ordinarily restricted to distances where the system may be reliably described in terms of the field and local dielectric functions, and therefore the limiting case of two touching hemispheres forming a spherical shell may not normally be taken. However, Kenneth and Klich's model is applicable to the case of a perfectly conducting material $\varepsilon \to \infty$ adopted by Boyer. In such an idealised model, there is no reference to a real material response function, and no other scale of relevance. We seem to be left with the bizarre situation in which the hemispheres composing a spherical shell, when touching in a spherical configuration, are apparently disposed to fall apart, but when they are pulled slightly apart, are disposed to resume a spherical configuration; the Casimir energy flips from repulsive to attractive at the point of osculation.

A possible way forward (or around)

According to Gabriel Barton, the problem of the self-energy of the spherical shell is not well posed [16]. Considering the case of a weakly dielectric spherical ball, Barton computes a Casimir energy consisting of a sum of microscopic van der Waals interactions that neatly divides into terms that can be associated with the volume V and surface S energies of the configuration, and a geometric component referred to as the 'pure Casimir energy':

$$E = -\frac{V}{\lambda^4} + B\frac{S}{\lambda^3} + C\frac{\hbar c}{2r}, \tag{39.5}$$

where λ is the cut-off term in a standard regularisation*, as discussed in Section 7.1. Barton is content to concede that the pure Casimir energy becomes repulsive for the case of the spherical shell, but argues that, for the case of the self-energy of any single connected body, the physics is dominated by components of the energy that are typically discarded in ordinary regularisation and renormalisation procedures. The pure Casimir energy such procedures extract is relevant to interactions between well-separated bodies. In reality, however, this contribution is much smaller than the surface and volume terms. After all, no massive connected body can change its shape or size without modifying the relative displacement of adjacent atoms, resulting in short-range energy changes proportional to typical atomic excitation energies. The total Casimir energy is therefore *attractive* and depends in detail on the discrete structure of the solid.

However, Barton's analysis is limited to dilute materials in which the Casmir energy can be approximated as the sum of van der Waals interactions. As discussed in Section 16, however, Casimir interactions are properly *non-additive*. How might the analysis be modified by including multiple scattering? Furthermore, whilst Barton offers us a plausible account of why we would never *observe* a repulsive effect on a spherical shell, it still seems legitimate to raise questions about the changing sign of the 'pure Casimir energy' in such cases.

39.2   Divergence of the Casimir stress

Let's consider another conundrum in the field of Casimir physics. As we have seen, the calculation of the Casimir force requires a renormalisation procedure; the quantisation of the field results in an infinite contribution to the ground state energy of the system that must be removed before a finite, physical force can be extracted.

In the context of Lifshitz theory and its derivatives, which are useful for calculating more realistic cases, this is typically achieved by subtracting from the total Green function an auxiliary Green function associated with an infinite homogeneous medium. One can then compute a finite stress tensor for the system that depends on the dielectric functions of the material at imaginary frequencies, from which the

---

*Barton adopts a different scheme in which the lattice spacing λ performs the regularisation and affords a natural cut-off. This modifies the volume $\propto 1/\lambda^3$ and surface terms $\propto 1/\lambda^2$.

Casimir force can be derived. Both Casimir's and Lifshitz' methods give identical results in the limiting case of a cavity sandwiched between perfectly reflecting mirrors [17].

In more recent work, however, it was shown that Lifshitz' method failed to yield finite results when applied to an inhomogeneous dielectric in which the optical properties were varying continuously in space [18]. Let's see what happens when we attempt to renormalise the stress tensor for the general case of an inhomogeneous medium*.

The renormalised stress in the continuum limit

The usual expression for the stress tensor, when applied to a medium that is piece–wise defined along a single axis, is known to be finite[†]. For a region of width a where ε and μ are homogeneous, the value of the stress tensor at a point x can be written in terms of the reflection coefficients (as opposed to the Green functions) associated with sending q–polarized ($q = s, p$) plane waves to the right (r_{qR}) and to the left (r_{qL}) of this point [17, 20, 23]:

$$\sigma_{xx}(x) = 2\hbar c \sum_{q=s,p} \int_0^\infty \frac{d\kappa}{2\pi} \int_{R^2} \frac{d^2 k_\parallel}{(2\pi)^2} w \frac{r_{qL} r_{qR} e^{-2aw}}{1 - r_{qL} r_{qR} e^{-2aw}}, \qquad (39.6)$$

where $w = (n^2 \kappa^2 + k_\parallel^2)^{1/2}$, $k_\parallel = |k_\parallel|$, and n is the value of the refractive index in the homogeneous region surrounding x (cf. Eq. (21.9)). The reflection coefficients are functions of the imaginary frequency, $\omega = ic\kappa$, the (real) in-plane wave vector $k_\parallel$, and the material parameters of the media to the right and to the left of the homogeneous region.

The advantage of writing the stress tensor in this form is that the renormalisation procedure of Lifshitz theory is *automatically implemented* [17]. The contributions to the stress arise entirely from inhomogeneities in the system. Let's consider the behaviour of (39.6) in the limit as the piece–wise definition of the medium becomes a continuous function ($a \to 0$).

General argument for media inhomogeneous in one direction

Suppose we initially approximate the inhomogeneous medium as a series of N homogeneous strips of width a (see Fig. 6.3), only taking the limit of $N \to \infty$ and $a \to 0$ in the final step of the calculation. The transfer matrix technique can be used for such an analysis of the field. You can read more about this technique in [22–25], and about the details of the calculation outlined below in [19].

Applying the usual expressions for the reflection coefficients in terms of ratios of the elements of the transfer matrix, and considering the lth homogeneous slice in

---

*The results from this section are discussed in more detail in [19].

[†]This is so long as we do not ask for all of the components of the stress tensor as we approach the boundaries of the homogeneous regions.

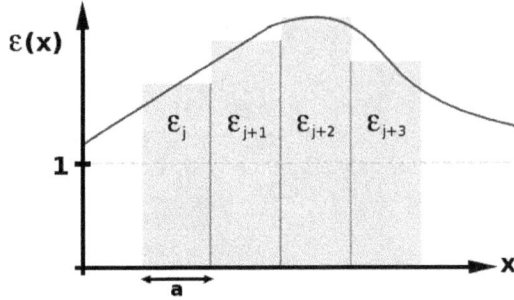

Figure 6.3: The medium is assumed inhomogeneous along x and is divided into N homogeneous slices of width a. The local value of the regularised stress tensor (39.6) is then investigated within the medium in the limit as $a \to 0$.

the stack, we find for the first polarisation

$$r_{1L}(x) = -\frac{\sum_{j=0}^{l-1} \frac{\Delta\mu_j}{\mu_j} e^{-2k_\parallel (l-j-1/2)a}}{\left(2 + \sum_{j=0}^{l-1} \frac{\Delta\mu_j}{\mu_j}\right)}, \qquad r_{1R}(x) = \frac{\sum_{j=l}^{N} \frac{\Delta\mu_j}{\mu_j} e^{-2k_\parallel (j-l+1/2)a}}{\left(2 + \sum_{j=l}^{N} \frac{\Delta\mu_j}{\mu_j}\right)}, \qquad (39.7)$$

where r_{1L} is the reflection coefficient associated with the left side of slice q, and r_{1R} is the reflection coefficient of the right. The reflection coefficients for the second polarisation are of an identical form but with the permeability replaced by the permittivity*. For the purposes of this analysis, we are restricting our attention to the regime of the integrand (39.6) where the in-plane wave vector is large in comparison with the refractive index multiplied by the 'frequency', $n_j\kappa/k_\parallel \ll 1$. The quantity w_j then becomes constant throughout the medium $w_j \sim k_\parallel$.

We will evaluate the integrand of (39.6) at a fixed κ, considering only a semi–infinite part of the integral over $k_\parallel$ $[K, \infty)$, where the reflection coefficients can be computed using (39.7) to within a reasonable approximation:

$$I = \sum_q \int_0^{2\pi} \frac{d\theta}{2\pi} \int_K^\infty \frac{k_\parallel^2 dk_\parallel}{2\pi} \frac{r_{qL} r_{qR} e^{-2k_\parallel a}}{1 - r_{qL} r_{qR} e^{-2k_\parallel a}} = \frac{1}{2\pi} \sum_{n=0}^\infty \sum_q I_{qn}.$$

In the final step we expand the denominator in a series of ascending powers of the reflection coefficients: we assume that $r_{qL} r_{qR} e^{-2k_\parallel a} < 1$ so that this sum converges for all a and all $k_\parallel$, interchange the order of integration and summation, and introduce the quantities

$$I_{qn} = \int_K^\infty k_\parallel^2 dk_\parallel (r_{qL} r_{qR})^{n+1} e^{-2(n+1)k_\parallel a}. \qquad (39.8)$$

---

*An additional factor of $a/2$ within the exponentials has been introduced such that the point x_l is at the centre of the lth slice.

The expressions given in (39.7) are now inserted into (39.8), then the integral over $k_\parallel$ in all these terms can be evaluated, yielding for instance

$$I_{10} = -\frac{\sum_{j=0}^{l-1}\sum_{k=l}^{N}\frac{\Delta\mu_j\Delta\mu_k}{\mu_j\mu_k}\left[\frac{K^2}{2(k-j+1)a}+\frac{2K}{4(k-j+1)^2a^2}+\frac{2}{8(k-j+1)^3a^3}\right]e^{-2K(k-j+1)a}}{\left(2+\sum_{j=0}^{l-1}\frac{\Delta\mu_j}{\mu_j}\right)\left(2+\sum_{j=l}^{N}\frac{\Delta\mu_j}{\mu_j}\right)}.$$

(39.9)

The size of this contribution to the stress (39.9) can be increased to an arbitrarily large value by decreasing the width of the slicing a. To see this, consider the continuum limit $(x_l \to x)$. In this limit, (39.9) becomes

$$I_{10} \to -\frac{\int_0^x dx_1 \int_x^L dx_2 \frac{d\ln[\mu(x_1)]}{dx_1}\frac{d\ln[\mu(x_2)]}{dx_2}\left[\frac{K^2}{2(x_2-x_1)}+\frac{2K}{4(x_2-x_1)^2}+\frac{2}{8(x_2-x_1)^3}\right]e^{-2K(x_2-x_1)}}{\left(2+\ln[\mu(x)]\right)\left(2-\ln[\mu(x)]\right)}.$$

(39.10)

This expression clearly diverges as $1/x^3$ in regions where $\mu(x)$ is not constant. For the second polarisation, I_{20} diverges in regions where $\varepsilon(x)$ is not constant. It therefore seems that there is no finite continuum limit of the regularised stress tensor (39.6) in these regions: the stress tensor contains infinite contributions*. As the remainder of the integral over $\mathbf{k}_\parallel$ is finite, we conclude that the whole integral diverges as $a \to 0$. Consequently (39.6) diverges everywhere within an inhomogeneous medium described by ε and μ that are continuous functions of position, independently of how these quantities depend on imaginary frequency. This is rather surprising.

Reflections

From this argument it seems clear at least that a calculation of (39.6) for a piecewise definition of an inhomogeneous medium does not represent an approximation to the continuous case. The expression for the stress tensor commonly employed in calculations of the Casimir force is not finite anywhere within an inhomogeneous medium. We might wonder how finite results ought to be extracted from this formalism. The advantage of the usual renormalisation procedure is that it removes an infinite quantity that does not depend on the inhomogeneity of the medium and is therefore of no relevance to the force. Conversely, here we have a divergent contribution that is due to the inhomogeneity of the medium. This might justly cause us to question the generality of the usual renormalisation procedure employed in Lifshitz theory.

The astute reader may have surmised that the divergence results from the reflection coefficients (39.7) failing to go to zero fast enough as $k_\parallel \to \infty$, in the limit where $a \to 0$. One might be tempted then to terminate the integral over $k_\parallel$ at some finite cut–off. However, without a good physical argument the value of this cut–off would be arbitrary. This problem does not seem to be widely appreciated in the literature. Why does it happen, and how can we fix it?

---

*It is important to emphasise that including the additional terms in the series (39.8), as well as corrections, does not affect this result. These contributions diverge in a similar manner, but represent higher powers of the derivatives of ε and μ—terms that vary quite independently as the spatial dependence of ε and μ is changed, and therefore cannot be expected to cancel in general.

39.3   Maxwell's fish eye and renormalisation

We began this section by questioning whether the predicted repulsive Casimir force on a spherical shell could be an artifact of the simple model used by Boyer. Another reason to reconsider this problem is the following: the bare stress of the quantum vacuum is always infinite and this infinity is removed by regularisation and renormalisation procedures. The most plausible method involves considering the relative stress between (or inside) macroscopic bodies. However, an infinitely thin sphere neither represents an extended macroscopic body nor multiple bodies. Suppose the physically relevant vacuum stress of an extended spherical shell tends to infinity in the limit where the shell becomes infinitely thin and infinitely conducting. In this case the renormalisation might remove this physically meaningful infinity, producing a finite result that may well have the wrong sign*. From our previous discussion, it appears that renormalisation in Casimir theory is not so well understood after all.

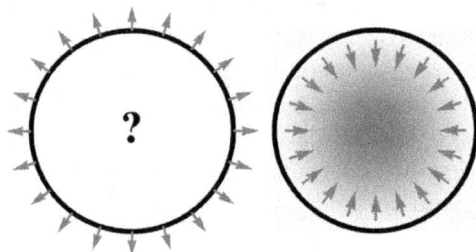

Figure 6.4: Boyer claimed that the Casimir force on a spherical shell is repulsive. In this example, we assume the shell to be filled with a medium (right) and find an attractive Casimir stress in the material. The shades of grey indicate the profile of the medium.

Let's consider a modification of Casimir's model. Imagine the spherical shell – still infinitely conducting and infinitely thin – is no longer hollow, but filled with an impedance-matched medium $\varepsilon = \mu$ of gradually varying electric permittivity ε and magnetic permeability μ, with the refractive index profile of Maxwell's fish eye:

$$n(r) = \frac{2}{1+r^2}. \tag{39.11}$$

For the boundary, we introduce a perfect mirror at $r = 1$. In this way we have extended the shell to a macroscopic body where the Casimir stress gradually builds up, whilst keeping the idealised boundary intact. The Casimir stress tensor $\sigma(r)$ must be radially symmetric. However, Maxwell's fish eye represents an inhomogeneous medium, and we have already encountered problems trying to renormalise the Casimir stress in precisely such cases. But perhaps we can get around this problem:

---

*According to [26] 'Lifshitz theory shows that the self-force is... inwardly directed and infinite'.

The geometry of the fish eye

From Fermat's principle, we know that light rays travel extremal optical paths with path lengths s measured by the refractive index n:

$$\mathrm{d}s = n \, \mathrm{d}l. \tag{39.12}$$

In other words, light travels along geodesics. We can think of optical materials as modifying the measure of space for light and implementing effective geometries. It can be shown that a medium with the properties of Maxwell's fish eye implements the stereographic projection of the surface of a hypersphere to 3d space (see Fig. 6.5).

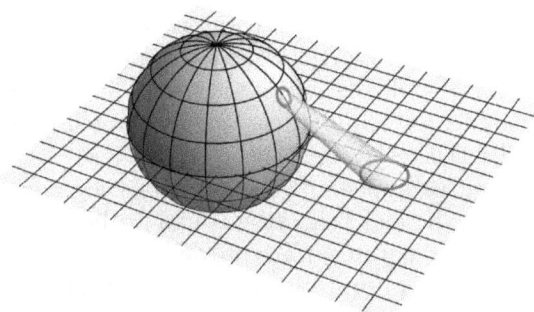

Figure 6.5: In 2d, the fish eye induces the stereographic projection of the surface of a sphere onto a plane; in 3d, it performs the stereographic projection of the hypersphere. *Image credit: Geometry and Light* [27].

The Green function of the fish eye with a mirror

The Green function of the bare fish eye satisfying Maxwell's equations and determining the behaviour of electromagnetic waves in the medium has been derived in [28]. The derivation is rather lengthy and involved and we will not recite it here, but it is documented fully and accessibly in [27]. The Green function may be succinctly expressed in terms of scalar waves propagating on the surface of the hypersphere. The effect of adding the mirror to our model can be described by an adaptation of the method of images [12]. The mirror lies on the equator (a 2-dimensional surface for the 4-dimensional hypersphere). Our method is to subtract from the incident Green function G_i the electromagnetic wave generated by the image source on the hypersphere:

$$G(r) = G_i(r) - J G_i(r^{-1}), \tag{39.13}$$

where J is the Jacobian matrix. This reflected field is simply the mirror image of the original field*. Since reflection results in a phase shift of π, we subtract the reflected field from the original one.

Renormalising the fish eye

Having obtained the Green function, we can in principle calculate the Casimir stress $(21.3)^\dagger$. As noted earlier, the standard procedure in Lifshitz theory is to renormalise the physical quantities by subtracting from the Green function $G(r)$ an auxiliary Green function $G_0(r)$ corresponding to an infinite homogeneous medium sharing the optical properties of the system at the point of measurement r:

$$\tilde{G}(r) = G(r) - G_0(r). \tag{39.14}$$

This physical contribution to the Casimir stress is understood to arise due to inhomogeneities in the surrounding material, so the Casimir force is 'renormalised' to zero for a homogeneous space. Unfortunately, as we have found (see Section 39.2), this method does not remove all the divergences from an inhomogeneous medium. Moreover, modifications to the standard procedure for renormalising the Casimir force must be physically well-motivated.

However, in this case we appear to occupy a privileged position: the Green function of an infinitely extended fish eye medium without a mirror corresponds to the Green function on the entire surface of the hypersphere, which is a uniform space. It can only produce a uniform vacuum stress σ_0 that does not contribute to the Casimir-force density $\nabla \cdot \sigma$. There is no Casimir force in the bare fish eye. It follows that the renormalised Green function

$$G(r) - G_0(r) = G_i(r) - G_0(r) - JG_i(r^{-1}) \tag{39.15}$$

has a redundant component $G_i - G_0$ that we can ignore; we need only focus on the reflected part of the radiation field.

The Casimir stress in Maxwell's fish eye

We thus consider only $-J\,G_i(r^{-1})$ in the total Green function (39.13) in the calculation of the stress tensor, which is described in more detail in [30]. In doing so, we find that

$$\sigma(r) = -\frac{\hbar c}{\pi^2 n \, (1 - r^2)^4} \mathbf{1}_3. \tag{39.16}$$

This is a remarkable result: using an alternative renormalisation motivated by geometric considerations, we have found a simple, exact expression for the Casimir

---

*In stereographic projection, reflection at the equator transforms $r \to r^{-1}$.

$\dagger$In this calculation, we use the Minkowski-like stress tensor: the electric contribution to the stress is weighted by ε, and the magnetic contribution by $1/\mu$. However, there has been some controversy about which stress tensor to use in media. The accepted use of the Minkowski-like stress tensor for computing Casimir forces in the Lifshitz theory [29] was challenged in [31], resulting in some debate [32–34]. A more recent argument for the accepted result can be found in [35], derived in the context of the theory of macroscopic quantum electrodynamics [36] described in Chapter 4.

stress in a system composed of spherical mirrors and filled with the inhomogeneous material of Maxwell's fish eye – the spherical analogue of a Casimir cavity filled with an inhomogeneous medium. This *bona fide* stress is negative and falls monotonically, so the Casimir-force density $\nabla \cdot \sigma$ is always attractive in this model. The result is easily generalised to the case of a system of arbitrary radius a:

$$\sigma(r) = -\frac{\hbar c}{\pi^2 a^4 n \left(1 - r^2/a^2\right)^4} 1_3. \tag{39.17}$$

Close to the mirror $r \to a$ the stress and the force tend to infinity, however. At the mirror itself $r = a$ the stress is infinite; it is still not possible to predict the actual force on the mirror (see Fig. 6.6). Is this residual infinity an artifact of using a perfect mirror – something that does not exist in nature? Could the approach we adopted here for obtaining the Casimir stress offer any general insight into the problem of renormalising the stress in inhomogeneous media?

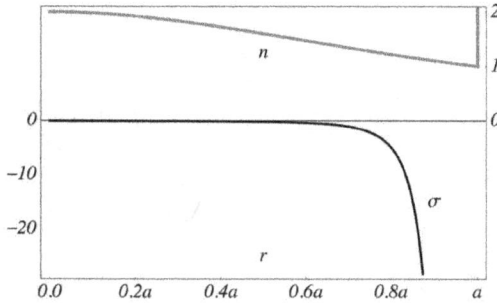

Figure 6.6: Index profile $n(r)$ (grey curve) of the medium inside the shell and the resulting vacuum stress $\sigma(r)$ (black curve, in units of $\hbar c/a^4$). As $r \to a$, the stress $\sigma \to \infty$.

39.4   The Casimir effect and cosmology

Dark energy

The Casimir effect is a physically observable manifestation of the quantum vacuum that becomes especially important in the nano-world. It seems there is another manifestation of the quantum vacuum, however, on a truly grand scale – in cosmology. In the modern era of cosmology (when cosmologists are no longer "often in error but seldom in doubt"*) there is overwhelming quantitative evidence for a strange effect: in addition to the familiar attractive gravitational force of matter there is a repulsive gravitational force due to what has been called *dark energy* [37]. Admittedly, 85% of matter seems to be *dark matter* of unknown origin [38], unless the laws of gravity are changed [39], but the hardest riddle of modern cosmology

---

*L. D. Landau.

is this mysterious dark energy that makes up nearly 70% of the total energy of the universe. Mathematically, it appears in Einstein's field equation of gravity [40],

$$R_{\alpha\beta} - \frac{R}{2} g_{\alpha\beta} + \Lambda\, g_{\alpha\beta} = \frac{8\pi G}{c^4} T_{\alpha\beta}, \qquad (39.18)$$

where $g_{\alpha\beta}$ is the metric tensor, $R_{\alpha\beta}$ the Ricci curvature tensor, R the curvature scalar, Λ the cosmological constant, G Newton's gravitational constant and $T_{\alpha\beta}$ the energy-momentum tensor. The tensor $T_{\alpha\beta}$ consists of the energy density $\mathcal{H}$, the momentum density $\boldsymbol{p}$ and the stress tensor σ as the 4×4 matrix

$$T_{\alpha\beta} = \begin{pmatrix} \mathcal{H} & -c\,\boldsymbol{p} \\ -c\,\boldsymbol{p} & -\sigma \end{pmatrix} \qquad (39.19)$$

written in a local Galileian frame [40]. The energy density $\mathcal{H}$ contains the rest-mass density ρc^2 that normally dominates gravity, far exceeding the other contributions to the energy-momentum tensor. We may turn the cosmological constant Λ into the dark energy if we simply write the term $\Lambda\, g_{\alpha\beta}$ on the right-hand side of Einstein's equation (39.18):

$$R_{\alpha\beta} - \frac{R}{2} g_{\alpha\beta} = \frac{8\pi G}{c^4} \left(T_{\alpha\beta} - \frac{c^4}{8\pi G} \Lambda\, g_{\alpha\beta} \right). \qquad (39.20)$$

We see that the energy density gets a contribution proportional to Λ with the opposite sign of the rest-mass density, leading to a repulsive gravitational force. One can exactly solve the Einstein equation (39.20) in the case of vanishing ordinary energy-momentum tensor $T_{\alpha\beta}$. The solution is known as the de Sitter space [40] and describes an exponential expansion of the universe driven by the cosmological constant, *i.e.* by the dark energy. Astronomical evidence shows that we are currently in such a phase, where the dark energy dominates cosmological evolution.

Renormalising dark energy

What can dark energy be? It seems to fill the universe, just like the quantum vacuum. The quantum vacuum, however, as we have seen in Section 4.1, would have an infinite energy density and hence would cause an infinite gravitational force, unless it is renormalised. In the theory of the Casimir effect we have encountered renormalisation, as well as some of its difficulties. Unfortunately cosmology is particularly unforgiving when it comes to the problem of renormalisation: in the Casimir effect, we can simply subtract from the total vacuum energy the part that does not produce a mechanical stress, but in Einstein's equation (39.20) *all* of the energy contributes to gravity.

One might introduce a cut-off to the vacuum energy that corresponds to wavelengths shorter than the Planck length, where the laws of physics are believed to be unknown, but such calculations have failed spectacularly – by 120 orders of magnitude [37]. On the other hand, we have seen in Section 39.2 that even the renormalisation of the Casimir effect fails for inhomogeneous planar media. Clearly, we need to understand better how to renormalise vacuum energies, and the chances that we will succeed, if we begin with what we know best, are exceedingly favourable: the part of physics the Casimir effect belongs to (quantum electromagnetism in media) is well-founded and reasonably well-understood. Many theoretical cross-checks – and most importantly, *experimental* tests – are possible at this level.

Understanding the Casimir effect in down-to-Earth settings through laboratory trials and thought-experiments may give us some real insight into the mystery of dark energy in cosmology. After all, Nature seems to form a unity, where each part is connected to the others in subtle (but not malicious) ways. Physics has shown us that we can understand Nature, at least to some extent, by drawing mathematical conclusions from wildly extrapolated empirical evidence. And that is the greatest mystery of all [41].

§40. Bibliography

[1] R. S. Decca, D. López, H. B. Chan, E. Fischbach, D. E. Krause, and J. R. Jamell, "Constraining new forces in the Casimir regime using the isoelectronic technique", *Phys. Rev. Lett.* **94**, 240401 (2005).

[2] R. S. Decca, D. López, E. Fischbach, G. L. Klimchitskaya, D. E. Krause, and V. M. Mostepanenko, "Tests of new physics from precise measurements of the Casimir pressure between two gold-coated plates", *Phys. Rev. D* **75**, 077101 (2007).

[3] C.-C. Chang, A. A. Banishev, R. Castillo-Garza, G. L. Klimchitskaya, V. M. Mostepanenko, and U. Mohideen, "Gradient of the Casimir force between Au surfaces of a sphere and a plate measured using an atomic force microscope in a frequency-shift technique", *Phys. Rev. B* **85**, 165443 (2012).

[4] R. Castillo-Garza and U. Mohideen, "Variable-temperature device for precision Casimir-force-gradient measurement", *Rev. Sci. Instrum.* **84**, 025110 (2013).

[5] R. Castillo-Garza, J. Xu, G. L. Klimchitskaya, V. M. Mostepanenko, and U. Mohideen, "Casimir interaction at liquid nitrogen temperature: comparison between experiment and theory", *Phys. Rev. B* **88**, 075402 (2013).

[6] A. A. Banishev, G. L. Klimchitskaya, V. M. Mostepanenko, and U. Mohideen, "Casimir interaction between two magnetic metals in comparison with nonmagnetic test bodies", *Phys. Rev. B* **88**, 155410 (2013); Erratum *Phys. Rev. B* **89**, 159901 (2014)

[7] A. O. Sushkov, W. J. Kim, D. A. R. Dalvit, and S. K. Lamoreaux, "Observation of the thermal Casimir force", *Nature Phys.* **7**, 230 (2011).

[8] D. García-Sánchez, K. Y. Fong, H. Bhaskaran, S. Lamoreaux, and H. X. Tang, "Casimir force and *in situ* surface potential measurements on nanomembranes", *Phys. Rev. Lett.* **109**, 027202 (2012).

[9] R. O. Behunin, Y. Zeng, D. A. R. Dalvit, and S. Reynaud, "Electrostatic patch effects in Casimir-force experiments performed in the sphere-plane geometry", *Phys. Rev. A* **86**, 052509 (2012).

[10] R. O. Behunin, D. A. R. Dalvit, R. S. Decca, C. Genet, I. W. Jung, A. Lambrecht, A. Liscio, D. López, S. Reynaud, G. Schnoering, G. Voisin, and Y. Zeng, "Kelvin probe force microscopy of metallic surfaces used in Casimir force measurements", arXiv:1407.3741 (2014).

[11] H. B. G. Casimir, "Introductory remarks on quantum electrodynamics", *Physica (Utrecht)* **19**, 846 (1953).

[12] J. D. Jackson, *Classical Electrodynamics* (Wiley, New York, 1998).

[13] T. H. Boyer, "Quantum electromagnetic zero-point energy of a conducting spherical shell and the Casimir model for a charged particle", *Phys. Rev.* **174**, 1764 (1968).

[14] K. A. Milton, L. L. DeRaad Jr., and J. Schwinger, "Casimir self-stress on a perfectly conducting spherical shell", *Ann. Phys.* **115**, 388 (1978).

[15] O. Kenneth and I. Klich, "Opposites Attract: a theorem about the Casimir force", *Phys. Rev. Lett.* **97**, 160401 (2006).

[16] G. Barton, "Perturbative Casimir energies of dispersive spheres, cubes and cylinders", *J. Phys. A: Math. Gen.* **34**, 4083 (2001).

[17] U. Leonhardt, *Essential Quantum Optics*, Cambridge University Press (2010).

[18] T. G. Philbin, C. Xiong, and U. Leonhardt, "Casimir stress in an inhomogeneous medium", *Ann. Phys.* **325**, 579 (2009).

[19] W. M. R. Simpson, S. A. R. Horsley, and U. Leonhardt, "Divergence of Casimir stress in inhomogeneous media", *Phys. Rev. A* **87**, 043806 (2013).

[20] E. M. Lifshitz. "The theory of molecular attractive forces between solids", *Sov. Phys. JETP* **2**, 73 (1956).

[21] C. Genet, A. Lambrecht, and S. Reynaud, "Casimir force and the quantum theory of lossy optical cavities", *Phys. Rev. A* **67**, 043811 (2003).

[22] M. Born and E. Wolf, *Principles of Optics* (Cambridge University Press, 1999).

[23] C. Genet, A. Lambrecht, and S. Reynaud, "Casimir force and the quantum theory of lossy optical cavities", *Phys. Rev. A* **67**, 043811 (2003).

[24] I. A. Shelykh and V. K. Ivanov, "Differential equation for the transfer matrix", *Int. J. Theor. Phys.* **43**, 477 (2004).

[25] M. Artoni, G. C. La Rocca, and F. Bassani, "Resonantly absorbing one-dimensional photonic crystals", *Phys. Rev. E* **72**, 046604 (2005).

[26] S. A. R. Horsley, "Canonical quantization of the electromagnetic field interacting with a moving dielectric medium", *Phys. Rev. A* **86**, 023830 (2012).

[27] U. Leonhardt and T. G. Philbin, *Geometry and Light: The Science of Invisibility* (Dover, New York, 2010).

[28] U. Leonhardt and T. G. Philbin, "Perfect imaging with positive refraction in three dimensions", *Phys. Rev. A*, **81**, 011804 (2010).

[29] A. D. McLachlan, "Van der Waals forces between an atom and a surface", *Mol. Phys.* **7**, 381 (1964).

[30] U. Leonhardt and W. M. R. Simpson, "Exact solution for the Casimir stress in a spherically symmetric medium", *Phys. Rev. D* **84**, 081701(R) (2011).

[31] C. Raabe and D.-G. Welsch, Reply to "Comment on 'Casimir force acting on magnetodielectric bodies embedded in media'". *Phys. Rev. A* **80**, 067801 (2009).

[32] L. P. Pitaevskii, Comment on "Casimir force acting on magnetodielectric bodies embedded in media". *Phys. Rev. A* **73**, 047801 (2006).

[33] I. Brevik and S. A. Ellingsen, Comment on "Casimir force acting on magnetodielectric bodies embedded in media". *Phys. Rev. A* **79**, 027801 (2009).

[34] C. Raabe and D.-G. Welsch, "Casimir force acting on magnetodielectric bodies embedded in media", *Phys. Rev. A* **71**, 013814 (2005).

[35] T. G. Philbin, "Casimir effect from macroscopic quantum electrodynamics", *New. J. Phys.* **13**, 063026 (2011).

[36] T. G. Philbin, "Canonical quantization of macroscopic electromagnetism", *New J. Phys.* **12**, 123008 (2010).

[37] G. Brumfiel, "Unseen universe: a constant problem", *Nature* **448**, 245 (2007).

[38] J. Hogan, "Unseen universe: welcome to the dark side", *Nature* **448**, 240 (2007).

[39] M. Milgrom, "MD or DM? Modified dynamics at low accelerations vs dark matter", arXiv:1101.5122 (2010).

[40] C. W. Misner, K. S. Thorne, and J. A. Wheeler, *Gravitation* (Freeman, New York, 1999).

[41] E. P. Wigner, "The unreasonable effectiveness of mathematics in the natural sciences", Richard Courant lecture in mathematical sciences delivered at New York University, May 11, 1959. *Comm. Pure Appl. Math.* **13**, (1960).

Further reading

A number of textbooks and articles provide more extended insights into the topics featuring in this introduction.

Quantum optics in media

- U. Leonhardt, *Essential Quantum Optics: from Quantum Measurements to Black Holes* (Cambridge University Press, 2010).
  A basic introduction to quantum optics that also treats media.

- W. Vogel and D.-G Welsch, *Quantum Optics* (Wiley-VCH; 3rd, Revised and Extended Edition, 2006).
  Covers both the basic concepts of quantum optics and also more advanced aspects of the field.

- J. C. Garrison and R. Y. Chiao, *Quantum Optics* (Oxford Univ. Press, 2008).

Macroscopic quantum electrodynamics

- C. Cohen-Tanoudji, J. Dupont-Roc, G. Grynberg, *Photons & Atoms: Introduction to Quantum Electrodynamics* (Wiley, New York, 1989).
  A comprehensive and rigorous intoduction to quantum electrodynamics.

- D. P. Craig and T. Thirunamachandran, *Molecular Quantum Electrodynamics: An Introduction to Radiation Molecule Interactions* (Dover, New York, 1998).
  A concise introduction to quantum electrodynamics with a focus on molecular interactions.

- A. Salam, *Molecular Quantum Electrodynamics: Long-Range Intermolecular Molecule Interactions* (Wiley, New Jersey, 2010).
  Quantum electrodynamics with a modern version of molecular interactions using state-sequence diagrams.

- G. Compagno, R. Passante, F. Persico, *Atom–Field Interactions and Dressed Atoms* (Cambridge University Press, Cambridge, 1995).

Quantum electrodynamics with an alternative view on interatomic interactions using a dressed-atom approach.

- S. Y. Buhmann, *Dispersion Forces I: Macroscopic Quantum Electrodynamics and Ground-State Casimir, Casimir–Polder and van der Waals Forces* (Springer, Heidelberg, 2012).
  Macroscopic quantum electrodynamics in absorbing media.

Quantum vacuum effects

- P. W. Milonni, *The Quantum Vacuum: An Introduction to Quantum Electrodynamics* (Academic Press, San Diego, 1994).
  An accessible and comprehensive overview over vacuum effects with historical references.

Casimir forces

- D. Dalvit, P. Milonni, D. Roberts, F. da Rosa, *Casimir Physics* (Springer, Berlin, 2011).
  An comprehensive collection of articles on different aspects of the Casimir force written by experts of the field.

- M. Bordag, G. L. Klimchitskaya, U. Mohideen, V. M. Mostepanenko, *Advances in the Casimir effect* (Oxford University Press, Oxford, 2009).
  An exhaustive and detailed description of the Casimir force in theory and experiment.

- K. A. Milton, *The Casimir Effect: Physical Manifestiations of Zero-Point Energy* (World Scientific, New Jersey, 2001).
  An introduction to Casimir forces with an emphasis on field theory.

Casimir–Polder forces

- S. Y. Buhmann, *Dispersion Forces II: Many-Body Effects, Excited Atoms, Finite Temperature and Quantum Friction* (Springer, Heidelberg, 2012).
  An overview over Casimir–Polder forces and related effects.

General Casimir Literature

- A collection of bibliographic resources are maintained by Prof. James Babb from Harvard University. The URLs are https://www.cfa.harvard.edu/ babb/ casimir-bib.html and http://www.mendeley.com/groups/558181/casimir-effects/. It has a collection of introductory level articles, *Scientific American* level articles and monographs. There is also a selection of research articles, which is unfortunately somewhat outdated.

- There are two excellent Resource Letters in American Jornal of Physics. "Resource Letter VWCPF-1: Van der Waals and Casimir-Polder forces", K. A. Milton, *Am. J. Phys.* **79**, 697 (2011) and "Resource Letter CF-1: Casimir Force", S. K. Lamoreaux, *Am. J. Phys.* **67** 850 (1999).

- A wealth of original research articles can be found in the various *Proceedings* of the workshops on *Quantum Field Theory Under the Influence of External Conditions*.

- S. K. Lamoreaux, "The Casimir force: background, experiments, and applications." *Rep. Prog. Phys.* **68**, 201 (2005).

- S. K. Lamoreaux, "Casimir forces: Still surprising after 60 years." *Physics Today*, Feb. 2007, pp. 40-45.

- P. W. Milonni and M.-L. Shih, "Casimir forces." *Contemp. Phys.* **33**, 313 (1992).

Measurements of Weak Forces

- V.B. Braginsky and A.B. Manukin, *Measurements of Weak Forces in Physics Experiments* (The University of Chicago Press, Chicago, IL, 1977).
  A fairly comprehensive work on the effect of noise and different approaches to measuring forces in classical systems.

- V.B. Braginsky, V.P. Mitrofanov, and V.I. Panov, *Systems with Small Dissipation* (The University of Chicago Press, Chicago, IL, 1985).
  An extension of the previous work: it not only considers mechanical oscillators, it also deals with electromagnetic and superconducting resonators.

Regularisation techniques

As mentioned in Section 7, a range of alternative techniques have been developed for removing divergences from the vacuum energy in order to obtain a finite Casimir energy. Some of them, point-splitting regularisation, dimensional regularisation and zeta function regularisation, are introduced in this Appendix.

§A. POINT-SPLITTING REGULARISATION

The commutator (4.2) of the electromagnetic fields diverges in the *coincidence limit* $r \to r'$. This indicates that infinities arise whenever we evaluate expectation values of field-operator products taken at equal position arguments. We have done exactly that when evaluating the total vacuum energy

$$E_0 = \int \mathrm{d}^3 r \left\langle \frac{\varepsilon_0}{2} \, \hat{\boldsymbol{E}}^2(\boldsymbol{r}) + \frac{1}{2\mu_0} \, \hat{\boldsymbol{B}}^2(\boldsymbol{r}) \right\rangle. \tag{A.1}$$

In order to remove the associated singularities, one can instead evaluate field-operator products with slightly displaced position arguments:

$$E_0 \mapsto \int \mathrm{d}^3 r \left\langle \frac{\varepsilon_0}{2} \, \hat{\boldsymbol{E}}(\boldsymbol{r}) \cdot \hat{\boldsymbol{E}}(\boldsymbol{r} + \Delta\boldsymbol{r}) + \frac{1}{2\mu_0} \, \hat{\boldsymbol{B}}(\boldsymbol{r}) \cdot \hat{\boldsymbol{B}}(\boldsymbol{r} + \Delta\boldsymbol{r}) \right\rangle. \tag{A.2}$$

This is called *point-splitting regularisation.*

What is the effect of this splitting on the Casimir energy? Originally, the contribution of each mode to the vacuum energy is proportional to $\int \mathrm{d}^3 r \, \boldsymbol{E}_j(\boldsymbol{r}) \cdot \boldsymbol{E}_{j'}^*(\boldsymbol{r})$. The point-splitting modifies these contributions to $\int \mathrm{d}^3 r \, \boldsymbol{E}_j(\boldsymbol{r}) \cdot \boldsymbol{E}_{j'}^*(\boldsymbol{r} + \Delta\boldsymbol{r})$. Applying this to the modes (3.37) of a cuboid cavity and restricting our attention to one dimension, $\Delta\boldsymbol{r} = \Delta x \boldsymbol{e}_x$, this amounts to a change

$$\int_0^{L_x} \mathrm{d}x \, \cos^2(\pi n_x x / L_x) \mapsto \int_0^{L_x} \mathrm{d}x \, \cos(\pi n_x x / L_x) \cos[\pi n_x (x + \Delta x)/L_x]. \tag{A.3}$$

The effect of this change on different modes is illustrated in Fig. A.1. For modes whose wavelength $\lambda_{n_x} = 2L_x/n_x$ is much larger than the point-splitting Δx, one obtains a tiny reduction of the vacuum-energy contribution, Fig. A.1(a). For modes with wavelengths comparable or smaller than the point-splitting, the contribution

(a) $n_x = 4$

$\boldsymbol{E}_{n\lambda}(\boldsymbol{r}) \cdot \boldsymbol{E}_{n\lambda}^*(\boldsymbol{r} + \Delta\boldsymbol{r})$

(b) $n_x = 14$

$\boldsymbol{E}_{n\lambda}(\boldsymbol{r}) \cdot \boldsymbol{E}_{n\lambda}^*(\boldsymbol{r} + \Delta\boldsymbol{r})$

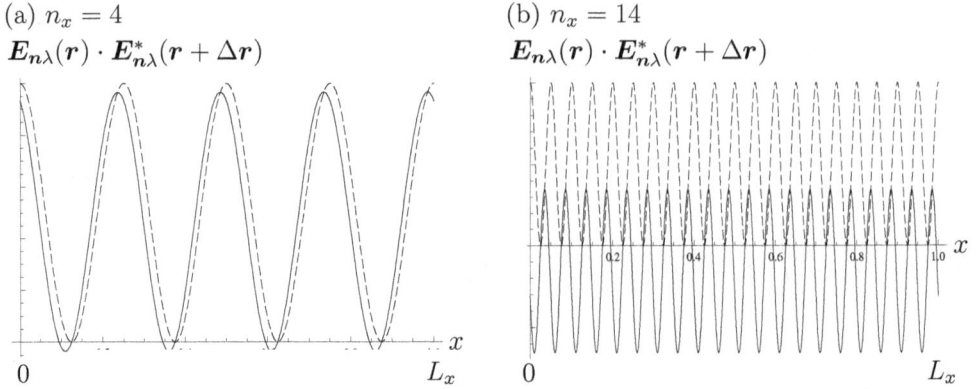

Figure A.1:   Point-splitting: we compare the amplitude $\boldsymbol{E}_{n\lambda}(\boldsymbol{r}) \cdot \boldsymbol{E}_{n\lambda}^*(\boldsymbol{r})$ of a given mode in the coincidence limit (dashed line) and the same amplitude $\boldsymbol{E}_{n\lambda}(\boldsymbol{r}) \cdot \boldsymbol{E}_{n\lambda}^*(\boldsymbol{r} + \Delta\boldsymbol{r})$ with point splitting $\Delta x = 0.03\, L_x$ (solid line). (a) A mode $n_x = 4$ whose wave length is much larger than the point-splitting is only weakly affected. (b) A mode $n_x = 14$ with wave length comparable to the splitting is strongly affected.

can be drastically altered and even turned negative, as shown by Fig. A.1(b). The effect can be quantified by evaluating the two integrals above. One finds that the point-splitting leads to a change

$$1 \mapsto \cos(\pi n_x \Delta x / L_x). \tag{A.4}$$

Generalised to three dimensions, the point-splitting alters the vacuum energy (7.2) to

$$E_0 \mapsto \frac{\hbar c}{2} \sum_{n\in N^3} \sum_\lambda \sqrt{\left(\frac{\pi n_x}{L_x}\right)^2 + \left(\frac{\pi n_y}{L_y}\right)^2 + \left(\frac{\pi n_z}{L_z}\right)^2}$$
$$\times \cos(\pi n_x \Delta x / L_x) \cos(\pi n_y \Delta y / L_y) \cos(\pi n_z \Delta z / L_z). \tag{A.5}$$

The cosine functions rapidly oscillate as a function of the mode index for modes with $n_i \gg L_i/\Delta x_i$. This leads to an effective cut-off of high-frequency modes, as required for obtaining a finite Casimir energy.

The point-splitting regularisation has the same physical meaning as the cut-off regularisation discussed in Section 7: the introduction of a finite point-splitting Δr implies a course-grained approximation to space where we cannot resolve distances smaller than this value. Similarly, the restriction of the electromagnetic field to modes with frequencies smaller than ω as imposed by the cut-off means that only distances greater than the minimum wavelength $\Delta r \geq \lambda = 2\pi c/\omega$ can be resolved.

An alternative point-splitting strategy involves the introduction of a finite time difference Δt between the two field operators in the vacuum energy (A.1). This leads to temporally oscillating terms $e^{i\omega_i \Delta t}$. They again have the effect of suppressing high-frequency modes. More efficient exponential convergence is achieved by introducing imaginary displacements in space $i\Delta\boldsymbol{r}$ or time $i\Delta t$. The price for this mathematical convenience is the more obscure physical meaning of the point-splitting.

§B. DIMENSIONAL REGULARISATION

Dimensional regularisation has a similarly esoteric physical meaning. The starting point is the observation that the Casimir energy (7.6) depends on the number of dimensions parallel to the plates

$$E_0(L) = \hbar c A \int \frac{d^2 k_\parallel}{(2\pi)^2} \sum_{n=0,1,\ldots}' \sqrt{k_\parallel^2 + \left(\frac{\pi n}{L}\right)^2}. \tag{B.1}$$

So what if this number was not $d = 2$ as in the physical case, but allowed to take arbitrary values? Treating the number of transverse dimensions as a continuous, complex-valued variable d, we have

$$E_0(L, d) = \hbar c A \int \frac{d^d k_\parallel}{(2\pi)^d} \sum_{n=0}^{\infty}{}' \sqrt{k_\parallel^2 + \left(\frac{\pi n}{L}\right)^2}. \tag{B.2}$$

We will evaluate this expression following the steps outlined by Milton [57]. In order to solve the $k_\parallel$-integral, we use a trick first introduced by *Schwinger* [58]: starting from the definition of the *Gamma function*

$$\Gamma(z) = \int_0^\infty d\tau\, \tau^{z-1} e^{-\tau} \tag{B.3}$$

he deduced the identity

$$\frac{1}{k^a} = \frac{1}{\Gamma(a)} \int_0^\infty d\tau\, \tau^{a-1} e^{-k\tau}. \tag{B.4}$$

This is known as the *proper time representation*, because it can be interpreted as the propagation in proper time τ of a photon. Applying it to the wave number k and noting that $\Gamma(a) = \Gamma(-1/2) = -2\sqrt{\pi}$, the Casimir energy in d dimensions takes the form

$$E_0(L, d) = -\frac{\hbar c A}{2\sqrt{\pi}} \int \frac{d^d k_\parallel}{(2\pi)^d} \sum_{n=0}^{\infty}{}' \int_0^\infty d\tau\, \tau^{-3/2} e^{-\tau(k_\parallel^2 + \pi^2 n^2/L^2)}. \tag{B.5}$$

Exercise B.1

Use substitution of variables to derive Schwinger's proper time representation. Use your result to obtain the above representation of the Casimir energy.

The Gaussian integral in d dimensions can then be carried out using

$$\int_{-\infty}^{\infty} dx\, e^{-x^2} = \sqrt{\pi}, \tag{B.6}$$

resulting in

$$E_0(L, d) = -\frac{\hbar c A}{2^{d+1}\pi^{(d+1)/2}} \sum_{n=0}^{\infty}{}' \int_0^{\infty} d\tau\, \tau^{-(d+3)/2} e^{-\tau\pi^2 n^2/L^2}. \tag{B.7}$$

Recalling the above definition of the Gamma function, the τ-integral leads to

$$E_0(L, d) = -\frac{\hbar c A}{2^{d+1}\pi^{(d+1)/2}} \left(\frac{\pi}{L}\right)^{d+1} \Gamma\left(-\frac{d+1}{2}\right) \sum_{n=1}^{\infty} \frac{1}{n^{-d-1}}, \tag{B.8}$$

where we have discarded the divergent, but L-independent $n = 0$ term. For values $\mathrm{Re}\, d < -2$, the sum over n converges and it is given by the *Riemann zeta function*

$$\zeta(z) = \sum_{n=1}^{\infty} \frac{1}{n^z}. \tag{B.9}$$

Using the reflection property

$$\pi^{-z/2}\Gamma\left(\frac{z}{2}\right)\zeta(z) = \pi^{-(1-z)/2}\Gamma\left(\frac{1-z}{2}\right)\zeta(1-z), \tag{B.10}$$

we finally obtain a explicit result for the Casimir energy:

$$E_0(L, d) = -\frac{\hbar c A \zeta(d+2)}{2^{d+1}\pi^{d/2+1}L^{d+1}} \Gamma\left(\frac{d}{2}+1\right). \tag{B.11}$$

Exercise B.2

Starting from Eq. (B.5), perform the $k_{\parallel}$- and τ-integrals as well as the sum over n to derive the above Casimir energy.

The result is valid only for the above, unphysical values for d. However, the expression on the right hand side of (B.11) is the unique infinitely differentiable complex function which is well-defined for arbitrary complex values of $d \neq 1$ and agrees with the Casimir energy for $\mathrm{Re}\, d < -2$. What happens if we evaluate this expression at our physically relevant value $d = 2$? Noting that $\zeta(4) = \pi^4/90$ and $\Gamma(2) = 1$, we find

$$E_0(L, d = 2) = -\frac{\pi^2 \hbar c A}{720 L^3}. \tag{B.12}$$

This is exactly the Casimir energy (7.12) as obtained from the cut-off regularisation. By introducing the number of transverse dimensions as a complex-valued parameter, we have thus been able to evaluate the Casimir energy for a range of unphysical parameter values. Extending the solution to our physical value of interest via analytic continuation, we have recovered the correct, finite Casimir energy.

How does this magic work? Let us have a closer look at the mathematics behind analytic continuation. The Riemann zeta function lies at the heart of this regularisation scheme. This series is defined as

$$\zeta(z) = \lim_{N\to\infty} \sum_{n=1}^{N} \frac{1}{n^z} \tag{B.13}$$

whenever this limit exists. When the series diverges, we can only study its asymptotic behaviour for large N,

$$f_z(N) = \sum_{n=1}^{N} \frac{1}{n^z}. \tag{B.14}$$

To render this a smooth function of N, one commonly introduces a cut-off function η as defined by Eq. (7.14):

$$f_z(N) \mapsto \sum_{n=1}^{N} \eta(n/N) \frac{1}{n^z}. \tag{B.15}$$

For $\mathrm{Re}\, z < -1$, one can show that

$$\sum_{n=1}^{N} \eta(n/N) \frac{1}{n^z} \to \zeta(z) + c_\eta(z) N^{1-z} \quad \text{for } N \to \infty, \tag{B.16}$$

where $\zeta(z)$ is the value of the zeta function obtained by analytic continuation and $c_\eta(z)$ is a cut-off dependent coefficient. In other words: the value for $\zeta(z)$ from analytic continuation is the constant part of the regularised series in the limit $N \to \infty$. Analytic continuation hence implements exactly the two steps involved in the cut-off regularisation: first, a regularisation via cut-off function η; secondly, a renormalisation whereby contributions from high-n modes are discarded. We observed in Section 7 that the low-n modes give the main contribution to the Casimir energy. We have hence made plausible why analytic continuation reproduces the Casimir force found via cut-off regularisation and renormalisation.

§C. Zeta-function regularisation

The central idea of dimensional regularisation was the introduction of a free complex parameter. The closely related *zeta-function regularisation* relies on a simpler form of the regularising parameter s. In this case, we generalise the Casimir energy (4.1) to

$$E_0 = \langle \hat{H}_F \rangle = \langle 0 | \hat{H}_F | 0 \rangle = \sum_j \tfrac{1}{2} \hbar \omega_j^{1-2s} \tag{C.1}$$

where the physically relevant result corresponds to the case $s = 0$. Applying this to the cuboid cavity (7.7), we have

$$E_0(L, s) = \frac{\hbar c A}{2\pi} \int_0^\infty \mathrm{d}k_\parallel \, k_\parallel \sum_{n=0}^{\infty}{}' \left[k_\parallel^2 + \left(\frac{\pi n}{L} \right)^2 \right]^{\frac{1}{2} - s}. \tag{C.2}$$

Neglecting the L-independent $n = 0$ term and introducing a variable $x = k_\parallel L/(n\pi)$, the Casimir energy takes the form

$$E_0(L, s) = \frac{\hbar c A}{2\pi} \left(\frac{\pi}{L} \right)^{3-2s} \int_0^\infty \mathrm{d}x \, x \left(1 + x^2\right)^{\frac{1}{2} - s} \sum_{n=1}^{\infty} \frac{1}{n^{2s-3}}. \tag{C.3}$$

For $\mathrm{Re}\, s > 2$, the sum over n converges and is given by the Riemann zeta function (B.9), hence the name zeta-function regularisation. The x-integral also converges in this parameter range,

$$\int_0^\infty \mathrm{d}x\, x\left(1 + x^2\right)^{1/2-s} = \frac{1}{2s - 3} \,. \tag{C.4}$$

Combining these results, we find the expression

$$E_0(L, s) = \frac{\hbar c A}{2\pi} \left(\frac{\pi}{L}\right)^{3-2s} \frac{\zeta(2s - 3)}{2s - 3} \tag{C.5}$$

for the Casimir energy. As in the case of dimensional regularisation, we use analytic continuation to extrapolate our $\mathrm{Re}\, s > 2$ result to the physically relevant case $s = 0$ where the zeta function takes the value $\zeta(-3) = 1/120$. We once more recover the Casimir energy

$$E_0(L, s = 0) = -\frac{\pi^2 \hbar c A}{720 L^3} \,. \tag{C.6}$$

When calculating Casimir forces, we may hence choose among a whole range of alternative regularisation techniques. The frequency cut-off regularisation at one end of this range has the advantage of involving physically meaningful limits whose applicability can be explicitly checked for each geometry and material of interest. The zeta-function regularisation at the other extreme is mathematically more elegant and often much easier to evaluate. However, greater care needs to be taken when judging its applicability as the limits involved are implicit in the mathematics of divergent sums.

Index

About the authors

THE EDITORS

Prof Ulf Leonhardt is a professor of physics at the
Weizmann Institute of Science. He completed his PhD
at the Humboldt University Berlin, receiving the Tiburtius
Prize of the Senate of Berlin and the Otto Hahn Medal
of the Max Planck Society. In 2006 he received a Scientific American 50 Award, in 2008 a Royal Society Wolf-
son Research Merit Award, in 2009 a Theo Murphy Blue
Skies Award of the Royal Society. In 2012 he received
a Thousand Talents Award of China. His research interests span many fields, including Casimir phenomena,
quantum optics, Hawking radiation and transformation optics.

Dr William Simpson is a research fellow at the Weizmann Institute of Science and a research associate at The
University of St. Andrews. He completed his PhD in theoretical physics at the University of St. Andrews, having obtained a SUPA prize doctoral studentship, and was
awarded the Springer thesis prize. He was the chairman
and organiser of the Foundations of Casimir Physics symposium in Stockholm in 2013, and the 'PhysPhil' Philosophy of Physics conferences in St Andrews (2012,4).
He has publications in physics, philosophy and theol-
ogy.

THE CONTRIBUTORS

Dr Stefan Yoshi Buhmann is an Emmy Noether Fellow
and Junior Research Group Leader in the Institute of Physics
at the University of Freiburg. He completed his PhD in
theoretical physics at the University of Jena. Subsequently,
he worked as a Feodor Lynen Fellow of the Alexander von
Humboldt Foundation in the Quantum Optics and Laser Science Group at Imperial College London. He is the author
of numerous papers on Casimir and related phenomena in
macroscopic quantum electrodynamics. He has also written a two volume series on Dispersion Forces, published by
Springer.

Dr Ephraim Shahmoon is a research fellow at the Weizmann Institute of Science, where he also completed his PhD in physics. His research focuses on quantum optics and electrodynamics, including dipolar and Casimir interactions in confined geometries, nonlinear quantum optics, electromagnetic-induced transparency and quantum open systems.

Prof Stefan Scheel is a Professor of Theoretical Physics in the Faculty of Mathematics and Natural Sciences at the University of Rostock. He obtained his PhD in 2001 at the University of Jena, receiving the Best Dissertation Award. In 2002 he received a Feodor Lynen Award of the Alexander von Humboldt foundation. He was awarded an EPSRC Advanced Research Fellowship at Imperial College London in 2004, becoming a Lecturer in 2005 and a Reader in 2010. His research interests include QED in linear and nonlinear dielectric materials, ultracold trapped neutral atoms and Rydberg physics, dispersion forces and quantum tomographic reconstruction.

Dr Simon Horsley is a lecturer in theoretical physics at the University of Exeter. He obtained his PhD at The University of York, and was awarded an EPSRC Postdoctoral Fellowship for research at the University of St. Andrews. His research interests include the physics of metamaterials, transformation optics, light in moving dielectric media, and the Casimir effect.

Prof Ricardo Decca is Professor of Physics and Co-Director of the Nanoscale Imaging Center at IUPUI (Indiana University–Purdue University). He received his PhD in Instituto Balseiro, Argentina. His research interests include the spectroscopic investigation of quantum systems, the function and morphology of biomembranes, and the investigation of forces (eg. gravity) at the submicron range.